Studies in Fuzziness and Soft Computing 291

For further volumes:
http://www.springer.com/series/2941

Ronald R. Yager, Ali M. Abbasov, Marek Z. Reformat,
and Shahnaz N. Shahbazova (Eds.)

Soft Computing:
State of the Art Theory
and Novel Applications

Springer

Editors
Ronald R. Yager
Machine Intelligence Institute
Iona College
New Rochelle
NY
USA

Shahnaz N. Shahbazova
Department of Information Technology
and Programming
Azerbaijan Technical University
Baku
Azerbaijan

Ali M. Abbasov
Ministry of Communications and
Information Technologies of the
Republic of Azerbaijan
Baku
Azerbaijan

Marek Z. Reformat
Department of Electrical
and Computer Engineering
University of Alberta
Edmonton, Alberta
Canada

ISSN 1434-9922 e-ISSN 1860-0808
ISBN 978-3-642-34921-8 e-ISBN 978-3-642-34922-5
DOI 10.1007/978-3-642-34922-5
Springer Heidelberg New York Dordrecht London

Library of Congress Control Number: 2012951833

Printed on acid-free paper

Springer is part of Springer Science+Business Media (www.springer.com)

Contents

Future Novel Directions

Fuzzy Sets and Relations

Rules Construction and Analysis

Information Boom: New Trends and Expectations

Ali M. Abbasov[1,2]

[1] Minister of Communications and Information Technologies of the Republic of Azerbaijan
[2] Ministry of Communications and Information Technologies of the Republic of Azerbaijan
33 Zarifa Aliyeva str., Baku, Azerbaijan, AZ1000
abbasov@mincom.gov.az

Abstract. This report presents new approaches to managing information resources to address the problems that result from rapid information growth. It provides a current-state analysis and shows how nanoelectronics can enable the use of physical resources, such as high-capacity storage elements, new sensors, high-speed computing, and data transmission networks.

The report investigates new computing platforms, emerging information models and architectures, and their potential impact on information management. Furthermore, it shows that unlike the traditional number-based form of computing, computing based on perceptions may open up new horizons in the field of information management by making a shift from quantitative to qualitative information measurement in real-world situations. It suggests that the quantum features of nanoelements can enable the implementation of the necessary operations in this shift from quantitative to qualitative information processing.

Finally, it describes the substitution of Boolean Algebra by Lotfi Zadeh's fuzzy sets and fuzzy logic, which makes it possible to investigate new approaches to the management of information resources.

1 Introduction

Today, the never-ending expansion of information has become a global concern, and the term "information overload" is listed as a problem and challenge for the 21st century. Although this phrase was first used in 1970 by Alvin Toffler, an overwhelming growth of information resources due to unprecedented development of the Internet in recent years has transformed information overload into a pressing global problem [1, 2].

Figures and estimations vary widely. Even a very rough estimation of the overall volume of information is much too complicated and any attempt at precise solutions seems to be unreasonable due to the huge speed of escalation.

For instance, data regarding a simple occurrence in nature may be of immense volume depending on the information unit and the physical features of the objects in it. Nevertheless, not to be distracted by this issue, we will merely focus on digital information.

The current upsurge of digital information seriously surpasses the technical capacity needed for its collection, storage, and processing [3]. Even if in the near future the Internet and other specific networks suffer only traffic jams and information and

R.R. Yager et al. (Eds.): *Soft Computing: State of the Art Theory*, STUDFUZZ 291, pp. 1–12.
DOI: 10.1007/978-3-642-34922-5_1

knowledge loss, it is possible to expect information wars and severe global economic crises, as well as anthropogenic disasters in the long run; therefore, discovery of new approaches to information management is a pressing issue. Moreover, if we consider that in the current world information serves as a primary source of acquiring knowledge and an important aid in making decisions, then information management becomes a global issue in the 21st century.

Information management can be analyzed in four stages, namely, information collection, storage, processing, and transmission. Current research shows that nanoelectronics can provide opportunities for creating high-capacity storage elements, new sensors, high-speed computing, and data transmission networks, and so on. Emerging computing platforms, information models, and architecture can provide new tools for information management. However, it should be mentioned that the development of these new computing technologies faces certain constraints. An additional concern is that algebraic computation, in spite of its use in promoting the development of electronics, has not been replaced with anything else since the first computer was introduced 60 years ago. This is because numeric, quantity-based. Measurement of information has always been the focus in computing models. Therefore, perhaps it is time to apply qualitative models to computing platforms and physical elements, given the fact that quantum features of the nanoelements are able to aid in the necessary manipulations. This goal constitutes the major objective of this presentation.

2 Modern Problems of Information Overload

Information is of immense importance in the current world. However, an overload of information causes serious problems in the collection, storage, processing, and transmission stages of information management.

There is a considerable increase in the speed of gathering of electronic information (both analogue and digital); it is caused by a rapid growth of the Internet, social and sensor networks as well as the wider use of digital cameras, TVs, observation systems, mobile phones, medical scanners, and so on. The Internet is primarily responsible for this increase; for example, in the year 2010 alone, 107 trillion e-mails were sent, that is, 294 billion e-mails per day, and the number of Web pages exceeded 280 million. There are more than 2 billion photos uploaded every month on Facebook [4].

All the above serves as a basic sample of the rapid growth of the digital world. Moreover, analyses have revealed that the digital world is growing even faster than it was first envisioned. The annual increase in speed of the digital world is estimated as 60 to 70%. In 2009, this increase was estimated at 0.8 zettabyte (ZB), whereas in 2010 it surged to 1.2 ZB [5]; it is expected to reach 35 ZB in 2020.

Another source of concern regards the storage of information. Because storage of unstructured data greatly exceeds that of structured data in the world, it becomes particularly important to back the data up properly and make it available for use. If we look at the Large Hadron Collider, we would see that every second it generates 40 terabytes of information in CERN. Today, even a cow generates 200 megabytes of information every year.

Transmission of information is another challenge in a rapidly growing information period. The Internet traffic volume rapidly increases; its annual growth is about 34%. Of major impact are the files with video content as they cover 30% of the current Internet traffic. It is expected that the Internet video demand, as well as IP-TV, will grow to 90% in the near future. Today, demand for information transmission severely exceeds the capabilities of the transmission channels. As a result, information gathers in one part of the world without being able to be transmitted to other parts. In the end, it causes a digital gap between the regions and leads to serious problems. Given the fact that products and services are shifting from tangible into intangible, this would mean that on one side of the world there is going to be vast storage of information, as well as services and digital products, whereas on the other side of the world, an increased demand for them. In its turn, this misbalance between supply of services/products and demand for them might cause severe economic crises in the coming 20 to 25 years. Moreover, taking into account that those who are "in need" are 80% of the population, it is more likely to predict the enormity of the economic crisis.

Another problem is the processing of information. Currently, computing technologies are facing challenges that cannot be overcome by traditional computers. The problems include information search, cryptography, nuclear studies, and electronic translations. Even though different computing systems and supercomputers like "Tianjin," which has 2.5 petaflops power, have been developed and parallel/flow/conveyer models are in place, thousands of such supercomputers are unable to manipulate the required volume of operations. However, new computing technologies like Cloud, Grid, and others are emerging and they are increasing the efficiency of computing technology several folds, thus helping to overcome the current problems. Nevertheless, computing technology has not been revolutionized and there is a lot to be done to ensure smooth information management.

In sum, we observe that the problem of information overload exists at all levels of information management and traditional means are outdated. Therefore, it is vital to focus on new approaches and analyze them accordingly.

3 Expectations and Proposals

Today, we are witnessing a revolutionary shift from semiconductors to nano-technologies. There are already nanostructures that receive and store information; scientific theory regarding their manipulation on the level of molecules, atom, and photons has become a reality. The ability of atoms to speed up and slow down in the stable situation opens up new perspectives in the area of information management, as in the measurement conducted with Qbits [7,8]. As an example, if we create a scheme of 300 atoms, then we can get 2^{300} bits or approximately 10^{80} digits, which exceed the number of elementary particles in the universe. Moreover, if we add the quantum features of the atoms to this, then we may observe a revolution in this sphere in the near future. Consequently, although theoretically, a scheme of 2^{300} bit capacity may simultaneously calculate 10^{80}. In its turn, this would lead to the advancement in information processing as well as to the development of quantum computers. As a result, these suggested changes might speed up the shift from traditional computing platforms to the more progressive ones. It should also be noted that, in case this scheme is created, it would become possible to apply Zadeh's fuzzy logic and algebra to it. If this happens, then computing might be done by perceptions rather than by numbers.

In this case, the quantum uncertainty Qbits can be represented as a classical fuzzy variable, the value of which lies within the interval [0, 1]. Developing the analogy of quantum effects with the properties of fuzzy variables, we can pass onto the higher level of abstraction - the linguistic variables. [9, 10]

Theoretically, by using nanoelectronics the computational algorithms of fuzzy logic can be carried out at the level of physical elements, which we cannot imagine.

Some studies have already been done in this regard and a number of achievements, for example, the application of algorithms in quantum computers with the purpose of manipulating complex operations, are to be considered as successes [11, 12, 13, 14].

In [11], D'hooghe, Pykacz, and Zapatrin study the possibility of performing fuzzy set operations on a quantum computer. It is shown that due to the famous quantum parallelism, quantum computers can operate "globally" on whole membership functions of fuzzy numbers, instead of calculating them "point by point" as classic computers do, which leads to a considerable decrease in the number of operations involved in storing and calculating fuzzy numbers.

In general, the transition from crisp to fuzzy logic and use of quantum calculations for this is certainly a big step toward creation of artificial intelligence.

The ever-expanding information scope can also play a positive role in creating artificial intelligence. As we have not as yet achieved considerable advances in the study of the human mind, statistical approaches can be of help in understanding the core processes taking place in the depths of the human mind. We can precisely model the processes of the human brain, such as incoming information to the brain (incoming information, for instance in the form of speech – words) and generation of adequate actions by the human brain (outgoing information, for instance, reciprocal speech – also words), such as the process of transforming information from incoming to outgoing in the case of translation in the human brain. In this case, mapping words to a set of words happens as because of computations over incoming words, which results in a few or a completely different set of words - the answer.

Accordingly, by having an almost unlimited amount of textual information, we can probably determine the nature and patterns of computing with words as proposed by Zadeh. All of these will allow us to better understand the structure of human brain and decrease the time to creating artificial intelligence, which is one of the ultimate goals of world science.

Rocha and Massad [14] have begun work in this new direction. This work reflects an original and astounding new understanding of the brain based on novel achievements in fuzziness and quantum information theory. By blending neuroscience, soft computing, quantum theory, and recent developments in mathematics, the actual knowledge about the brain function is formalized into a coherent theoretical framework. This monograph demonstrates how the physiology of the neuron can be understood based on the fundamentals of fuzzy formal language and introduces the basics of quantum computation and quantum information to the brain.

Theoretically we argue that computing based on fuzzy logic, with the help of nanotechnologies, is possible. However, information storage and smooth transmission is not enough for the overall success of information management; there are numerous challenges, namely, the analyses of information, language translation, and perception of propositions, as well as matters concerning artificial intellect, that require new approaches to information technology. Still, it is vital to reduce the impact of information quantity through different nonstatistical measures to information, such as its validity, usefulness, and value [15].

Not all information on the Web is authentic. Some of this information is "trash information." Other data are ably disguised misinformation, which is not only invaluable in terms of knowledge, but which also decreases the probability of correct problem solving of important managerial problems and as a result hinders individuals from making adequate decisions. This situation pushes us toward creation of more powerful techniques to manage the ever-emerging information boom; all these techniques could be elaborated based upon fuzzy logic. . To separate useful (for a specific demand) information from the general flow of information, it is essential to develop intelligent tools of "filtering" using various criteria of fuzzy character.

Owing to fuzzy logic it has become possible to define new opportunities for efficiency.

In its turn, this requires a new approach toward information volume. The unit used for measurement of the information may vary depending on the nature and purpose of the application. In other words, “bits” are used to measure the electric condition of the electronic elements. However, the previously discussed “information load” is primarily about the texts, photos, speeches, and information with video content, and it covers different fields of human activities.

In this regard, computing with words (CW) may play an important role, and if it becomes possible to express information in natural languages by propositions, then it seems better to use “word” instead of “bit” as an information measurement unit. Accordingly, the quantity of words in any given language could be equal to logical algebra, and the grammar of a language, including syntactic as well as morphological rules, could be considered as tools for calculating, i.e., computing with words. As Zadeh stated, this could be “approaching the words with computation” [10, 16-20]. Computing with words is inspired by the remarkable human capability of performing a wide variety of mental tasks without quantitative measurements and provides a basis for an important generalization of probability theory, namely, perception-based probability theory, which is creating the new methods of design and analysis to deal with real-world problems [16].

There are two major imperatives for CW. CW is a necessity when the available information is too imprecise to justify the use of numbers, as well as when there is a tolerance for imprecision that can be exploited to achieve tractability, robustness, low-solution cost, and better rapport with reality. Exploitation of the tolerance for imprecision is an issue of central importance in CW. In CW, a word is viewed as a label of a granule, that is, a fuzzy set of points drawn together by similarity, with the fuzzy set playing the role of a fuzzy constraint on a variable. In coming years, CW is likely to evolve into a basic methodology in its own right with wide-ranging ramifications on both basic and applied levels [17].

Traditional approaches to natural language understanding are based on classic, Aristotelian, bivalent logic. So far, the use of traditional approaches has been met with limited success. The principal problem is that, basically, natural languages are systems for describing perceptions, and as such, are intrinsically imprecise in ways that place them beyond the reach of bivalent logic and probability theory [18].

Unlike traditional numbers, CW can be expressed both in abstract and various algebras. CW can be based both on crypt and fuzzy sets. The undisputed pioneer in this sphere is Professor Lotfi Zadeh. The development of a computing platform based on linguistic variables, with the help of fuzzy logic, has been spearheaded by him and his academic team. CW also covers "if-then," crypt and fuzzy rules, linguistic phrases, and other similar models. In other words, to implement linguistic research tasks (e.g., information search, automatic translation, speech recognition), a syntactic and semantic computing of the words in natural language must be done. Here the most challenging issue is the creation of a linguistic algebra (translation, understanding of the text, question-answering system, synthesis of speech, compressed data) because it requires creation of relevant CW models for every operation. Recently, we have started to apply the CW model in the automatic translation of the Azerbaijani language and Azerbaijani speech recognition [21]. It should also be stressed that CW has a disadvantage due to its strict dependence on languages.

For example, a model of calculus of linguistically quantified propositions is used for the automatic text document categorization problem. CW using an ontological model of natural language is used as well [23]. In general, there are dozens of academic research and papers dedicated to CW, although we are still far from having a single, perfect and completed theory. Zadeh's generalized theory of uncertainty-GTU must be considered a landmark in this field.

It should be pointed out that it is possible to use traditional computing algebra (Boolean algebra) to actualize the logic based on linguistic algebra. However, it is possible to create multilevel logical elements based on quantum specification of the nanoelements and here we can benefit from Zadeh's fuzzy logic. In this case, there is a question: If we are talking about the traditional equivocation and statist condition of information, then the measurement unit of uncertainty is entropy. But if we use words as an information unit, then what will be the measurement of equivocation in this case? Presumably, this is also going to be an indefinite measurement as was done with entropy.

The value of information may play an important role in this process, as we are discussing the linguistic model. Moreover, if we develop from the equivocation of information and accept the value of information, it would make it easier to understand the process. On the other hand, the dictionary of frequently used words in natural languages could help us. Recently, we prepared a list of frequently used words in the Azerbaijani language and we are using it in automatic translation and understanding of the language [21]. These dictionaries vary depending on the field of study and the time. However, during a certain period of development in a particular field of study, distribution of collocations remain asymmetric and about 20% of words hold 80% of frequency. In other words, if we shift to the linguistic network model, then we would get a scale-free network. There are methods of defining "value" junctions in the scale-free network [26,27].

Another specific characteristic of the fuzzy or linguistic model is that it can be used in information transmission. Today, the speed of transmission is confined to Shannon's information theory, Kotelnikov's discretization model, Boolean algebra, and short frequency length of light spectrum in optical lifts. If we apply the linguistic model to information and use quantum electronic techniques as a physical element, we are likely to achieve crucial changes in information transmission. There have been successful studies done in the sphere of transmission of elementary particles and photon's transmission in quantum conditions in optical lifts and in the establishment of quantum networks [3, 5, 27]. Use of quantum channels in transmission may unprecedentedly increase information flow speed.

4 Introduction to Computing with Words

Zadeh [28, 29, 30] with the introduction of fuzzy subsets and the related developments of computing with words [19, 20, 31, 32, 33, 34, 35] has provided a tool to enable the automated manipulation of human knowledge. The starting point for Zadeh's paradigm of computing with words (CW) is a collection of propositions expressed in a natural language. The fundamental position taken by Zadeh is that the knowledge contained in a natural language proposition can be viewed as a constraint on one or more of the implicit variables. The first step in the CW methodology consists of a translation (or explication) of these propositions into a formal computer manipulatable language which he calls **G**eneralized **C**onstraint **L**anguage (GCL). The second step in the process is a goal directed manipulation of these propositions. This step, which Zadeh calls granular computing, can be seen as a kind of inference process. This inference process is based on a constraint propagation mechanism. The result of this second step is a proposition in GCL providing a constraint on a variable of interest. The final step is a process of retranslation, here we convert a statement in GCL into an appropriate statement in natural language. Figure #1 provides a schematic view of the process of CW.

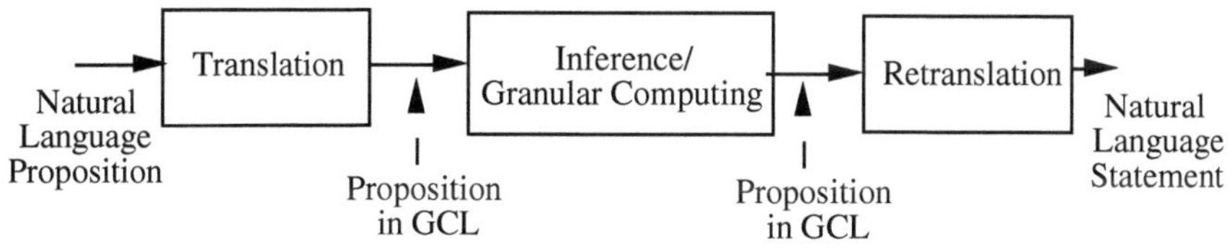

Fig. 1. Computing with Words Paradigm

Each of these three steps presents us with considerable opportunities for research and development. In step one we need to provide ways for expressing many types of natural language statements within the general constraint language. The manipulation of the propositions in the inference step is very important and clearly needs work. One promising direction in this inference process is what Zadeh calls protoform based reasoning [35]. Step three, retranslation, involves the process of converting a statement in GCL into a proposition in natural language

As indicated an important part of the paradigm of computing with words is translation step, as we indicated this involves translating human knowledge into a formal computer manipulatable language. Much of human knowledge involves a combination of possibilistic and probabilistic information. In order to represent this type of information Zadeh [36] has recently introduced the idea of a Z-number associated with an uncertain variable V. Yager [37] has studied these in considerable detail. A Z-number is an ordered pair of fuzzy numbers, (A, B). Here A is a fuzzy subset of the domain X of the variable V and B is generally a fuzzy subset of the unit interval. A Z-number is used to represent a datum of about the uncertain variable of the type where A represents a value of the variable and B represents an idea of certainty or other closely related concept such as sureness, confidence, reliability, strength of truth or probability. A Z-number is closely related to the idea of linguistic summary [38, 39[. Zadeh [361] refers to the ordered triple, (V, A, B) as a Z-valuation and indicates this is equal to the assignment statement V *is* (A, B). Here the Z-number is providing information about the value of the variable V. Generally this Z-valuation is indicating the V takes the value A with certainty equal B. Some examples of these Z-valuations are

(Age Mary, Young, likely)
(Income John, bout 100k, very likely)
(Income Bill, about 200K, not likely)
(Enemy number of soldiers, about 300, pretty sure)
(Weight Bill heavy, confident)

Thus Z-valuations are providing some information about the value of the associated variable. A number of questions arise regarding these objects such as the representation of the information contained in a Z-valuation, the manner in which we can manipulate this information and our ability to combine or fuse multiple pieces of information of this type. Many of the answers are dependent upon the nature of the underling uncertainty associated with the variable.

In [1] Zadeh focused on the situation in which it is assumed that the undying uncertainty associated with V is probabilistic, in this case V is a random variable. We see that these z-numbers can provide a very powerful tool for representing and manipulating many different types of human knowledge, These Z-numbers will play a very important role in the representation of the types of information that is currently being mined on the internet.

5 Fuzzy Modeling of Social Networks

Social relational networks, such as Facebook and LinkedIn, are rapidly becoming a powerful source of human interaction and communication. They are breaking down borders and connecting people from one end of the world to another. They played a central role in the so–called "Arab Spring." They contain a great body of valuable information and clearly are useful to both governments and business for discovering new developing trends. Recently Yager [40, 41] has begun using fuzzy logic and related technologies to extend our capabilities to analyze and model social relational networks. His objective is to provide tools that can enable analyst to interact with formal models of social networks using human concepts and ideas. Just as we are able to query a database about its contents, he wants to be able to query a formal model of a social network for information about its contents. Since human beings predominantly use linguistic terms in which to communicate, reason and understand Yager noted we are faced with the task of trying to build bridges between human conceptualization and the formal mathematical representation of the social network. Consider for example a concept such as "leader". An analyst may be able to express, in linguistic terms, using a network relevant vocabulary, properties of a leader. Our task then becomes translating this linguistic description into a mathematical formalism that allows us to determine how true it is that a particular node is a leader.

In his work he began looking at the possibility of using fuzzy set methodologies and more generally granular computing to provide the necessary bridge between the human analyst and the formal model of the social network.

The interest in focusing on fuzzy technology is based on the confluence of two important factors. One of these is that fuzzy set theory and particularly Zadeh's paradigm of computer with words was especially developed for the task of representing human linguistic concepts in terms of a mathematical object, a fuzzy subset. Fuzzy logic has a large repertoire of operations that allows for the combination of these sets in ways that mimic the logic of human reasoning and deduction. The second important factor is the nature of the formal mathematical model of social networks. The standard formal model used to represent a social network is a mathematical structure called a relationship. Using this structure the members of the network constitute a set of elements, the connections in a network are represented as pairs of elements and the network is viewed as the set of all these pairs. The key observation here is that the standard form of network representatives is in terms of set theory. The fact that the underlying representation of the social network is in set theoretic terms makes it to well suited to a marriage with the fuzzy set approach. In figure 2 we show Yager's Paradigm for Intelligent Social Network Analysis.

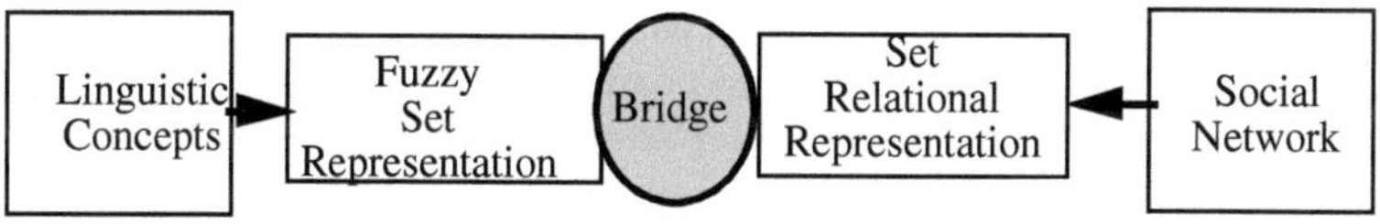

Fig. 2. Paradigm for Intelligent Social Network Analysis (PISNA)

6 Conclusion

Today, one of the global problems is the ever-increasing availability of information, the so-called phenomenon of "information overload." "Information overload" can be viewed as a kind of natural disaster of the 21st century; it presents problems, threats, and demands that must be faced by our modern information society. After analyzing the current status of the world information environment as well as looking at methods and hardware/software tools, we can conclude that the use of quantum effects for the creation of new calculating systems and algorithms developed on entirely different principles of data processing will offer new choices in the management of information resources that were not possible before.

The development of nanotechnology and nanoelectronics enables previously impossible alternatives such as the practical implementation of no quantitative algebraic operations directly on the physical elements using the quantum properties of elementary particles. This makes it possible to qualitatively improve the management of information resources in large ways at all levels of its implementation. Additionally, fuzzy algebra with elements of words and linguistic operations based on them, the so-called computing with words paradigm, may fundamentally change the direction of computing and information sciences.

Thus, we may be on the threshold of the birth of a new paradigm, Zadeh science, in contrast to the hard numeric science.

As governments, business and individuals become more and more dependent on the internet with its vast amounts of data available they will find fuzzy set related technologies an invaluable tool for converting this data into the kinds useful knowledge needed to make intelligent decisions for a better future.

References

1. Reichardt, R.: Digital reference overload: thoughts on how to deal. Internet Reference Services Quarterly 11(2), 105–112 (2006)
2. Cisco Visual Networking Index: Forecast and Methodology (2009-2014)
3. Data, data everywhere. Special report: Managing information. The Economist, February 27-March 5, pp. 1-16 (2010)
4. Internet 2010 in numbers, `http://royal.pingdom.com/2011/01/12/internet-2010-in-numbers/`
5. Gantz, J., Reinsel, D.: The Digital Universe Decade-Are You Ready? IDC-Analyze the Future, 16 p. (May 2009) (This explosive growth means that by 2020, our Digital Universe will be 44 times as big as it was in 2009)
6. The Expanding Digital Universe. An IDC White Paper (2010), `http://www.emc.com/collateral/analyst-reports/expanding-digital-idc-white-paper.pdf`
7. Goser, K., Glösekötter, P., Dienstuhl, J.: Nanoelectronics and nanosystems: from transistors to molecular and quantum devices, p. 284. Springer (2004)
8. Mermin, N.D.: From Cbits to Qbits: Teaching computer scientists quantum mechanics. Am. J. Phys. 71, 23–30 (2003)
9. Deutsch, D.: Quantum Theory, the Church-Turing Principle and the Universal Quantum Computer. Proc. Roy. Soc., Lond. A400(1818), 97–117 (1985)
10. Zadeh, L.A.: Fuzzy logic = computing with words. IEEE Transactions on Fuzzy Systems 4(2), 103–111 (1996)
11. D'hooghe, B., Pykacz, J., Zapatrin, R.: Quantum Computation of Fuzzy Numbers. International Journal of Theoretical Physics 43(6), 1423–1432 (2004)
12. Kanfmann, A.: Introduction to the Theory of Fuzzy Subsets, vol. 1. Academic Press, New York (1975)
13. Shor, P.W.: Algorithms of quantum computation: Discrete logarithms and factoring. In: Proceedings of the 35th Annual Symposium of the Foundations of Computer Sciences, pp. 124–134. IEEE Computer Society Press (1994)
14. Rocha, A.F., Massad, E.: The Brain: Fuzzy Arithmetic to Quantum Computing (Studies in Fuzziness and Soft Computing), vol. 165, p. 234. Springer (2004)

15. Shankin, G.P.: The value information. The Theory and Applications–Filomatis, 128 p. (2004)
16. Zadeh, L.A.: Computing with words (CW) and its application to decision support and systems analysis. In: 2003 IEEE International Symposium on Intelligent Signal Processing, pp. 1–2 (2003)
17. Zadeh, L.A.: The concept of a generalized constraint - a bridge from natural languages to mathematics. In: 2005 IEEE International Workshop on Annual Meeting of the North American Intelligent Signal Processing Fuzzy Information Processing Society, NAFIPS 2005, pp. 1–6 (2005)
18. Zadeh, L.A.: Computing with Words and perceptions—A paradigm shift. In: IEEE International Conference on Information Reuse & Integration, IRI 2009, pp. viii–x (2009)
19. Zadeh, L.A.: From computing with numbers to computing with words. From manipulation of measurements to manipulation of perceptions. IEEE Transactions on Circuits and Systems I: Fundamental Theory and Applications 46(1), 105–119 (1999)
20. Zadeh, L.A., Kacprzyk, J. (eds.): Computing With Words in Information/Intelligent Systems 1 & 2. Physica-Verlag, New York (1999)
21. Abbasov, A.M., Fatullayev, A.B.: Forming the set of recognition units for the speech recognition system for the Azerbaijani language. Applied and Computational Mathematics 6(2), 181–191 (2007)
22. Zadrozny, S., Kacprzyk, J.: computing with words for text processing: An approach to the text categorization. Information Sciences 176, 415–437 (2006)
23. Havens, C.T., Kellen, M.J.: Popescn ontological Self-Organizing Map. IEEE Transaction of Fuzzy Systems 18(3) (2010)
24. Zadeh, L.A.: Toward a generalized theory of uncertainty (GTU) – an outline. Information Sciences 172(1-2), 1–40 (2005)
25. Barabasi, A.L., Bonabeau, E.: Scale-Free Networks. Scientific American, 50–59 (May 2003)
26. Barabasi, A.-L.: Scale-Free Networks: A Decade and Beyond. Science, 412–414 (July 24, 2009)
27. Kimble, H.J.: The quantum Internet. Nature 453, 1023–1030 (2008)
28. Zadeh, L.A.: Fuzzy logic and approximate reasoning. Synthese 30, 407–428 (1975)
29. Zadeh, L.A.: A theory of approximate reasoning. In: Hayes, J., Michie, D., Mikulich, L.I. (eds.) Machine Intelligence, vol. 9, pp. 149–194. Halstead Press, New York (1979)
30. Yager, R.R., Ovchinnikov, S., Tong, R., Nguyen, H.: Fuzzy Sets and Applications: Selected Papers by L. A. Zadeh. John Wiley & Sons, New York (1987)
31. Zadeh, L.A.: Fuzzy logic = computing with words. IEEE Transactions on Fuzzy Systems 4, 103–111 (1996)
32. Zadeh, L.A.: Outline of a computational theory of perceptions based on computing with words. In: Sinha, N.K., Gupta, M.M. (eds.) Soft Computing and Intelligent Systems, pp. 3–22. Academic Press, Boston (1999)
33. Zadeh, L.A.: A new direction in AI - toward a computational theory of perceptions. AI Magazine 22(1), 73–84 (2001)
34. Zadeh, L.A.: Toward a logic of perceptions based on fuzzy logic. In: Novak, W., Perfilieva, I. (eds.) Discovering the World with Fuzzy Logic, pp. 4–28. Physica-Verlag, Heidelberg (2001)
35. Zadeh, L.A.: Toward a perception-based theory of probabilistic reasoning with imprecise probabilities. Journal of Statistical Planning and Inference 105, 233–264 (2002)
36. Zadeh, L.A.: A note on Z-numbers. Information Science 181, 2923–2932 (2011)

37. Yager, R.R.: On Z-valuations using Zadeh's Z-numbers. International Journal of Intelligent Systems 27, 259–278 (2012)
38. Yager, R.R.: A new approach to the summarization of data. Information Sciences 28, 69–86 (1982)
39. Yager, R.R.: On linguistic summaries of data. In: Proceedings of IJCAI Workshop on Knowledge Discovery in Databases, Detroit, pp. 378–389 (1989)
40. Yager, R.R.: Intelligent social network analysis using granular computing. International Journal of Intelligent Systems 23, 1197–1210 (2008)
41. Yager, R.R.: Concept representation and database structures in fuzzy social relational networks. IEEE Transactions on Systems, Man and Cybernetics: Part A 40, 413–419 (2010)

Using Alternative Contexts in Concept Hierarchies to Inspire Creativity

Frederick E. Petry[1] and Ronald R. Yager[2]

[1] MC&G Code 7440.5
Naval Research Laboratory
Stennis Space Center MS 39529 USA
fpetry@nrlssc.navy.mil
[2] Machine Intelligence Institute
Iona College
New Rochelle NY USA
yager@panix.com

Abstract. In this paper we examine issues involving measures of creativity for data generalization using hierarchies. In particular we consider consensus and specificity measures for the partitions that result using crisp concept hierarchies. We note that fuzzy hierarchies do not produce partitions of data in general so some approaches to considering "partitionness' is described.

Keywords: creativity, concept hierarchies, congurence, specificity, partitions.

1 Introduction

Often it is said that the essential aspect of human intelligence is creativity. It is very difficult to interpret what is meant by creativity as the term is used in domains ranging from the arts to the sciences. However since the area of artificial intelligence would strive ultimately to approach human intelligence it becomes essential to try to introduce some qualitative / quantitative measures of creativity [1]. We have previously considered a framework utilizing a knowledge base with the components: of multiple concept hierarchies, inference rules and scenarios [2]. This was intended to assist in the reformulation of data to aid analysts in intelligence assessments. Multiple hierarchies were used to generalize input data and develop common concepts that may be found in the data. Underlying hierarchies are similarity relations that provide contexts for semantic interpretations. With these multiple hierarchies the system can emulate the creative process of viewing information from various perspectives.

In this paper we wish to examine some aspects of creativity as related to data generalization. Specifically we consider the technique of attribute-oriented induction of data which we have studied in previous work [3]. Data generalization provides more granular categories that enhance creative discovery. We examine and assess some measures of the creative, especially, novel aspects of information that is generated by the generalization process **G** and assess the value of different hierarchies.

R.R. Yager et al. (Eds.): *Soft Computing: State of the Art Theory*, STUDFUZZ 291, pp. 13–23.
DOI: 10.1007/978-3-642-34922-5_2

2 Background

2.1 Creativity

Generalization construed broadly is a central facet of intelligent behavior, an inductive process going from the specific to the general. Here we focus on a data generalization process **G** as found in the data mining area. Relevant concept hierarchies are used to reduce the specific data in a database into small set of general concepts by an induction process.

There have been a number of approaches to evaluating machine creativity and we discuss here some aspects relevant to data generalization [4]. Usually it is desired to use domain independent criteria to be as broadly applicable as possible. A creative act can be thought of in two stages – generation and evaluation. The basis for the evaluation of creativity can be viewed as an assessment of the output of a generation process after factoring out the input to the process.

The input to the process can be considered as the implicit and explicit knowledge termed the inspiring set I by Ritchie [5] If we denote by R the results of the generation, then the items to be considered as creative must lie in R/I, i.e. R-I. For the data generalization process **G** we are considering that $I = D \cup H_i$, where D is the data in the database and $H_i \in \{H_1, H_2, \ldots H_n\}$ is one hierarchy of the set of hierarchies that may be used for generalization. $R_i = \mathbf{G}(I)$ therefore is the result of the generalization process on D using H_i. We will consider results in which the data and hierarchies may be either crisp or fuzzy.

Often it may become difficult to exactly specify the input I so strong and weak versions of I have been introduced [4]. I_S contains those values specifically known to the generalization process **G**, so a creative item must be completely new. Often the influence of other information on the process is difficult to quantify so I_W is introduced, containing items that are known to have influenced the generalization. Since this information may be difficult to identify exactly, it may be desirable to consider I_W as a fuzzy set. In this paper we consider the input to **G**, $I = D \cup H_i$, as a strong I_S. What we will investigate in this paper are approaches to the assessment of creativity by measuring the novelty of some given results in R.

2.2 Generalization

Generalization is a broad concept that has been used in several contexts. One is the idea of data summarization, a process of grouping of data, enabling transformation of similar item sets, stored originally in a database at the low (primitive) level, into more abstract conceptual representations. Summarization of data is typically performed with utilization of concept hierarchies [6,7], which in ordinary databases are considered to be a part of background knowledge In fuzzy set theory an important consideration is the treatment of data from a linguistic viewpoint. From this an approach has been developed that uses linguistically quantified propositions to summarize the content of a database, by providing a general characterization of the analyzed data [8-11]. There have also been several approaches to the use of fuzzy hierarchies for data generalization [12-14]. Fuzzy gradual rules for data summarization have also been considered [15].

In a previous research effort [3] we developed an approach to data summarization that involves aspects of generalization and compression. The use of concept hierarchies, ontologies, to provide categories to be utilized in this process has been well established [16]. At a meta-level this summarization task can be seen as being driven by the following imperative:

Using a given concept hierarchy, provide a valid, succinct description of the contents of the database which accounts for most of the data in a manner that is useful to the user.

Now consider an example of data generalization letting D= {Oakland, San Jose, …., Sacramento} be a set of cities. However for a particular application, this data may be at too low a level, i.e. too specific.

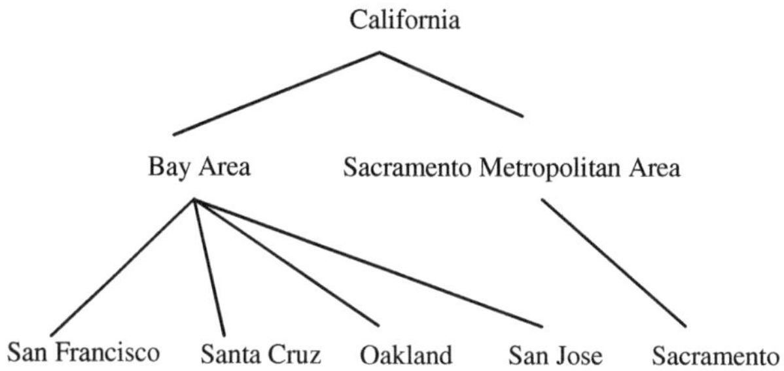

Fig. 1. Example Concept Hierarchy for Cities in California

Figure 1 illustrates part of a concept hierarchy H_1 for an attribute Location, describing US cities based on the geographical location. This concept hierarchy represents some of the domain background knowledge we have a priori.

By ascending the hierarchy, for the attribute Location in the set D, the values San_ Francisco, Santa_Cruz, Oakland, and San_Jose are generalized to the higher level category Bay_Area, while the value Sacramento is generalized to Sacramento_ Metropolitan_Area. Thus R_1 = G (D, H_1) = {Bay_Area, Sacramento_ Metropolitan_ Area. }.

As we have discussed depending on a semantic context there may be other hierarchy for the data being generalized. These may represent another application for the data or another context that is desired to be related to the original one. For the domain of cities we have discussed, another context might be the classification of the city based on population compared to the geographical context of Figure 1. This is illustrated by H_2 below in Figure 2.

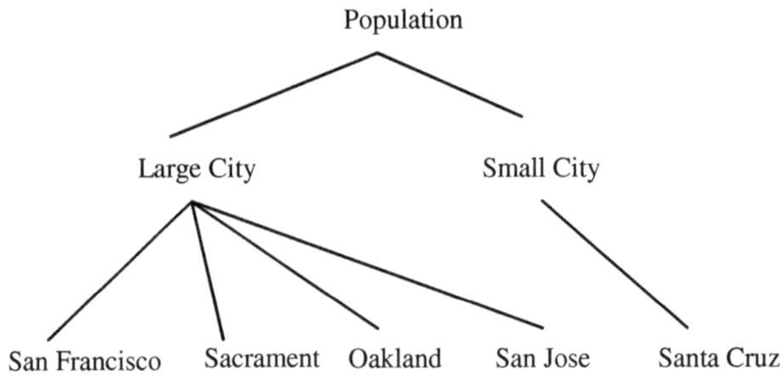

Fig. 2. Concept Hierarchy Based on Population Size

2.3 Hierarchies and Data Partitions

We discuss concept hierarchies associated with a specific attribute. Let A be an attribute and let D (A) be the domain of possible data values of A. A concept hierarchy consists of a number of levels each of which is a partitioning of the space D (A). Furthermore this partitioning becomes coarser and coarser as we go up the hierarchy. The lowest possible level of a hierarchy consists of a partitioning by the individual elements of D (A) and the highest level possible is the whole domain D (A).

Formally each level of a crisp concept hierarchy H is an equivalence relationship. Thus at level k of the concept hierarchy we have a relationship, L_k: D(A) $\times$D(A) $\rightarrow$ {0, 1}, with properties:

1. Reflexive: $L_k(x, x) = 1$, 2. Symmetric: $L_k(x, y) = L_k(y, x)$ and 3. Transitive: If $L_k(x, y) = L_k(y, z) = 1$ then $L_k(x, z) = 1$ for $x,y,z \in D(A)$.

The semantics of this relationship is that $L_k(x, y) = 1$ indicates these two elements, x and y, are essentially the same.

As is well known such an equivalence relationship partitions the space D(A) into n_k disjoint subsets of D(A). These subsets we denote by $E_{k|i}$, the ith equivalence class for the partition of level k. So for x, y $\in E_{k|i}$ we have $L_k(x, y) = 1$.

The increased coarseness of partitioning as we ascend the concept hierarchy is reflected in the requirement that if k > j then for all pairs x and y we have $L_k(x, y) \geq L_j(x, y)$. Essentially this requires that if x and y are in the same class for level j of the hierarchy, they are in the same class in any higher-level k. This implies that if k > j then for any equivalence class $E_{j|h}$ at level j there exists an equivalence class $E_{k|i}$ at level k such that $E_{j|h} \subseteq E_{k|i}$.

So at each level k, the concept hierarchy is a partition of the set of possible data values D(A) into n_k categories (equivalence classes) :

$$E_{k|1}, E_{k|2}, \ldots, E_{k|n_k}$$

If we have m levels then the concept hierarchy is a collection of m partitions of the space D(A). In particular the concept hierarchy consists of

Partition 1:. $E_{1|i}$ for $i = 1$ to n_1

Partition 2:. $E_{2|i}$ for $i = 1$ to n_2

Partition m:. $E_{m|i}$ for $i = 1$ to n_m

We should note that while formally each category $E_{k|i}$ corresponds to a subset of the data space D(A), typically the category has an associated name $C_{k|i}$ which essentially describes the elements in $E_{k|i}$. In general we make no distinction between these two uses of $E_{k|i}$, as a subset of D(A) and as a denotation of the subset. However when actually generalizing data for a given attribute in a tuple, we use the representative name, $C_{k|i}$, in the generalized tuple.

3 Creativity Evaluation Approaches

In discussion of assessment of the creativity of data generalization based on the input I we want to consider all possible results R_1, R_2, …over the set of hierarchies $\{H_1, H_2, \ldots H_n\}$ There may be a number of motivations to consider the generalization results based on the semantic context of various hierarchies. However one major motivation that might be considered is novelty of the results from various hierarchies. As previously discussed we know that one viewpoint of the generalization results R_i are the partitions of the input data D induced by the hierarchies. It is a comparison and characterization of these partitions we consider next.

3.1 Consensus of Partitions

One aspect of the novelty of a given data generalization can be based on the how different the original data generalized from different hierarchies appears to be. We consider the idea of a consensus of generalized data that has been introduced by Yager [18] in terms of the concept of congruence.

One approach is to introduce a measure of similarity, congruence, between two partitions using the underlying equivalence relations. Here we now consider formulating a congruence measure from the perspective of the partitions themselves.

Assume we have two partitions of the set D,

$$P_1 = A_1, \ldots, A_q$$

$$P_2 = B_1, \ldots, B_p$$

where $D = \bigcup_{j=1}^{q} A_j$ and $A_i \cap A_j = \varnothing$ for $i \neq j$ and $D = \bigcup_{j=1}^{p} B_j$ and $B_i \cap B_j = \varnothing$ for $i \neq j$.

Without loss of generality we shall assume $q = p$. If $q > p$ we can augment the partition P_2 by adding $q - p$ subsets, $B_{P+1} = B_{P+2} = ...= B_q = \varnothing$. Thus in the following we assume the two partitions have the same number of classes, q.

We now introduce an operation called a pairing of P_1 and P_2, denoted $g(P_1, P_2)$, which associates with each subset A_i of P_1 a unique partner B_j from P_2. Formally if $Q = \{1, 2, ..., q\}$ then a pairing is a mapping $g: Q \rightarrow Q$ that is bijective, one to one and onto. Essentially g is a permutation of Q. We then have that a pairing $g(P_1, P_2)$ is a collection of q pairs, $(A_j, B_{g(j)})$.

We shall now associate with each pairing a score, $Score(g(P_1, P_2))$, defined as follows. Denoting $C_{g.j} = A_j \cap B_{g(j)}$ for j= 1 to q we obtain

$$Score(g(P_1, P_2)) = (\sum_{j=1}^{q} Card\ (C_{g.j}\ ))/\ Card\ (D)$$

Example: Now we consider the partitions induced by the hierarchies in Figures 1 and 2 for which D = [San Francisco, Santa Cruz, Oakland, San Jose, Sacramento}. For the first hierarchy H_1 we have P_1 consisting of: A_1 = [San Francisco, Santa Cruz, Oakland, San Jose}, A_2 = {Sacramento}. Then for the second hierarchy H_2 the partition P_2 is B_1 = [San Francisco, Sacramento, Oakland, San Jose}, and B_2 = {Santa Cruz}. In this case there are two pairings.

One pairing is $g(j) = j$ in which case we get the pairs (A_1, B_1), (A_2, B_2). From this

$C_{g.1} = A_1 \cap B_1$ = { San Francisco, Oakland, San Jose }

$C_{g.2} = A_2 \cap B_2 = \varnothing$

In this case $Score(g(P_1, P_2)) = 3/5$.

The other pairing is $g(1) = 2$, $g(2) = 1$ and here our pairs are

(A_1, B_2), (A_2, B_1).

In this case

$C_{g.1} = A_1 \cap B_2$ = {Santa Cruz}

$C_{g.2} = A_2 \cap B_1$ = {Sacramento}

In this case $Score(g(P_1, P_2)) = 2/5$

We now shall use this to obtain a measure of congruence, $Cong_2(P_1, P_2)$. Let G be the set of all pairings, $g \in G$. We define

$$Cong_2(P_1, P_2) = \underset{g \in G}{Max}\ Score(g(P_1, P_2))$$

Thus this measure of congruence is the score of the largest pairing.

We see for any pairing g, $0 \le \sum_{j=1}^{q} \text{Card}(C_{g.j}) \le \text{Card}(D)$. From this it follows that $0 \le \text{Cong}_2(P_1, P_2) \le 1$.

More precisely since for any two partitions we can always find a pairing g in which $\sum_{j=1}^{q} \text{Card}(C_{g.j}) \ge 1$ we see that

$$\frac{1}{\text{Card}(D)} \le \text{Cong}_2(P_1, P_2) \le 1$$

So this measure allows us to compare partitions produced by generalization using different hierarchies. Thus we can say the generalization result R based on one hierarchy versus another is more or less novel. This approach can be used for a simple form of a fuzzy database using a membership for each tuple in the relation. When this data is generalized the fuzzy membership affects only the count in the generalized data [17]. The actual partitions in this case still remain crisp set and so these approaches apply.

3.2 Specificity of Partitions

The concepts of specificity and its complement non-specificity have been investigated in considerable detail particularly by Yager [19] and Klir [20]. It is closely related to the idea of generality.

Let $A \subseteq D$ and $|D| = n$. Based on Yager's [21,22] work on specificity we can use the following definition for non–specificity

$$NS(A) = \frac{|A| - 1}{n - 1}$$

This measure takes values in the unit interval. It attains its maximal value of one for the case when A = D and its minimum value of zero for the case when A is a singleton.

So this measure of non-specificity can be used to provide a measure of generality. The larger the set A (the more non-specific), the more general the concept the set is representing.

Next consider the extension of the concept of non-specificity from sets to partitions. We are interested in quantifying how general the classes comprising a partitioning are. Assume P: $\langle A_1, ..., A_q \rangle$ is a partition of D arising from a generalization using some hierarchy H_i. The calculation of the non-specificity of P, NS(P), will use the measure of non-specificity of the individual classes in P:

$$NS(P) = \sum_{i=1}^{q} (|A_i| / n)\, NS(A_i)$$

This is a weighted average of the non-specificities of the component classes in P. The weights are determined by the number of elements in the class.

This definition is independent of the indexing of the classes, each class is treated in the same manner. We also note that it is idempotent, if for all i, $NS(A_i) = a$ then $NS(P) = a$. Thus if all the classes have the same degree of non-specificity then this is the non-specificity of the partition as a whole. We also note that since each of $NS(A_i) \in [0, 1]$ then $NS(P) \in [0, 1]$. Let us look at some notable special cases. Consider the case where $P = P^*$, that is $q = 1$ and $A_1 = D$. We see that $NS(A_1) = 1$ and $|A_1| = n$ and therefore $NS(P^*) = 1$. Thus the non–specificity of a partition consisting of just the whole space is one. At the opposite extreme is the case when the classes are just singletons P_*, here we have n classes with $A_i = \{d_i\}$. In the case $NS(A_i) = 0$ and hence from the idempotency we get $NS(P) = 0$.

The measure of specificity provides another way to compare and evaluate generalization by multiple hierarchies. So the non-specificity measure provides a way of expressing the novelty of the possible generalization results. Consider the partitions P_1 and P_2 used in the example of congruence. Their sets have the same cardinality and we have $|A_1| = 4$, $|A_2| = 1$. So

$$NS(P_1) = NS(P_2) = [\,(4/4) * (3/4) + (1/4) * 0\,] = 3/4.$$

Now if we have another hierarchy H_3 categorizing the cities based on some new criteria e.g. nearness to ocean, that leads to a new partition P_3 = {San Francisco, Santa Cruz}; {San Jose, Oakland, Sacramento}. For this partition $|A_1| = 2$, $|A_2| = 3$. So

$$NS(P_3) = [\,(2/4) * (1/4) + (3/4) * (2/4)\,] = 1/2.$$

Thus we can say that this partition is more general and represents a more novel generalization of the data.

4 Generalization in Fuzzy Hierarchies

Fuzzy hierarchies enable the expression of partial ISA relationships with membership values as fraction numbers between two incident concept nodes [23]. For fuzzy hierarchies, a concept is regarded as a partial specification of its upper concept with the corresponding membership degree μ in the [0, 1] interval. If $\mu = 1$, there is a complete specification as in crisp concept hierarchies

Now we address how generalization proceeds in a fuzzy hierarchy. Figure 3 is a fuzzy hierarchy FH_1 corresponding to Figure 1 which we used before in the generalization example. So Santa Cruz may be considered as being in the Bay Area only to some degree as it is over a coastal range on the coast from cities that are more "certainly" in the Bay Area. The other point to note is that a concept may be related to more than one higher level concept to some degree as illustrated by Sacramento.

To reflect the different membership values when terms are generalized we require the result to be a fuzzy set. A straight forward approach is to formulate the memberships of the higher level concepts in the tree for the result as the average of the memberships of the lower level concepts being generalized. For example let D= {San Francisco, San Jose, Sacramento}. The generalization of this data is then R = {((1.0 + 0.9 + 0.3)/3) / Bay Area, ((1.0/1.0))/ Sacramento } = { 0.73 / Bay Area , 1.0 / Sacramento}. Note that this is consistent with the crisp hierarchy where the memberships are implicitly one.

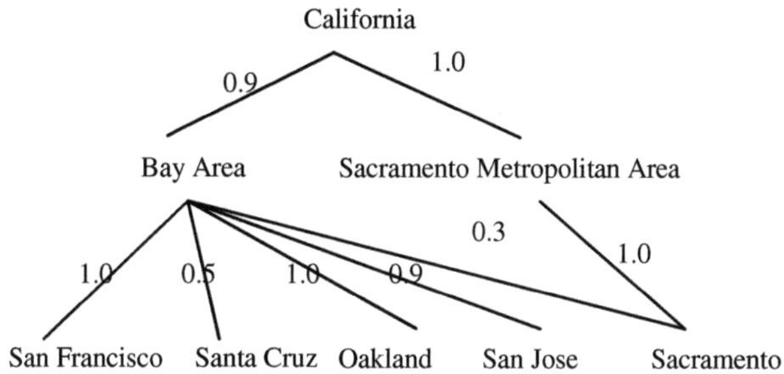

Fig. 3. Fuzzy Concept Hierarchy

Next we need to describe a fuzzy hierarchy as was done for a crisp concept hierarchy in terms of partitions. For a fuzzy hierarchy at each level k we have a defining fuzzy relationship

$$Z_k: X \times X \rightarrow [0, 1]$$

Such a fuzzy relationship naturally leads to fuzzy equivalence classes [13] of concepts at each level k of the corresponding hierarchy. However the sets of these fuzzy equivalence classes are fuzzy sets and as a consequence do not form a formal set partitioning of a domain as for crisp equivalence classes. So rather than a partitioning of the domain X, we have a set decomposition. At each level this is $F_{k|1}, F_{k|2}, ..$ where in general

$$F_{k|i} \cap F_{k|j} \neq \emptyset$$

This implies that there may not be a unique concept at level k to which a value at level k-1 generalizes. Graphically we can illustrate this in Figure 4

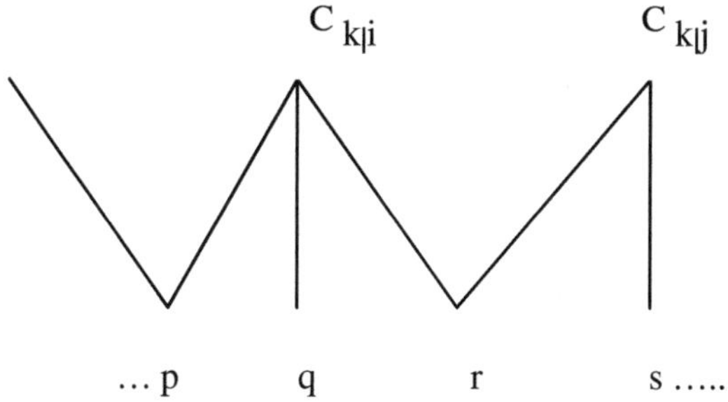

Fig. 4. Fuzzy Hierarchy Notation Example

So $F_{k|i}$ corresponding to $C_{k|i}$ is the set { ...p,q,r,...} and $F_{k|j}$ representing $C_{k|j}$ is { ...r,s,t...}. An overlapping value such as r has a degree of membership $\mu_{k|i}(r)$ in $F_{k|i}$ and $\mu_{k|j}$ (r) in $F_{k|j}$ as determined by Z_k.

Clearly our previous approach to assessing the result of the generalization cannot apply directly here as we do not have partitions. Wee have been considering some general measures of categorization [24] that could be applied. Consider a collection **F** of fuzzy sets, **F** = $\{A_1, A_2, \ldots A_N\}$ over a domain D. We are now interested in how closely the set of fuzzy subsets in **F** resembles a partition of D. Two issues are the coverage of the domain X and the degree of overlapping of the subsets. It is this latter case of the set overlapping we want to represent. One approach we have utilized is to consider for each $d \in D$ how much d is "spread out" over the sets of F. So let

$$\Delta d_d = [\max_{i=1,\ldots N} \mu_{A_i}(d)_j - \text{2nd Largest}_{j=1,\ldots N;\, j\neq i} \mu_{A_j}(d)_j]$$

Clearly if for some k, $d \in A_k$ only, $\Delta_d = 1$ since d being just one fuzzy subset is least "spread out." $\Delta_d < 1$ implies that d is spread over multiple A_i 's. There are other more involved measures that we are considering but for this paper the above is illustrative of the approach. Measures such as these could combine with the previous measures for partitions to be applied to the results of generalizations from fuzzy hierarchies.

5 Conclusion

We have discussed creativity measures that can be used for the results of data generalization. The data partitions that result from the generalization using different hierarchies are compared to assess their novelty using the congruence and specificity measures.

Acknowledgment. We would like to thank the Naval Research Laboratory's Base Program, Program Element No. 0602435N for sponsoring this research.

References

[1] Boden, M.: The Creative Mind: Myths and Mechanisms. Weidenfield and Nicholson, London (1990)

[2] Petry, F., Yager, R.: A Framework for Detecting Impending Critical Scenarios Using the Enhancement of Intelligence. Int. Jour. of Computational Intelligence Research 3(4), 287–295 (2007)

[3] Yager, R., Petry, F.: A Multicriteria Approach to Data Summarization Using Concept Ontologies. IEEE Trans. on Fuzzy Syst. 14(6), 767–780 (2006)

[4] Pease, A., Winterstein, D., Colton, S.: Evaluating Machine Creativiity. In: Proc. of ICCBR 2001 – Workshop on Creative Systems, Vancouver, CA, pp. 56–61 (2001)

[5] Ritchie, G.: Assessing Creativity. In: Proc. of AISB 2001 Symp. on AI and Creativity in Arts and Science, pp. 3–11 (2001)

[6] Han, J., Cai, Y., Cercone, N.: Knowledge discovery in databases: An attribute-oriented approach. In: Proceedings of 18th VLDB Conf., pp. 547–559 (1992)
[7] Han, J.: Mining Knowledge at Multiple Concept Levels. In: Proc. 4th Int'l Conf. on Information and Knowledge Management, pp. 19–24. ACM Press, New York (1995)
[8] Yager, R.: On linguistic summaries of data. In: Piatesky-Shapiro, G., Frawley (eds.) Knowledge Discovery in Databases, pp. 347–363. MIT Press, Boston (1991)
[9] Kacprzyk, J.: Fuzzy logic for linguistic summarization of databases. In: Proc. 8th Int'l Conf. on Fuzzy Systems, Seoul, Korea, pp. 813–818 (1999)
[10] Dubois, D., Prade, H.: Fuzzy sets in data summaries - outline of a new approach. In: Proc. 8th Int'l Conf. on Information Processing and Management of Uncertainty in Knowledge-Based Systems, Madrid, pp. 1035–1040 (2000)
[11] Feng, L., Dillon, T.: Using Fuzzy Linguistic Representations to Provide Explanatory Semantics for Data Warehouses. IEEE Transactions on Knowledge and Data Engineering 15(1), 86–102 (2003)
[12] Lee, D., Kim, M.: Database summarization using fuzzy ISA hierarchies. IEEE Transactions on Systems, Man, and Cybernetics - Part B 27(1), 68–78 (1997)
[13] Raschia, G., Mouaddib, N.: SAINTETIQ:a fuzzy set-based approach to database summarization. Fuzzy Sets and Systems 129, 37–162 (2002)
[14] Angryk, R., Petry, F.: Data Mining Fuzzy Databases Using Attribute-Oriented Generalization. In: Proc. IEEE Int. Conf. Data Mining Workshop on Foundations and New Directions in Data Mining, Melbourne, FL, pp. 8–15 (2003)
[15] Cubero, J., Medina, J., Pons, O., Vila, M.: Data Summarization in Relational Databases Through Fuzzy Dependencies. Information Sciences 121(3-4), 233–270 (1999)
[16] Han, J., Kamber, M.: Data Mining: Concepts and Techniques, 2nd edn. Morgan Kaufmann, San Francisco (2006)
[17] Petry, F., Zhao, L.: Data Mining by Attribute Generalization with Fuzzy Hierarchies in Fuzzy Databases. Fuzzy Sets and Systems 160(15), 2206–2223 (2009)
[18] Yager, R.: Some Measures Relating Partitions Useful for Computatuional Intellgience. Int. Jour. Computational Intelligence Systems 1(1), 1–18 (2008)
[19] Yager, R.R.: On Measures of Specificity. In: Kaynak, O., Zadeh, L.A., Turksen, B., Rudas, I.J. (eds.) Computational Intelligence: Soft Computing and Fuzzy-Neuro Integration with Applications, pp. 94–113. Springer, Berlin (1998)
[20] Klir, G.J.: Uncertainty and Information. John Wiley, New York (2006)
[21] Yager, R.R.: Entropy and specificity in a mathematical theory of evidence. Int. J. of General Systems 9, 249–260 (1983)
[22] Yager, R.R.: Specificity measures of possibility distributions. In: Proceedings of the Tenth NAFIPS Meeting. U. of Missouri, Columbia, pp. 240–241 (1991)
[23] Angryk, R., Petry, F.: Consistent Fuzzy Concept Hierarchies for Attribute Generalization. In: Proc. IASTED Int. Conf on Information and Knowledge Sharing, Scottsdale AZ, pp. 158–163 (2003)
[24] Petry, F., Yager, R.: A Framework for Use of Imprecise Categorization in Developing Intelligent Systems. IEEE Trans. on Fuzzy Systems 18(2), 348–361 (2010)

The Essence of Fuzzy Set Qualitative Comparative Analysis (fsQCA)

Jerry M. Mendel

Signal & Image Processing Institute, Ming Hsieh Department of Electrical Engineering, University of Southern California, Los Angeles, CA 90089-2564

Abstract. Fuzzy Set Qualitative Comparative Analysis (fsQCA) is a methodology for obtaining linguistic summarizations from data that are associated with cases. It was developed by the eminent social scientist Prof. Charles C. Ragin, but has, as of this date, not been applied by engineers or computer scientists. Unlike more quantitative methods that are based on correlation, fsQCA seeks to establish logical connections between combinations of causal conditions and an outcome, the result being rules that summarize (describe) the sufficiency between subsets of all of the possible combinations of the causal conditions (or their complements) and the outcome. The rules are connected by the word OR to the output. Each rule is a possible path from the causal conditions to the outcome. This chapter provides a high-level introduction to fsQCA.

1 Introduction

Fuzzy Set Qualitative Comparative Analysis (fsQCA)[1] is a methodology for obtaining linguistic summarizations from data that are associated with cases. It was developed by the eminent social scientist Prof. Charles C. Ragin, but has, as of this date, not been applied by engineers or computer scientists. Unlike more quantitative methods that are based on correlation, fsQCA seeks to establish logical connections between combinations of causal conditions and an outcome, the result being rules that summarize (describe) the sufficiency between subsets of all of the possible combinations of the causal conditions (or their complements) and the outcome. The rules are connected by the word OR to the output. Each rule is a possible path from the causal conditions to the outcome, i.e. different causal combinations leading to the same outcome.

According to Ragin [20, p. 183)]: "The goal of Qualitative Comparative Analysis (QCA) is to derive a logically simplified statement describing the different combinations of conditions linked to an outcome." Each combination of conditions and same outcome is sometimes referred to as a *type* or a *typological configuration* [1] According to Rihoux and Ragin [23, p. 33 and p. 66]:

Crisp set Qualitative Comparative Analysis (csQCA) was the first QCA technique, developed in the late 1980s, by Professor Charles Ragin[2] and programmer Kriss

[1] This chapter is based on [11].

[2] He is now a professor of sociology and political science at the University of Arizona. In the 1980's he was a professor of sociology and political science at Northwestern University.

R.R. Yager et al. (Eds.): *Soft Computing: State of the Art Theory*, STUDFUZZ 291, pp. 25–37.
DOI: 10.1007/978-3-642-34922-5_3

Drass. Ragin's research in the field of historical sociology led him to search for tools for the treatment of complex sets of binary data that did not exist in the mainstream statistics literature. He adapted for his own research, with the help of Drass, Boolean algorithms that had been developed in the 1950s by electrical engineers to simplify switching circuits, most notably Quine[3] [17] and McCluskey [9]. In these so-called minimization algorithms, he had found an instrument for identifying patterns of multiple-conjunctural causation and a tool to "simplify complex data structures in a logical and holistic manner [18, p. viii]. … csQCA is based on Boolean algebra, which uses only binary data (0 or 1), and is based on a few simple logical operations[4] [union, intersection and complement]. … [In csQCA,] it is important to follow a sequence of steps, from the construction of a binary data table to the final 'minimal formulas.' … Two key challenges in this sequence, before running the minimization procedure, are: (1) implementing a useful and meaningful dichotomization of each variable, and (2) obtaining a 'truth table' (table of configuration) that is free of 'contradictory configurations.' … The key csQCA procedure is 'Boolean minimization.'

csQCA was extended by Ragin to fuzzy sets, because he realized that categorizing social science causes and effects as black or white was not realistic. Fuzzy sets let him get around this. According to [23, p. 120]:

fsQCA retains key aspects of the general QCA approach, while allowing the analysis of phenomena that vary by level or degree. … The fsQCA procedure … provides a bridge between fuzzy sets and conventional truth table analysis by constructing a Boolean truth table summarizing the results of multiple fuzzy-set analyses. … Fuzzy membership scores (i.e., the varying degree to which cases belong to sets) combine qualitative and quantitative assessments. … The key set theoretic relation in the study of causal complexity is the *subset relation*; cases can be precisely assessed in terms of their degree of consistency [subsethood] with the subset relation, usually with the goal of establishing that a combination of conditions is sufficient for a given outcome.

Both csQCA and fsQCA are set-theoretic methods. They differ from conventional quantitative variable-based methods (e.g., correlation and regression) in that they [1] "… do not disaggregate cases into independent, analytically separate aspects but instead treat configurations as different types of cases." Additionally, [1] "The basic intuition underlying QCA[5] is that cases are best understood as configurations of attributes resembling overall types and that a comparison across cases can allow the researcher to strip away attributes that are unrelated to the outcome in question."

This chapter provides a high-level introduction to fsQCA. For more quantitative descriptions of fsQCA see [13], [14].

2 The Basic Principles of fsQCA

The basic principles of fsQCA are: conjunctural causation, equifinal causation, limited diversity, causal asymmetry, the sufficiency of a causal combination is not

[3] Actually, Quine is a famous logician and is not an electrical engineer.

[4] Bracketed phrases, inserted by the present author, are meant to clarify quoted materials.

[5] It is quite common to refer to both csQCA and fsQCA as "QCA" letting the context determine which QCA it is. More recently, the phrase Configurational Comparative Methods is used to cover all QCA methods, e.g., [23].

black and white, and the same set of cases do not have to be used for different outcomes or for different objectives for the same outcome. In this section we elaborate on these basic principles.

1. *Conjunctural Causation*: It is usually not just one postulated causal condition that by itself causes a desired outcome. Instead, it is a combination of causal conditions that causes a desired outcome. fsQCA can determine such a combination of causal conditions. Not all of the postulated causal conditions may be in a causal combination that produces a desired outcome. fsQCA can strip away the unneeded causal conditions in each causal combination.
2. *Equifinal Causation*: There can be different combinations of causal conditions that produce a desired outcome. fsQCA can establish which causal combinations do this.
3. *Limited Diversity*: Usually there are not enough cases available to provide instances for each of the possible causal combinations. Substantive knowledge provided by expert(s) during thought experiments— **counterfactual analysis**— supplements case-data.
4. *Causal asymmetry*: Generally there is **causal asymmetry** between fsQCA for a desired outcome and fsQCA for the complement of that outcome. Generally it is not the complements of the causal combinations associated with the desired outcome that are associated with the complement of the desired outcome.
5. *The Sufficiency of a Causal Combination is Not Black and White*: Each winning causal combination is not 100% sufficient to be a cause of the desired outcome. fsQCA computes a fuzzy measure of sufficiency.
6. *The Same Set of Cases do not Have to be Used for Different Outcomes or for Different Objectives for the Same Outcome*: Identify the best possible instances of the phenomenon to be explained and then study those instances (cases) in great depth. Casing is outcome driven, i.e. you can have different choices of cases for different kinds of studies: (a) Study for which there is only one case; (b) Study when there are a set of cases for the same outcome; (c) Study for which there are both[6] negative and positive cases for the same outcome; (d) Study that uses the entire population (such a study seeks generalizations about the population). According to Ragin [21]: "It is wrong to label a study flawed simply because it omits negative cases, for there are many good reasons to study positive cases in isolation from negative cases." Choosing appropriate cases should be done first, and this choice does not have to be done once and for all, i.e. it can be modified during the entire fsQCA procedure.

3 fsQCA Overview

fsQCA begins[7] (Fig. 1) with your *substantive knowledge* (❀) about a problem. You specify a *desired outcome* (✿) (a separate fsQCA is run for each such outcome) and

[6] A *positive case* is one for which the desired outcome is strongly present and a *negative case* is one for which the desired outcome may not be present at all or is weakly present.

[7] This section is taken from [13] and [14].

then choose the *cases* (③) from which you hope to extract new knowledge about the potential causes for that outcome. Next you postulate a *set of k potential causes* (④) that you believe could have, either individually or in various combinations, led to the desired outcome. You might be wrong about postulating a cause and so you protect yourself against this by simultaneously considering each cause and its *complement*.

fsQCA connects the 2^k possible (candidate) causal combinations to the desired outcome as a simple if-then rule, namely "if *this causal combination*, then *the desired outcome*." Each causal combination contains exactly k terms (the causal condition or its complement) connected to each other by AND, to the desired outcome. All 2^k candidate rules are for the same desired outcome and are therefore connected by the word OR (⑤).

fsQCA now uses the *case-based data* to reduce the number of rules from 2^k candidate rules to a much small number of rules, and it simplifies the rules so that they usually contain causal combinations with fewer than k terms (⑥). The latter happens because all of the rules are for the same desired outcome; hence, they can be logically combined using set theory reduction techniques, and by doing this it frequently happens that some or many causal conditions are absorbed (so they disappear from the final causal combination).

There may not be enough cases (Ragin calls this "limited diversity") to provide evidence (or enough evidence) about all 2^k candidate causal combinations, so more substantive knowledge is obtained from domain experts (⑦). This additional substantive knowledge is incorporated into the fsQCA computations (⑧).

At the end of fsQCA one has a small collection of simplified if-then rules (⑨) that provide at least one simplified causal combination for a desired outcome (unless no such rule can be found). It is then possible to connect cases to each rule that are its *best instances* (⑩), and to compute the *coverage* ((11)) of the cases by each rule.

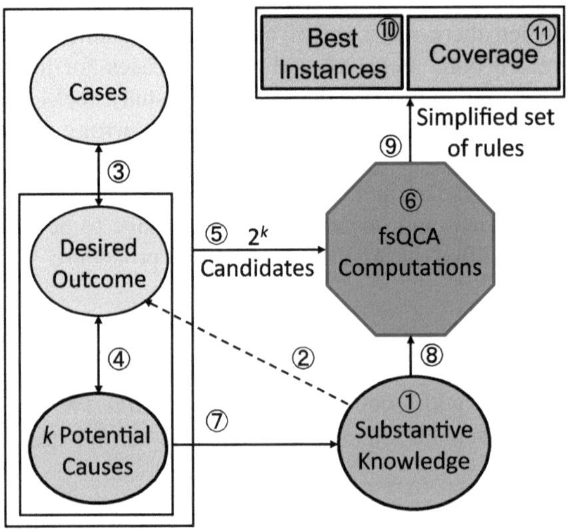

Fig. 1. fsQCA summarized

Fuzzy sets are used in some of the fsQCA steps because things are not always black and white; instead, they are a matter of degree.

The reader needs to be aware that Ragin presented his first ideas about crisp QCA in his 1987 book [18], but after he realized that things are not black and white in social science applications he extended (abandoned) crisp QCA to fsQCA. This was done in two versions. The first version is in his 2000 book [19] and the second version is in his 2008 book [20]. The major difference between the two versions is in the way he computes consistency. In his 2008 book [20, p. 48] he refers to the 2000 approach as "simplistic." It does not use the subsethood formula, but instead uses a crisp counting technique. I mention all of this because if a reader only reads [19] he/she will be implementing an out-of-date version of fsQCA.

4 The Steps of fsQCA

fsQCA seeks to establish logical connections between combinations of causal conditions and a desired outcome, the result being rules (typological configurations) that summarize[8] (describe) the sufficiency between subsets of all of the possible combinations of the causal conditions (or their complements) and the outcome. It is not a methodology that is derived through mathematics, e.g. as the solution to an optimization problem, although it uses mathematics. Our mathematical description of fsQCA does not appear in the existing literatures about fsQCA. It is needed, though, if engineers and computer scientists are to use fsQCA.

It has taken us close to two years to explain fsQCA as a collection of 13 steps[9] that provide one or more subsets of sufficient conditions between a collection of postulated causal conditions and a desired outcome. Why so long? Because Ragin does not explain all of the steps of fsQCA, nor does he quantify the steps in a way that engineers and computer scientists require so that they can implement fsQCA. Instead he provides software that is available on-line that guides a user through fsQCA without the user having to understand the mathematical underpinnings of fsQCA, at:

[http://www.u.arizona.edu/~cragin/fsQCA/software.shtml].

Beginning in October 2009, I was and continue to be in very close e-mail contact with Prof. Ragin about a multitude of fsQCA issues. Understanding all aspects of fsQCA turned out to be akin to peeling an onion. As I went from one step of fsQCA to the next, a new "ring" was uncovered that needed further explaining. Prof. Ragin was extremely patient and generous with his time and always responded to my

---

[8] Ragin does not think of fsQCA as linguistic summarization; he thinks of it as describing what's going on between a collection of causal conditions and an outcome. It is only in [23, page 15, Box 1.4] that "summarizing data" is acknowledged as one of the five types of uses of QCA techniques. Consequently, it now seems legitimate to use fsQCA for linguistic summarization. The other four uses for QCA are: check coherence of data, check hypotheses of existing theories, quickly test conjectures, and develop new theoretical arguments.

[9] To the best of our knowledge, prior to [13] and [14] fsQCA has never before been enumerated as a collection of 13 steps. In fact, some of the steps have never before been explained in the fsQCA literature.

enquiries quickly and in great detail. [10] reprints our e-mail dialogs, organizing them by category and sub-category, so that others may benefit from Prof. Ragin's insights and wisdom about fsQCA.

As mentioned above, we now describe fsQCA as a collection of 13 steps. The first five steps are preparatory to the next six computational steps, after which there are two summary steps. Because of space limitations, we only provide very high-level statements of the 13 steps.

4.1 Preparatory Steps

(1) Choose a desired outcome from the space of all desired outcomes for the application, $\mathcal{S}_O$, and its appropriate cases[10], $\mathcal{S}_{Cases}$, [e.g., *Low MPG* for a subset of automobiles (each auto is a case) that are in the UCI Repository].

(2) Choose k causal conditions, $\mathcal{S}_C$, and, if a condition is described by more than one term, treat each term as an independent causal condition (e.g., *low acceleration*, *high acceleration*, *light weight*, *heavy weight*, etc.).

(3) Treat the desired outcome and causal conditions as fuzzy sets, and determine membership functions (MFs) for all of them (many methods exist for doing this, none of whose details are needed for the rest of this article).

(4) Evaluate these MFs for all available cases, the results being *derived MFs*.

(5) Create 2^k candidate causal combinations (rules), $\mathcal{S}_F$, and view each as a possible corner in a 2^k-dimensional vector space.

4.2 Computational Steps

Steps 6-11 are computational steps and are summarized in Fig. 2, which we have found to be a relatively simple mnemonic way to remember these steps.

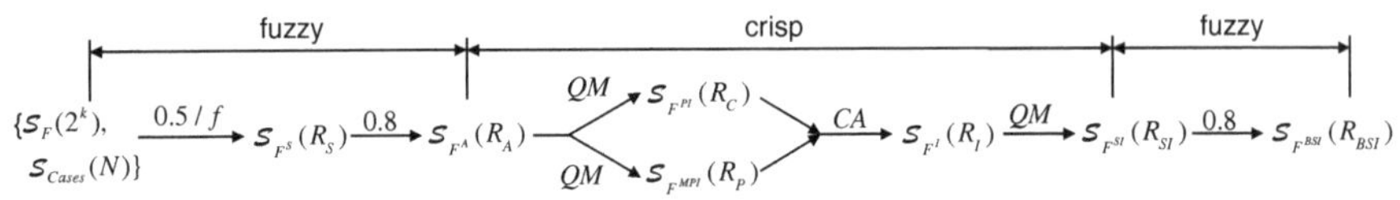

Fig. 2. Mnemonic summary of Steps 6-11 of fsQCA

Steps 6 and 7 use fuzzy sets to allow for uncertainties about the natures of the causal conditions and desired outcome:

---

[10] Cases have no natural ordering, but instead each case is identified by an integer, so that by knowing the integer one also knows the case. The integers x = 1, 2, …, N are used to represent the N cases, and in this way the cases are ordered. For a person to repeat someone else's fsQCAs, and compare their intricate details with someone else's intricate details, they need to know the ordering of the N; hence, it is assumed that this information is provided to them.

(6) Compute the MF of each of the 2^k candidate causal combinations in all of the available cases, and keep only the ones—the R_S surviving causal combinations (firing-level surviving rules), $\mathcal{S}_{F^S}$, —whose MF values are > 0.5, i.e., keep the causal combinations that are closer to corners and not the ones that are farther away from corners.

(7) Compute the consistencies (subsethoods[11]) of these R_S surviving causal combinations, and keep only those R_A causal combinations, $\mathcal{S}_{F^A}$, —the actual causal conditions (actual rules)—whose consistencies are > 0.80.

Steps 8-10 use crisp sets, because the R_A causal combinations are now treated as actual corners, since it has just been established these causal combinations do indeed exist, i.e. existence is treated as crisp. It is quite possible, however, that there are still too many rules, but now for a totally different reason than before. When the R_A actual rules are combined using the logical OR (disjunction) operation, then, because all of these rules share the same consequent there can be a lot of redundancies. Set theory reduction techniques can be used to obtain two kinds of minimal solutions; however, it is usually very difficult to perform these reductions by hand, and so they are automated by using the Quine-McCluskey (QM) Algorithm [17], [9], [16].

(8) Use the QM algorithm to obtain R_C prime implicants[12] (the complex solutions, $\mathcal{S}_{F^{PI}}$) and R_P minimal prime implicants[13] (the parsimonious solutions, $\mathcal{S}_{F^{MPI}}$).

(9) Use substantive knowledge from an expert to perform Counterfactual Analysis[14] (CA) on the complex solutions, constrained by the parsimonious solutions, to obtain the R_I intermediate solutions, $\mathcal{S}_{F^I}$. The complexity of an intermediate solution is supposed to be between complexities of the complex and parsimonious solutions, and, according to Ragin, the intermediate solutions are the most useful ones. Rules for CA are given in [13] and [14].

(10) Perform QM on the R_I intermediate solutions (they are also connected by OR) to obtain the R_{SI} simplified intermediate solutions, $\mathcal{S}_{F^{SI}}$.

---

[11] If a causal combination is sufficient for the desired outcome, then the rule actually exists; however, if it is not then the rule does not exist. So, this calculation of fsQCA establishes whether or not a rule exists. This calculation is a quantification of the fact that a causal combination is *sufficient* for an outcome if the outcome always occurs when the causal combination is present (however, the outcome may also occur when a different causal combination is present), i.e. the causal combination (the antecedents) is a *subset* of the outcome.

[12] A *prime implicant* is a combination of primitive Boolean expressions that differ on only one cause and have the same output.

[13] *Minimal prime implicants* (also called *essential prime implicants* by Ragin [18]) cover as many of the primitive Boolean expressions as possible with a logically minimal number of prime implicants. For an example of how minimal prime implicants are determined from prime implicants, see [18, pp. 95-98]. Other examples can be found in [16].

[14] CA [1], [22] offers a way to overcome the limitations of a lack of empirical instances, i.e. the problem of *limited diversity* [19, pp. 81 ff.], and involves thought experiments in which the substantive knowledge of a domain expert is used.

Step 11-13 uses fuzzy sets because the influence of a causal combination on a desired output is not 100%, i.e. it is fuzzy:

(11) Retain only those simplified intermediate solutions whose consistencies are approximately greater than 0.80, the R_{BSI} believable simplified intermediate solutions, $\mathcal{S}_{F^{BSI}}$.

4.3 Summarization Steps

(12) Connect the R_{BSI} believable simplified intermediate solutions (this can also be done for the complex and parsimonious solutions) with its best instances, $\mathcal{S}_{BeIn}(s)$, $s = 1,\ldots,R_{BSI}$. This is a very important step because it reconnects the fsQCA results back to cases.

(13) Compute the coverage of each solution. Coverage provides a measure of *generality* of a summary because it shows how many cases support the summary. It is an assessment of the degree to which a solution is supported by cases. In other words, coverage determines what percentage of cases covers a solution. Ragin [20, Ch. 3] mentions three kinds of coverage and Rihoux and Ragin [23, p. 64] define them as: (1) *solution coverage*, C_s, which is the proportion of cases that are covered by *all* of the terms; (2) *raw coverage*, C_r, which is the proportion of cases that are *simultaneously* covered by *each* term *one at a time*; and, (3) *unique coverage*, C_u, which is the proportion of cases that are uniquely covered by *a specific term* (no other terms cover those cases). Each measure of coverage provides a different insight into the believable simplified intermediate solutions. We usually only focus on raw coverage and solution coverage. Formulas for all three kinds of coverage are in [14].

The results of fsQCA for each desired outcome O are given by

$$\left\{F_s^{BSI}, \mathcal{S}_{BeIn}(s), \text{Coverage}(s), \text{Subsethood}(s)\right\}_{s=1}^{R_{BSI}}$$

Subsethood(s) is a number that is between 0.80 and 1 (usually, $\text{Subsethood}(s) \neq 1$). Some may say that it provides a measure of causality between a causal combination and the desired outcome.

5 Example

In order to illustrate fsQCA, we present a very simple example that is taken from some of Ragin's books [20, Ch. 7] and [23, Ch. 5]. The example is about the breakdown of democracy of European countries between World Wars 1 and 2, for which the desired outcome is O = *Breakdown of Democracy* and the three causal conditions are A = *developed* (country), B = *urban* (country) and C = *literate* (country). This example is a simplification of a more complete example in which there are two more causal conditions D = *industrial* (country) and E = *stable* (country) that appears in [13].

Table 1. Data- and fuzzy-membership-matrix (showing original variables and their fuzzy-set membership function scores)[a]

Case	Outcome		Condition and MF scores					
	o	*MF(O)*	*A*	*MF(A)*	*B*	*MF(B)*	*C*	*MF(C)*
1	-9	0.95	720	0.81	33.4	0.12	98	0.99
2	10	0.05	1098	0.99	60.5	0.89	94.4	0.98
3	7	0.11	586	0.58	69	0.98	95.9	0.98
4	-6	0.88	468	0.16	28.5	0.07	95	0.98
5	4	0.23	590	0.58	22	0.03	99.1	0.99
6	10	0.05	983	0.98	21.2	0.03	96.2	0.99
7	-9	0.95	795	0.89	56.5	0.79	98	0.99
8	-8	0.94	390	0.04	31.1	0.09	59.2	0.13
9	-1	0.58	424	0.07	36.3	0.16	85	0.88
10	8	0.08	662	0.72	25	0.05	95	0.98
11	-9	0.95	517	0.34	31.4	0.10	72.1	0.41
12	10	0.05	1008	0.98	78.8	1	99.9	0.99
13	-6	0.88	350	0.02	37	0.17	76.9	0.59
14	-9	0.95	320	0.01	15.3	0.02	38	0.01
15	-4	0.79	331	0.01	21.9	0.03	61.8	0.17
16	-8	0.94	367	0.03	43	0.30	55.6	0.09
17	10	0.05	897	0.95	34	0.13	99.9	0.99
18	10	0.05	1038	0.98	74	0.99	99.9	0.99

[a] This table is modeled after Table 5.2 in [23], and the numbers in it are the same as the ones in that table.

Table 2. Fuzzy set membership of cases in causal combinations[a]

Case	Membership in corners of vector space formed by causal conditions: Firing Levels							
	F_1 *abc*	F_2 *abC*	F_3 *aBc*	F_4 *aBC*	F_5 *Abc*	F_6 *AbC*	F_7 *ABc*	F_8 *ABC*
1	0.01	0.19	0.01	0.12	0.12	**0.81**	0.01	0.12
2	0.01	0.01	0.01	0.01	0.01	0.11	0.02	**0.89**
3	0.02	0.02	0.02	0.42	0.02	0.02	0.02	**0.58**
4	0.02	**0.84**	0.02	0.07	0.07	0.16	0.02	0.07
5	0.01	0.42	0.01	0.03	0.03	**0.58**	0.01	0.03
6	0.01	0.02	0.01	0.02	0.02	**0.97**	0.01	0.03
7	0.01	0.11	0.01	0.11	0.11	0.21	0.01	**0.79**
8	**0.87**	0.13	0.09	0.09	0.09	0.04	0.04	0.04
9	0.12	**0.84**	0.12	0.16	0.16	0.07	0.07	0.07
10	0.02	0.28	0.02	0.05	0.05	**0.72**	0.02	0.05
11	**0.59**	0.41	0.10	0.10	0.10	0.34	0.10	0.10
12	0	0	0.01	0.02	0	0	0.01	**0.98**
13	0.41	**0.59**	0.17	0.17	0.17	0.02	0.02	0.02
14	**0.98**	0.01	0.02	0.01	0.02	0.01	0.01	0.01
15	**0.83**	0.17	0.03	0.03	0.03	0.01	0.01	0.01
16	**0.70**	0.09	0.30	0.09	0.30	0.03	0.03	0.03
17	0.01	0.05	0.01	0.05	0.05	**0.87**	0.01	0.13
18	0.01	0.01	0.01	0.02	0.01	0.01	0.01	**0.98**
> 0.5	5	3	0	0	0	5	0	5

[a] This table is modeled after Table 5.6 in [23], and the numbers in it are the same as the ones in that table.

Using knowledge and techniques from social science, numerical values were obtained for *A*, *B* and *C* for 18 European countries that in Table 1 are called "Cases 1–18." Numerical values were initially obtained by Ragin for *o* = *Survival of Democracy*, which was assumed to be the complement of *Breakdown of Democracy*;

hence, $MF(O)$ was computed from $MF(o)$ as $1 - MF(o)$. S-shaped MFs were used for *Survival of Democracy*, *developed* (country), *urban* (country), and *literate* (country) using a method that is described in [20, Ch. 5]. Using these MFs, Ragin obtained the MF scores that are also given in Table 1. These MFs implement Step 4.

From this point on, *A*, *B* and *C* are viewed as generic causal conditions for a generic outcome *O*, because there are more actual causal conditions that are associated with *Breakdown of Democracy* than are shown in Table 1, and because tables for three causal conditions are easy to display.

For three causal conditions there are eight causal combinations, all of which are given in Table 2, along with their memberships. These memberships are the firing levels in Step 6, and are computed by using the minimum for conjunction, i.e.,

$$\mu_{F_1}(x) = \min\{\mu_a(x), \mu_b(x), \mu_c(x)\} = \min\{1 - \mu_A(x), 1 - \mu_B(x), 1 - \mu_C(x)\}.$$

The bold-faced numbers in Table 2 indicate memberships that are greater than 0.5. The numbers of such memberships are listed in the last row of the table for each of the eight causal combinations. Using a frequency threshold of three only four of the eight causal combinations survive, i. e. $R_S = 4$. Those firing-level surviving rules are summarized in Table 3.

Table 3. Distribution of cases across causal conditions and set-theoretic consistency of causal combinations[a]

Best Instances	*Causal Conditions* A	B	C	*Corresponding Vector Space Corner*	*Number of cases with > 0.5 membership*	*Set-theoretic Consistency*
8, 11, 14, 15, 16	0	0	0	*abc*	5	0.98
4, 9, 13	0	0	1	*abC*	3	0.84
1, 5, 6, 10, 17	1	0	1	*AbC*	5	0.44
2, 3, 7, 12, 18	1	1	1	*ABC*	5	0.34

[a] This table is modeled after a combination of Tables 5.7 and 5.8 in [23], and the numbers in it are the same as the ones in Table 5.7.

The first column of Table 3 is called "Best Instances." It lists the cases that are associated with each surviving causal combination. This is a very important column because it directly connects the fsQCA back to the original cases. The next three columns of this table are for the three causal conditions, and their entries are listed as 0 or 1, where a 0 occurs if the complement of the causal condition appears in the causal combination, and a 1 appears if the causal condition appears in the causal combination [e.g., $abC \rightarrow (0,0,1)$]. The next column in this table states the causal combination (the corresponding vector space corner) using set notation (e.g., abC). The last column in this table gives the count (from Table 2) of the number of MF entries that are > 0.5.

Next, the consistencies are computed using Kosko's [8] subsethood formula (Step 7). Note that these calculations use the derived MFs for all 18 cases. Results are summarized in the last column of Table 3.

Using a consistency threshold of 0.80 only two of the four rules become actual rules, i.e. $R_A = 2$. These are the first two rules in Table 3, *abc* and *abC*. Observe that *abC*, which has fewer cases with > 0.5 membership than do *AbC* or *ABC*, survives, whereas *AbC* and *ABC* do not.

The prime implicant (Step 8) for $abc + abC$ is easy to obtain, because $abc + abC = ab(c + C) = ab$. The minimal prime implicant (Step 8), found from the QM algorithm, is *a*. These solutions can be expressed linguistically, as:

$$\begin{cases} \text{Complex solution} & \text{IF } C_1 = a \text{ and } C_2 = b, \text{THEN } O \\ \text{Parsimonious solution} & \text{IF } C_1 = a, \text{THEN } O \end{cases}$$

In words, these solutions are:

Complex solution	*Not developed* and *not urban* (rural) is a sufficient causal combination for *Breakdown of Democracy*
Parsimonious solution	*Not developed* is a sufficient causal condition for *Breakdown of Democracy*

To complete fsQCA, CA has to be performed (Step 9). *Substantive knowledge* (made up by the author, but reasonable) is: The desired outcome could have occurred if *a* (not developed), *b* (not urban (rural)) or *c* (not literate) occurred.

CA is trivial for this example because: (1) the parsimonious solution is contained in the complex solution, and (2) the substantive knowledge is contained in the complex solution. Additionally, since the complexity of an intermediate solution is supposed to be between complexities of the complex and parsimonious solutions, and the former is described by two causal conditions, whereas the latter is described by one causal condition, the intermediate solution must either be the same as the complex or parsimonious solutions. In this case because substantive knowledge indicates the desired outcome could have been caused by the presence of *a* or *b*, the complex solution becomes the intermediate solution, namely *ab*, or, in words:

Not developed and *not urban* (rural) is a sufficient causal combination for *Breakdown of Democracy*.

Generally CA is not so trivial; more examples of how to perform CA, as well as rules for CA are in [13] and [14].

Consistency (Step 10): The set theoretic consistency of *ab* was computed to be 0.837, which is greater than 0.80, so this solution is retained.

Because of page limitations, discussions about how to compute the *Best Instances* and *Coverage* are deferred to [13] and [14]. Those discussions show that Cases 4, 8, 11, 13, 14, and 16 are best instances for *ab*, and that the coverage of *ab* is 0.736.

6 Conclusions

It is quite common these days for people who work in the general field of computational intelligence (CI), which includes fuzzy sets as one of its major pillars (the others being neural networks and evolutionary computing), to inquire how a CI technique can be used to solve problems in interdisciplinary or non-traditional (i.e., non-engineering or non-computer-science) fields. The expectations are there will be a flow from CI into these fields. Rarely, does the flow occur in the other direction. Charles Ragin's fsQCA is one of those remarkable exceptions and represents a flow from social science and political science into CI.

Because of space limitations many aspects of fsQCA that are crucial to its actual use have not been discussed, including: new theoretical results for fsQCA [15], results that could only have been obtained after a formal quantification of fsQCA had occurred, including a totally different way to perform the early steps of fsQCA, one that saves orders of magnitude of computation time; a new method for determining the MFs for causal conditions and outcome from data [7]; the detailed steps of Counterfactual Analysis [13] and [14]; comparisons of fsQCA as a linguistic summarizer [13] with the pioneering works on linguistic summarization by Yager [25]-[28], Kacprzyk and Yager [2], Kacprzyk and Zadrożny, e.g., [3]-[5], and Kacprzyk, Yager and Zadrożny [6], as well as the more recent work by Wu and Mendel [24]. The applications of fsQCA to some engineering and computer science problems will appear in other publications.

Additional works on extending fsQCA from type-1 fuzzy sets to interval type-2 fuzzy sets is underway, because we view fsQCA as a new *Computing With Words Engine* that can be incorporated in a *Perceptual Computer* [12], to aid people in making subjective judgments.

It is also very important to study the robustness of fsQCA to both the frequency cutoff and consistency threshold numbers. Perhaps it is possible to fuzzify both of these numbers and propagate their fuzziness through the fsQCA calculations.

Acknowledgments. The author is very pleased to acknowledge the generous help provided to him by Prof. Charles Ragin in explaining the many ramifications of fsQCA. He would also like to acknowledge help provided by Prof. Peer Fiss, who met with him a number of times to discuss some ramifications of fsQCA; and to his student Mohammad Korjani, who provided the consistency and coverage computations for the example.

References

[1] Fiss, P.C.: Building better causal theories: a fuzzy set approach to typologies in organization research. Academy of Management Journal (2010) (in press)

[2] Kacprzyk, J., Yager, R.R.: Linguistic summaries of data using fuzzy logic. International Journal of General Systems 30, 133–154 (2001)

[3] Kacprzyk, J., Zadrożny, S.: On combining intelligent querying and data mining using fuzzy logic concepts. In: Bordogna, G., Pasi, G. (eds.) Recent Research Issues on the Management of Fuzziness in Databases, pp. 67–81. Physica-Verlag, Heidelberg (2000)

[4] Kacprzyk, J., Zadrożny, S.: CWW in intelligent database querying: standalone and Internet-based applications. Information Sciences 34, 71–109 (2001)

[5] Kacprzyk, J., Zadrożny, S.: Computing with words is an implementable paradigm: fuzzy queries, linguistic data summaries and natural language generation. IEEE Trans. on Fuzzy Systems 18, 461–472 (2010)

[6] Kacprzyk, J., Yager, R.R., Zadrożny, S.: A fuzzy logic based approach to linguistic summaries of databases. International Journal of Applied Mathematics and Computer Science 10, 813–834 (2000)
[7] Korjani, M.M., Mendel, J.M.: Fuzzy set qualitative comparative analysis (fsQCA), Part 2: challenges. Submitted for Presentation at FUZZ-IEEE 2012, Brisbane, AU (2012)
[8] Kosko, B.: Neural Networks and Fuzzy Systems: A Dynamical Systems Approach to Modeling Machine Intelligence. Prentice-Hall, Englewood Cliffs (1992)
[9] McCluskey, E.J.: Introduction to the Theory of Switching Circuits. McGraw-Hill, New York (1966)
[10] Mendel, J.M.: fsQCA: Dialog Between Jerry M. Mendel and Charles C. Ragin, USC-SIPI Report 411 (March 2011), `http://www.compasss.org/pages/resources/emailmendel.pdf`
[11] Mendel, J.M.: On Charles Ragin's fuzzy set qualitative comparative analysis (fsQCA). Presented at the World Conference on Soft Computing, San Francisco (May 2011)
[12] Mendel, J.M., Wu, D.: Perceptual Computing: Aiding People in Making Subjective Judgments. John Wiley and IEEE Press, Hoboken (2010)
[13] Mendel, J.M., Korjani, M.: Charles Ragin's Fuzzy Set Qualitative Comparative Analysis (fsQCA) Used for Linguistic Summarizations (2011) (submitted for publication)
[14] Mendel, J.M., Korjani, M.: Fuzzy set qualitative comparative analysis (fsQCA), part 1: mathematical description. Submitted for Presentation at FUZZ-IEEE 2012, Brisbane, AU (2012)
[15] Mendel, J.M., Korjani, M.: Fast fuzzy set qualitative comparative analysis (fast fsQCA). Submitted for Presentation at NAFIPS 2012, Berkeley, CA (2012)
[16] Mendelson, E.: Schaum's Outline of Theory and Problems on Boolean Algebra and Switching Circuits. McGraw-Hill, New York (1970)
[17] Quine, W.V.: The problem of simplifying truth functions. American Mathematical Monthly 59, 521–531 (1952)
[18] Ragin, C.C.: The Comparative Method: Moving Beyond Qualitative and Quantitative Strategies. University of California Press, Berkeley (1987)
[19] Ragin, C.C.: Fuzzy-set Social Science. Univ. of Chicago Press, Chicago (2000)
[20] Ragin, C.C.: Redesigning Social Inquiry: Fuzzy Sets and Beyond. Univ. of Chicago Press, Chicago (2008)
[21] Ragin, C.C.: Reflections on casing and case-oriented research. In: Byrne, D., Ragin, C. (eds.) The SAGE Handbook of Case-Based Methods, ch. 31, pp. 522–534. SAGE Publications, Inc., Thousand Oaks (2009)
[22] Ragin, C.C., Sonnett, J.: Between complexity and parsimony: limited diversity, counterfactual cases and comparative analysis. In: Kropp, S., Minkenberg, M., (eds.) Vergleichen in der Politikwissenschaft. VS Verlag for Sozialwissencahften, Wiesbaden
[23] Rihoux, B., Ragin, C.C. (eds.): Configurational Comparative Methods: Qualitative Comparative Analysis (QCA) and Related Techniques. SAGE, Los Angeles (2009)
[24] Wu, D., Mendel, J.M.: Linguistic summarizations using IF-THEN rules and interval type-2 fuzzy sets. IEEE Trans. on Fuzzy Systems 19, 136–151 (2011)
[25] Yager, R.R.: A new approach to the summarization of data. Information Sciences 28, 69–86 (1982)
[26] Yager, R.R.: On linguistic summaries of data. In: Piatetsky-Shapiro, G., Frawley, B. (eds.) Knowledge Discovery in Databases, pp. 347–363. MIT Press, Cambridge (1991)
[27] Yager, R.R.: Linguistic summaries as a tool for database discovery. In: Proc. FUZZ-IEEE 1995/IFES 1995, Yokohama, Japan, pp. 79–82 (1995)
[28] Yager, R.R.: Database discovery using fuzzy sets. International Journal of Intelligent Systems 11, 691–712 (1996)

The Kendall Rank Correlation between Intuitionistic Fuzzy Sets: An Extended Analysis

Eulalia Szmidt[1,2], Janusz Kacprzyk[1,2], and Paweł Bujnowski[1]

[1] Systems Research Institute, Polish Academy of Sciences, ul. Newelska 6, 01–447 Warsaw, Poland
{szmidt,kacprzyk,bujnowski}@ibspan.waw.pl

[2] WIT – Warsaw School of Information Technology ul. Newelska 6, 01–447 Warsaw, Poland
{szmidt,kacprzyk}@ibspan.waw.pl

Abstract. This paper is a continuation and extension of our previous works on correlation coefficients between Atanassov's intuitionistic fuzzy sets (A-IFSs, for short), notably on the *Pearson coefficient r* and the *Spearman correlation coefficient* to measure the degree of association between A-IFSs. Here, we develop, and illustrate on examples, the concept of the *Kendall rank correlation* for A-IFSs which is another important measure of correlation.

1 Introduction

While building models of processes and systems, one of the most interesting and relevant aspect is to determine relationships between variables, notably between relevant ones, and then to measure their intensity. These issues have been for a long time a subject of interest and a huge interest effort in many research and scholarly communities, and a wide array of various coefficients have been developed over the years.

A notable example in this line of research is the Pearson correlation coefficient r, i.e., a measure of a linear relationship between the variables, which is one of the most frequently used tools in statistics (cf. Rodgers and Nicewander [18]). This correlation coefficient indicates how well the values of two variables move together in a linear way (Rodgers and Nicewander [18], Aczel [2]). The assumption that is necessary is that the distributions of the variables are normal. When this assumption is not valid or the data are in the form of ranks, we can use some other measures of the degree of association between two variables, notably the *Spearman rank correlation coefficient* r_s (e.g., Aczel [2]) or the *Kendall* τ *correlation coefficient* (Kendall [14]).

Although the *Spearman rank correlation coefficient* is said to be used more often (cf. Griffiths [9]), the *Kendall* τ *correlation coefficient* has an intuitively simple interpretation, and its algebraic structure is simpler than that of the Spearman coefficient (cf. Noether [17]). This has clearly implied that the *Kendall* τ *correlation coefficient* is a common tool that is employed in contemporary research as exemplified by Bracke et al. [7], Benson et al. [6], Barret et al. [5], Kurvers et al. [8], Moller et al. [15], etc.

As Zadeh has observed [43], [44], most of information relevant to probabilistic analyzes is imprecise, and there is imprecision and fuzziness not only in probabilities, but also in events, relations and properties. In this context, the probabilistic concepts should also be extended to fuzzy models and their generalizations.

R.R. Yager et al. (Eds.): *Soft Computing: State of the Art Theory*, STUDFUZZ 291, pp. 39–54.
DOI: 10.1007/978-3-642-34922-5_4

One of the most relevant and widely used extension of the basic concept of Zadeh's fuzzy sets is the concept of Atanassov's intuitionistic fuzzy set (A-IFS) in which, roughly speaking the degree of membership alone, as in Zadeh's fuzzy set, is replaced by a degree of membership, degree of non-membership and hesitation margin, with all summing up to 1. Therefore, a challenge and necessity is to extend properties and relations defined for information represented by means of fuzzy sets to those for information defined by means of the A-IFSs. The analysis of relationships between the A-IFSs is therefore a very important challenge and research task.

In our quest for doing the above, we have already discussed in detail the *Pearson correlation coefficient r* (Szmidt and Kacprzyk [37], and the *Spearman rank correlation coefficient* r_s in (Szmidt and Kacprzyk [38]). This paper is a continuation of our work [39] on the *Kendall* τ *correlation coefficient.*

We have proposed the concept of the *Kendall correlation coefficient* for the A-IFSs which is a generalization of the *Kendall correlation coefficient* for crisp sets as it fulfills the same properties. Moreover, it takes into account all three terms describing the A-IFS, i.e., the membership values, non-membership values, and hesitation margins. We show that each term plays an important role in data analysis and decision making. In this paper we further elaborate upon that proposal.

An important part of this paper is the presentation of a well known benchmark commonly used in the broadly perceived analysis of data, i.e. the Iris data set from the University of California, Irvine repository [46].

2 A Brief Introduction to Intuitionistic Fuzzy Sets

One of the possible generalizations of a fuzzy set in X (Zadeh [42]), given by

$$A' = \{ < x, \mu_{A'}(x) > | x \in X \} \tag{1}$$

where $\mu_{A'}(x) \in [0,1]$ is the membership function of the fuzzy set A', is Atanassov's intuitionistic fuzzy set A (Atanassov [3], [4]), namely:

$$A = \{ < x, \mu_A(x), \nu_A(x) > | x \in X \} \tag{2}$$

where: $\mu_A : X \to [0,1]$ and $\nu_A : X \to [0,1]$ such that $0 \le \mu_A(x) + \nu_A(x) \le 1$, and $\mu_A(x)$, $\nu_A(x) \in [0,1]$ denote the degree of membership and a degree of non-membership of $x \in A$, respectively, and the *hesitation margin* of $x \in A$ is:

$$\pi_A(x) = 1 - \mu_A(x) - \nu_A(x) \tag{3}$$

The $\pi_A(x)$ expresses a lack of knowledge of whether x belongs to A or not (Atanassov [4]); obviously, $0 \le \pi_A(x) \le 1$, for each $x \in X$;

The hesitation margin turns out to be important while considering the distances (Szmidt and Kacprzyk [21], [23], [30], entropy (Szmidt and Kacprzyk [25], [32]), similarity (Szmidt and Kacprzyk [33]) for the A-IFSs, etc. i.e., the measures that play a crucial role in virtually all information processing tasks. The hesitation margin is shown

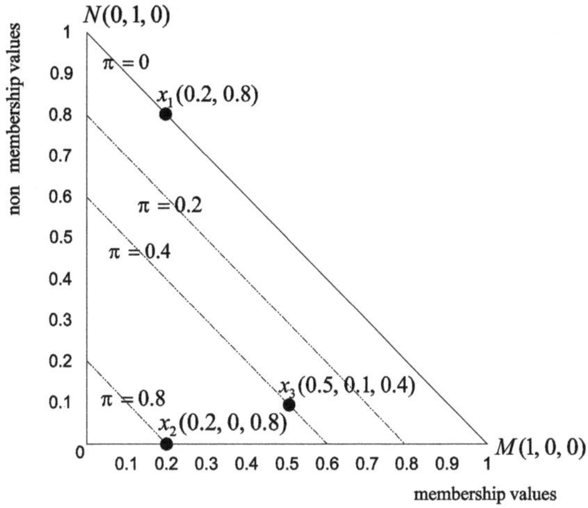

Fig. 1. Geometrical representation

to be indispensable also in the ranking of intuitionistic fuzzy alternatives as it indicates how reliable (sure) information represented by an alternative is (cf. Szmidt and Kacprzyk [34], [35]).

The use of A-IFSs instead of fuzzy sets implies the introduction of additional degrees of freedom (non-memberships and hesitation margins) into the set description. Such a generalization of fuzzy sets gives us an additional possibility to represent imperfect knowledge which may lead to describing many real problems in a more adequate way. This is confirmed by successful applications of A-IFSs to group decision making, negotiations, voting and other situations are presented in Szmidt and Kacprzyk [20], [22], [24], [26], [27], [28], [29], [31], [36], Szmidt and Kukier [40], [41].

2.1 A Geometrical Representation

One of possible geometrical representations of an intuitionistic fuzzy set is given in Fig. 1 (cf. Atanassov [4]). It is worth noticing that although we use a two-dimensional figure (which is more convenient to draw in our further considerations), we still adopt our approach (e.g., Szmidt and Kacprzyk [23], [30], [25], [32]), [33]) taking into account all three terms (membership, non-membership and hesitation margin values) describing an intuitionistic fuzzy set. Any element belonging to an intuitionistic fuzzy set may be represented inside an *MNO* triangle. In other words, the *MNO* triangle represents the surface where the coordinates of any element belonging to an A-IFS can be represented. Each point belonging to the *MNO* triangle is described by the three coordinates: (μ, ν, π). Points *M* and *N* represent the crisp elements. Point $M(1,0,0)$ represents elements fully belonging to an A-IFS as $\mu = 1$, and may be seen as the representation of the ideal positive element. Point $N(0,1,0)$ represents elements fully not belonging to an A-IFS as $\nu = 1$, i.e. can be viewed as the ideal negative element. Point $O(0,0,1)$ represents elements about which we are not able to say if they belong or not belong to an A-IFS (the intuitionistic fuzzy index $\pi = 1$). Such an interpretation is intuitively appealing and provides means for the representation of many aspects of imperfect information. Segment MN (where $\pi = 0$) represents elements belonging to the classic fuzzy sets ($\mu + \nu = 1$). For example, point $x_1(0.2, 0.8, 0)$ (Figure 1), like any element from segment MN represents an element of a fuzzy set. A line parallel to MN describes the elements with the same values of the hesitation margin. In Figure 1 we can see point $x_3(0.5, 0.1, 0.4)$ representing an element with the hesitation margin equal 0.4, and point $x_2(0.2, 0, 0.8)$ representing an element with the hesitation margin equal 0.8. The closer a line that is parallel to MN is to O, the higher the hesitation margin.

3 Correlation and the Kendall Coefficient between Crisp Sets

The correlation coefficient (Pearson's r) between two variables is a measure of the linear relationship between them.

The correlation coefficient is equal 1 in the case of a positive (increasing) linear relationship, -1 in the case of a negative (decreasing) linear relationship, and some value between -1 and 1 in all other cases. The closer the coefficient is to either -1 or 1, the stronger the correlation between the variables.

Let (X_1, Y_1), (X_2, Y_2), $\ldots$, (X_n, Y_n) be a random sample of size n from a joint probability density function $f_{X,Y}(x, y)$, let $\overline{X}$ and $\overline{Y}$ be the sample means of variables X and Y, respectively, then the sample correlation coefficient $r(X, Y)$ is given as (e.g., [18]):

$$r(A,B) = \frac{\sum_{i=1}^{n} (x_i - \overline{X})(y_i - \overline{Y})}{(\sum_{i=1}^{n} (x_i - \overline{X})^2 \sum_{i=1}^{n} (y_i - \overline{Y})^2)^{0.5}} \tag{4}$$

where: $\overline{X} = \frac{1}{n} \sum_{i=1}^{n} x_i, \quad \overline{Y} = \frac{1}{n} \sum_{i=1}^{n} y_i.$

When the assumption that the data distributions are normal is not valid or when data are in the form of ranks we may use the *Spearman rank correlation coefficient* or the *Kendall rank correlation coefficient*. In this paper we discuss the *Kendall rank correlation coefficient* which, due to – for instance – Noether [17] has an intuitively simple interpretation, and its algebraic structure is much simpler than that of the Spearman coefficient.

Definition 1. *(cf. e.g., Nelsen [1])* Let (x_1, y_1), (x_2, y_2), $\ldots$, (x_n, y_n) be a set of joint observations from two random variables X and Y, respectively, such that all the values of (x_i) and (y_i) are unique. Any pair of observations (x_i, y_i) and (x_j, y_j) are said to be concordant if the ranks for both elements agree: that is, if both $x_i > x_j$ and $y_i > y_j$ or if both $x_i < x_j$ and $y_i < y_j$. They are said to be discordant, if $x_i > x_j$ and $y_i < y_j$ or if $x_i < x_j$ and $y_i > y_j$. If $x_i = x_j$ or $y_i = y_j$, the pair is neither concordant, nor discordant; $i, j \in \{1, \ldots, n\}$.

For n observations, (i.e., $(i, j) \in \{1, \ldots, n\}^2$) the number of concordant C, discordant D, tied pars T in X, and tied pairs U in Y (if $x_i = x_j$, or $y_i = y_j$, $i, j \in \{1, \ldots, n\}$, the pair is said to be tied), is denoted:

$$C = |\{(i,j)|x_i < x_j \; and \; y_i < y_j\}| \tag{5}$$

$$D = |\{(i,j)|x_i < x_j \; and \; y_i > y_j\}| \tag{6}$$

$$T = |\{(i,j)|x_i = x_j\}| \tag{7}$$

$$U = |\{(i,j)|y_i = y_j\}| \tag{8}$$

The Kendall τ coefficient is defined as [14]:

$$\tau = \frac{C - D}{\frac{1}{2}n(n-1)} \tag{9}$$

where: C - number of concordant pairs; D - number of discordant pairs, and

$$|\tau| \leq 1$$

For the perfect agreement between the two rankings (i.e., all pairs are concordant), the coefficient has value 1. For the perfect disagreement between the two rankings (i.e., all pairs are discordant), the coefficient has value -1. All other arrangements yield the value of τ between -1 and 1 (increasing values imply an increasing agreement between the rankings). For completely independent rankings, the coefficient has value 0.

If two values of X or two values of Y with the same rank (i.e. ties) occur, the following formula is used [14]:

$$\tau_b = \frac{C - D}{\sqrt{\frac{1}{2}(n(n-1) - T}\sqrt{\frac{1}{2})n(n-1) - U}} \tag{10}$$

where: T - the number of ties in X (the number of pairs for which $x_i = x_j$; U - the number of ties in Y (the number of pairs for which $y_i = y_j$.

4 Correlation, and the Kendall Coefficient between A-IFSs

In Szmidt and Kacprzyk [37] we proposed a correlation coefficient for two A-IFSs, A and B, so that we could express not only a relative strength but also a positive or negative relationship between A and B. We took into account all three terms describing an A-IFSs (the membership values, non-membership values and the hesitation margins) because each of them influences the results (cf. [37]).

Suppose that we have a random sample $x_1, x_2, \ldots, x_n \in X$ with a sequence of data pairs $[(\mu_A(x_1), \nu_A(x_1), \pi_A(x_1)), (\mu_B(x_1), \nu_B(x_1), \pi_B(x_1))]$, $[(\mu_A(x_2), \nu_A(x_2), \pi_A(x_2)), (\mu_B(x_2), \nu_B(x_2), \pi_B(x_2))]$, $\ldots$, $[(\mu_A(x_n), \nu_A(x_n), \pi_A(x_n)), (\mu_B(x_n), \nu_B(x_n), \pi_B(x_n))]$ which correspond to the membership values, non-memberships values and hesitation margins of the A-IFSs A and B defined on X, then the correlation coefficient $r_{A-IFS}(A,B)$ is given by Definition 2.

Definition 2. The correlation coefficient $r_{A-IFS}(A,B)$ between two A-IFSs, A and B in X, is (Szmidt and Kacprzyk [37]):

$$r_{A-IFS}(A,B) = \frac{1}{3}(r_1(A,B) + r_2(A,B) + r_3(A,B)) \tag{11}$$

where

$$r_1(A,B) = \frac{\sum_{i=1}^{n}(\mu_A(x_i) - \overline{\mu_A})(\mu_B(x_i) - \overline{\mu_B})}{(\sum_{i=1}^{n}(\mu_A(x_i) - \overline{\mu_A})^2)^{0.5}(\sum_{i=1}^{n}(\mu_B(x_i) - \overline{\mu_B})^2)^{0.5}} \quad (12)$$

$$r_2(A,B) = \frac{\sum_{i=1}^{n}(\nu_A(x_i) - \overline{\nu_A})(\nu_B(x_i) - \overline{\nu_B})}{(\sum_{i=1}^{n}(\nu_A(x_i) - \overline{\nu_A})^2)^{0.5}(\sum_{i=1}^{n}(\nu_B(x_i) - \overline{\nu_B})^2)^{0.5}} \quad (13)$$

$$r_3(A,B) = \frac{\sum_{i=1}^{n}(\pi_A(x_i) - \overline{\pi_A})(\pi_B(x_i) - \overline{\pi_B})}{(\sum_{i=1}^{n}(\pi_A(x_i) - \overline{\pi_A})^2)^{0.5}(\sum_{i=1}^{n}(\pi_B(x_i) - \overline{\pi_B})^2)^{0.5}} \quad (14)$$

where: $\overline{\mu_A} = \frac{1}{n}\sum_{i=1}^{n}\mu_A(x_i)$, $\overline{\mu_B} = \frac{1}{n}\sum_{i=1}^{n}\mu_B(x_i)$, $\overline{\nu_A} = \frac{1}{n}\sum_{i=1}^{n}\nu_A(x_i)$, $\overline{\nu_B} = \frac{1}{n}\sum_{i=1}^{n}\nu_B(x_i)$, $\overline{\pi_A} = \frac{1}{n}\sum_{i=1}^{n}\pi_A(x_i)$, $\overline{\pi_B} = \frac{1}{n}\sum_{i=1}^{n}\pi_B(x_i)$,

The proposed correlation coefficient (11) depends on two factors: the amount of information expressed by the membership and non-membership degrees (12)–(13), and the reliability of information expressed by the hesitation margins (14).

Remark: Analogously as for the crisp and fuzzy data, $r_{A-IFS}(A,B)$ makes sense for the A-IFS type variables whose values vary. If, for instance, the temperature is constant and the amount of ice cream sold is the same, then it is impossible to conclude about their relationship (as, from the mathematical point of view, we avoid zero in the denominator).

The correlation coefficient $r_{A-IFS}(A,B)$ (11) fulfills the following properties:

1. $r_{A-IFS}(A,B) = r_{A-IFS}(B,A)$,
2. If $A = B$ then $r_{A-IFS}(A,B) = 1$,
3. $|r_{A-IFS}(A,B)| \leq 1$.

The above properties are not only fulfilled by the correlation coefficient $r_{A-IFS}(A,B)$ (11) but also by all of its three components (12)–(14).

Remark: It is should be emphasized that $r_{A-IFS}(A,B) = 1$ occurs not only for $A = B$ but also in the cases of a perfect linear correlation of the data (the same concerns each component (12)–(14)).

However, Definition 2 can not be used to measure the degree of association between A-IFSs when the assumption that the data distributions are normal is not valid or when data are in the form of ranks. In such a case we can use the *Spearman rank correlation coefficient* (cf. Szmidt and Kacprzyk [38]) or the *Kendall rank correlation coefficient* to be defined here for A-IFSs.

Definition 3. The Kendall rank correlation coefficient τ_{A-IFS} between two A-IFSs, A and B in X is defined as:

$$\tau_{A-IFS} = \frac{1}{3}(\tau_1 + \tau_2 + \tau_3) \quad (15)$$

where: τ_i, $i = 1,\ldots,3$ are the Kendall rank correlation coefficients between A and B with respect to their membership values, non-membership values, and hesitation margin values, given as:

$$\tau_1 = \frac{C_\mu - D_\mu}{\frac{1}{2}n(n-1)} \tag{16}$$

where: C_μ – the number of concordant pairs with respect to the membership values; D_μ – the number of discordant pairs with respect to the membership values, i.e.:

$$C_\mu = |\{(i,j)|\mu_A(x_i) < \mu_A(x_j) \text{ and } \mu_B(x_i) < \mu_B(x_j)\}| \tag{17}$$

$$D_\mu = |\{(i,j)|\mu_A(x_i) < \mu_A(x_j) \text{ and } \mu_B(x_i) > \mu_B(x_j)\}| \tag{18}$$

$$\tau_2 = \frac{C_\nu - D_\nu}{\frac{1}{2}n(n-1)} \tag{19}$$

where: C_ν – the number of concordant pairs with respect to the non-membership values; D_ν – the number of discordant pairs with respect to the non-membership values, i.e.:

$$C_\nu = |\{(i,j)|\nu_A(x_i) < \nu_A(x_j) \text{ and } \nu_B(x_i) < \nu_B(x_j)\}| \tag{20}$$

$$D_\nu = |\{(i,j)|\nu_A(x_i) < \nu_A(x_j) \text{ and } \nu_B(x_i) > \nu_B(x_j)\}| \tag{21}$$

and

$$\tau_3 = \frac{C_\pi - D_\pi}{\frac{1}{2}n(n-1)} \tag{22}$$

where: C_π – the number of concordant pairs with respect to the hesitation margins; D_π – the number of discordant pairs with respect to the hesitation margins, i.e.:

$$C_\pi = |\{(i,j)|\pi_A(x_i) < \pi_A(x_j) \text{ and } \pi_B(x_i) < \pi_B(x_j)\}| \tag{23}$$

$$D_\pi = |\{(i,j)|\pi_A(x_i) < \pi_A(x_j) \text{ and } \pi_B(x_i) > \pi_B(x_j)\}| \tag{24}$$

For the *Kendall rank correlation coefficient* τ (15) between the A-IFSs the same properties as for its crisp set counterpart are valid, i..e.:

1. $|\tau_{A-IFS}| \leq 1$

2. $\tau_{A-IFS}(A,B) = \tau_{A-IFS}(B,A)$

The interpretation of the obtained values is the same as in a case of the crisp sets. The perfect agreement of the rankings (i.e., the two rankings are the same) produces the value of the coefficient τ_{A-IFS} equal to 1. The perfect disagreement of the rankings (i.e., one ranking is the reverse of the other) produces the value of the coefficient τ_{A-IFS} equal to -1. All other arrangements produce the value of the coefficient τ_{A-IFS} which lies between -1 and 1, and increasing values imply increasing agreement between the rankings. For the completely independent rankings, the coefficient τ_{A-IFS} is equal to 0.

Each of the three components of the *Kendall rank correlation coefficient* (15) i.e., (16), (19), and (22) fulfill the above properties, too.

Remark: Definition 3 was introduced for the cases without ties. If ties occur, the counterparts of (10) replace the formulas (16), (19), and (22).

We will show now some small but illustrative examples.

Table 1. Example 1 - calculations of (16), (19), (22)

Calculation of the Kendall component:					
(16)		(19)		(22)	
pair	score	pair	score	pair	score
(0.5, 0.55)	+1	(0, 0.1)	+1	(0.05, 0.12)	+1
(0.5, 0.57)	+1	(0, 0.2)	+1	(0.05, 0.23)	+1
(0.5, 0.38)	-1	(0, 0.5)	+1	(0.05, 0.35)	+1
(0.5, 0.35)	-1	(0, 0.6)	+1	(0.05, 0.95)	+1
(0.55, 0.57)	+1	(0.1, 0.2)	+1	(0.12, 0.23)	+1
(0.55, 0.38)	-1	(0.1, 0.5)	+1	(0.12, 0.35)	+1
(0.55, 0.35)	-1	(0.1, 0.6)	+1	(0.12, 0.95)	+1
(0.57, 0.38)	-1	(0.2, 0.5)	+1	(0.23, 0.35)	+1
(0.57, 0.35)	-1	(0.2, 0.6)	+1	(0.23, 0.95)	+1
(0.38, 0.35)	-1	(0.5, 0.6)	+1	(0.35, 0.95)	+1

Example 1. An expert clinician, while administering any medication, should make a decision based on the context of the individual patient and his/her own past experience of the expected effect (e.g. Helgason and Jobe [10], [11]). The effects may be positive (expressed by a membership value), negative (expressed by a non-membership value), and difficult to foresee (expressed by a hesitation margin), for a specific patient. Suppose that two new medicines A and B are tested, and their effects on 5 patients are the following (Figure 2):

$$\begin{aligned} A = &\{(x_1, 0.05, 0.2, 0.75), (x_2, 0.15, 0.35, 0.5), (x_3, 0.25, 0.38, 0.37), \\ &(x_4, 0.4, 0.4, 0.2), (x_5, 0.5, 0.45, 0.05)\} \quad (25) \\ B = &\{(x_1, 0.5, 0, 0.95), (x_2, 0.55, 0.1, 0.35), (x_3, 0.57, 0.2, 0.23), \\ &(x_4, 0.38, 0.5, 0.12), (x_5, 0.35, 0.6, 0.05)\} \quad (26) \end{aligned}$$

where, for example, the positive effects of medicine A on the first patient (x_1) are expressed by the membership value equal to 0.05, the negative effects are expressed by the non-membership value equal to 0.2, and effects difficult to predict are expressed by the hesitation margin equal to 0.75; etc.

To calculate τ_{A-IFS}, we go through three steps using the three terms describing an A-IFS (membership values, non-membership values, and the hesitation margins) which are responsible for the three components (τ_1, τ_2, τ_3) of τ_{A-IFS} (15).

1. The membership values of the elements in A are: $0.05, 0.15, 0.25, 0.4, 0.5$ which means that they are already ordered, and the corresponding membership values of the elements in B are: $0.5, 0.55, 0.57, 0.38, 0.35$ (they first increase, and then decrease). Details of the calculations of τ_1 due to (17)–(18) are given in Table 1. It is worth mentioning that we calculate τ_1 using actual observations without first converting them to the ranks: we consider all possible pairs of the membership values of the elements in B – for each element we consider only the pairs build from elements occurring after the element considered, eg., for 0.57 we consider two pairs: (0.57, 0.38) and (0.57, 0.35), whereas for 0.38 we consider only one

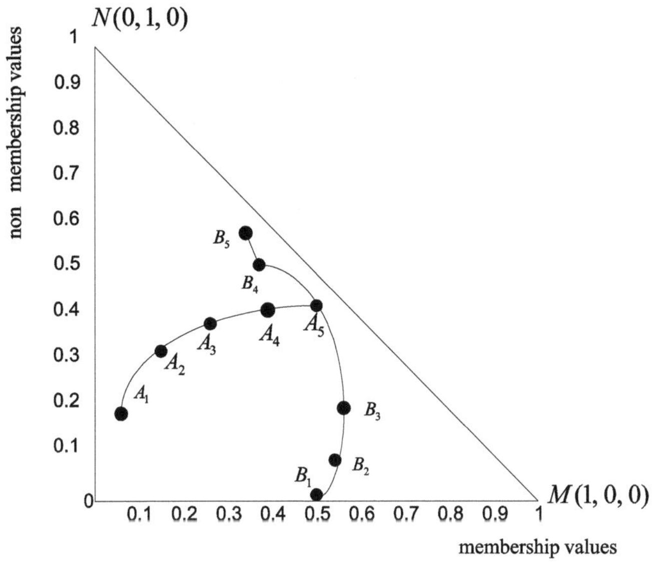

Fig. 2. Data from Example 1

pair: $(0.38, 0.35)$. If the second element in a pair is bigger than the first element, the score is equal 1, if the second element is smaller than the first element, the score is equal to -1 (see (Table 1)) As we have $C_5^2 = 10$ pairs, and the obtained score (Table 1, second column) is equal $3 - 7 = -4$, in result τ_1 (16) is equal to -0.4.

2. The non-membership values of the elements in A: $0.2, 0.35, 0.38, 0.4, 0.45$ are also ordered (increasing), and the non-membership values of the elements in B are: $0, 0.1, 0.2, 0.5, 0.6$ (also increasing). In result, performing the same calculations as previously (details in Table 1, third and fourth columns), we obtain τ_2 (19) equal to 1.
3. The hesitation margins of the elements in A, i.e.: $0.75, 0.5, 0.37, 0.2, 0.05$ decrease, and the hesitation margins of the elements in B i.e.: 0.95, 0.35, 0.23, 0.12, 0.05, decrease, too. First we arrange the hesitation margins of the elements of A in an increasing order, and obtain: $0.05, 0.2, 0.37, 0.5, 0.75$, with the respective order of the elements of B: $0.05, 0.12, 0.23, 0.35, 0.95$. Further calculations are like previously – details are in Table 1 (fifth and sixth columns). In effect τ_3 (22) is equal to 1.

Therefore, finally, from (15) we obtain $\tau_{A-IFS}(A,B) = \frac{1}{3}(-0.4 + 1 + 1) = 0.53$.

If we exclude from considerations the hesitation margins, and take into account two components (16) and (19) only, as is done in some works, we obtain $\tau_{A-IFS}(A,B) = \frac{1}{2}(-0.4 + 1) = 0.3$. But from the point of view of an expert clinician all three components may seem interesting. Let us notice that in this example the data are such that $\tau_3 = 1$ influences substantially the final result. If we consider a relationship between A and B in the categories of the positive effects and negative effects only, we obtain an average relationship equal to 0.3 which suggests that the relationship between A and B is not strong. On the other hand, τ_3 (22) equal to 1 suggests that both medicines are strongly associated as far as unpredictable effects are concerned. It may be an important information from a medical point of view.

It is worth emphasizing that for practical purposes (e.g., in decision making) it seems rather useful to know the component τ_3 (22) responsible for the association of A and B in respect to lack of knowledge represented by the variables considered. If, for example, the data represent reactions of patients to a new medicine, it seems unavoidable to carefully examine just the part (22) of the Kendall coefficient (15) as it may happen that a new treatment/medicine increases unpredictable reactions. In such situations it may be important not only to assess all components separately but even to give them different weights in (15).

Certainly, we can find cases for which τ_3 (22) does not influence the Kendall coefficient τ_{A-IFS} (15) in a sense of the final result (an obtained number). But such situations are exceptional, not a rule.

Example 2. Consider data in Figure 3:

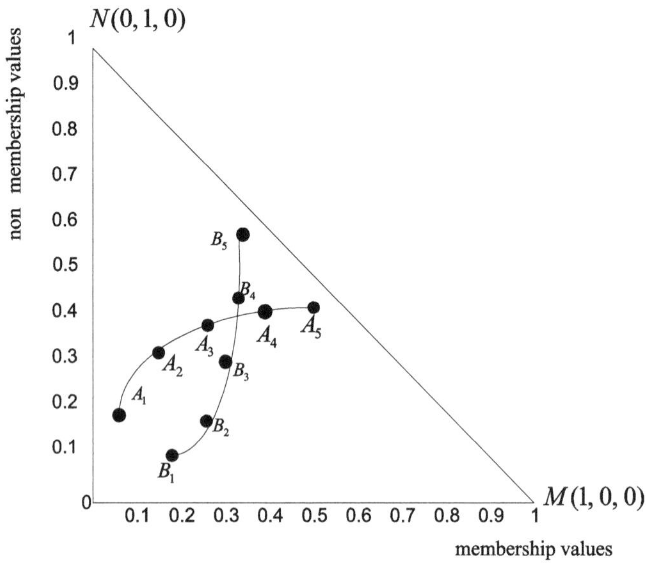

Fig. 3. Data from Example 2

$$
\begin{aligned}
A = &\{(x_1, 0.05, 0.2, 0.75), (x_2, 0.15, 0.35, 0.5), (x_3, 0.25, 0.38, 0.37),\\
&(x_4, 0.4, 0.4, 0.2), (x_5, 0.5, 0.45, 0.05)\} \qquad (27)\\
B = &\{(x_1, 0.18, 0.1, 0.72), (x_2, 0.25, 0.2, 0.55), (x_3, 0.3, 0.3, 0.4),\\
&(x_4, 0.32, 0.45, 0.23), (x_5, 0.34, 0.6, 0.06)\} \qquad (28)
\end{aligned}
$$

It is easy to verify that for the above data we obtain $\tau_1 = 1$, $\tau_2 = 1$, and $\tau_3 = 1$ (respectively from (16), (19), (22)). In other words, in spite of taking into account τ_3 (22) or

Table 2. The values of the first Kendall correlation component (16) between each pair of the attributes for the Iris Setosa data

Attribute	sepal length	sepal width	petal length	petal width
sepal length	1	0, 23	0,71	0,72
sepal width	-	1	0,4	0,37
petal length	-	-	1	0.84
petal width	-	-	-	1

Table 3. The values of the second Kendall correlation component (19) between each pair of the attributes for the Iris Setosa data

Attribute	sepal length	sepal width	petal length	petal width
sepal length	1	NA	0,59	0,63
sepal width	-	1	0,27	0,24
petal length	-	-	1	0.84
petal width	-	-	-	1

Table 4. The values of the Kendall correlation (15) (total, i.e. including the first and second component) in between each pair of the attributes for the Iris Setosa data

Attribute	sepal length	sepal width	petal length	petal width
sepal length	1	-0,03	0,65	0,67
sepal width	-	1	0,34	0,31
petal length	-	-	1	0.84
petal width	-	-	-	1

not, we obtain the same result for τ_{A-IFS} (15), as in this case $\tau_{A-IFS}(A,B) = \frac{1}{2}(1+1) = \frac{1}{3}(1+1+1) = 1$. But, as it was shown earlier, such a situation may rather be considered as a coincidence not a rule.

We have also examined the Kendall correlation coefficient using the Iris data [46] expressed in terms of the A-IFSs. Iris data consists of 3 classes with 50 instances each. Each class refers to a type of the iris plant (Iris Setosa, Iris Versicolor, Iris Virginica). There are 4 attributes: sepal length, sepal width, petal length, petal width.

We have used the algorithm based on the mass assignment theory proposed by Szmidt and Baldwin [19] to describe the data in terms of the A-IFSs, i.e., to assign the parameters of an A-IFS model which describes each attribute in terms of membership values, non-membership values, and hesitation margin values. Having description of the attributes in terms of A-IFSs, we have calculated the three components of (15) for each pair of the attributes. The results are in (Tables 2–13).

The results concerning the Iris Setosa data are shown in Tables 2 – 4. We can observe substantial correlation of petal length and petal width in respect with the components (16) – Table 2 and (19) – Table 3. The values of the Kendall correlation component (22) between each pair (except of sepal length and sepal width) of the attributes for the Iris Setosa data are not statistically significant (at the 0.01 level) so they are not presented.

Table 5. The values of the first Kendall correlation component (16) between each pair of the attributes for the Iris Versicolor data

Attribute	sepal length	sepal width	petal length	petal width
sepal length	1	NA	-0,12	NA
sepal width	-	1	0,33	0,36
petal length	-	-	1	0.62
petal width	-	-	-	1

Table 6. The values of the second Kendall correlation component (19) between each pair of the attributes for the Iris Versicolor data

Attribute	sepal length	sepal width	petal length	petal width
sepal length	1	0,17	0,42	0,35
sepal width	-	1	0,41	0,41
petal length	-	-	1	0.7
petal width	-	-	-	1

Table 7. The values of the third Kendall correlation component (22) between each pair of the attributes for the Iris Versicolor data

Attribute	sepal length	sepal width	petal length	petal width
sepal length	1	-0,17	0,15	0,18
sepal width	-	1	0,28	0,23
petal length	-	-	1	0.61
petal width	-	-	-	1

Table 8. The values of the total Kendall correlation (15) between each pair of the attributes for the Iris Versicolor data

Attribute	sepal length	sepal width	petal length	petal width
sepal length	1	0,004	0,15	0,27
sepal width	-	1	0,34	0,33
petal length	-	-	1	0.64
petal width	-	-	-	1

Table 9. The values of the first Kendall correlation component (16) between each pair of the attributes for the Iris Virginica data

Attribute	sepal length	sepal width	petal length	petal width
sepal length	1	-0,21	0,72	0,64
sepal width	-	1	-0,14	-0,16
petal length	-	-	1	0.86
petal width	-	-	-	1

Table 10. The values of the second Kendall correlation component (19) between each pair of the attributes for the Iris Virginica data

Attribute	sepal length	sepal width	petal length	petal width
sepal length	1	0,28	0,36	0,32
sepal width	-	1	0,49	0,52
petal length	-	-	1	0.89
petal width	-	-	-	1

Table 11. The values of the third Kendall correlation component (22) between each pair of the attributes for the Iris Virginica data

Attribute	sepal length	sepal width	petal length	petal width
sepal length	1	NA	0,47	0,41
sepal width	-	1	0,23	0,2
petal length	-	-	1	0.61
petal width	-	-	-	1

Table 12. The values of the total Kendall correlation (15) between each pair of the attributes for the Iris Virginica data

Attribute	sepal length	sepal width	petal length	petal width
sepal length	1	-0,04	0,61	0,55
sepal width	-	1	0,13	0,10
petal length	-	-	1	0.78
petal width	-	-	-	1

Table 13. The values of the total Kendall correlation (15) between each pair of the attributes for all Iris data

Attribute	sepal length	sepal width	petal length	petal width
sepal length	1	-0,03	0,47	0,5
sepal width	-	1	0,27	0,25
petal length	-	-	1	0.76
petal width	-	-	-	1

The Kendall correlation coefficient (15) – Table 4 is the biggest between petal length and petal width, and does not count only between sepal length and sepal width.

The results concerning the Iris Versicolor data are in Tables 5 – 8. The values of the Kendall correlation components (16), (19), and (22) are in Tables 5 – 7. Again, petal length and petal width are highly correlated (all three components (16), (19), and (22) are substantial and statistically significant). This trend is also reflected by (15) for Iris Versicolor – Table 8.

The Kendall correlation coefficient components (16), (19), and (22) between the Iris Virginica attributes are in Tables 9 – 12. Again, Kendall correlation between petal width and petal length is the highest in respect with all three components (16), (19), (22)), and "confirmed" in a case of (15), too. Sepal width is rather weakly correlated with all other attributes in respect with (16) – Table 9. On the other hand, sepal width is correlated to a stronger extent with other attributes with respect to the component (19) – Table10 but it does not have such a strong influence (because of the other two components) on (15) – Table 12.

Finally, in Table 13 there are shown the values of the (total) Kendall correlation (15) between each pair of the attributes for all the Iris data. We may observe that some results are still the same as for the separate components of the data, e.g., a strong correlation between petal length and petal width. On the other hand, e.g., sepal width is moderately correlated with other attributes (petal length and petal width) whereas for some situations previously analyzed, it was small. In other words, it seems important to consider a detailed structure of the data examined. Next, all three components of the Kendall correlation coefficient (15) are important and worth considering separately in the sense of providing much richer insight.

5 Conclusions

A new concept of the *Kendall correlation coefficient* for the A-IFSs was proposed. The coefficient is a generalization of the *Kendall correlation coefficient* as defined for the crisp sets, i.e., it fulfills the same properties, and reduces to its well known form for the crisp sets. It is worth emphasizing that all three terms describing the A-IFS were taken into account (the membership values, non-membership values and hesitation margins). Each term plays an important role in data analysis and decision making, by providing a deeper insight, so that each of them should be reflected while assessing the relationship between the A-IFSs.

References

1. Nelsen, R.B.: Kendall tau metric. In: Hazewinkel, M. (ed.) Encyclopedia of Mathematics. Springer (2001) ISBN 978-1556080104
2. Aczel, A.D.: Complete business statistics. Richard D. Irvin, Inc. (1998)
3. Atanassov, K.: Intuitionistic Fuzzy Sets. VII ITKR Session. Sofia (Centr. Sci.-Techn. Libr. of Bulg. Acad. of Sci., 1697/84) (1983) (in Bulgarian)
4. Atanassov, K.: Intuitionistic Fuzzy Sets: Theory and Applications. Springer (1999)
5. Barrett, P.M., McGowan, A.J., Page, V.: Dinosaur diversity and the rock record. Proc. Royal Society B 276, 2667–2674 (2009)
6. Benson, R.B., Butler, R.J., Lindgren, J., Smith, A.S.: Mesozoic marine tetrapod diversity: mass extinctions and temporal heterogeneity in geological megabiases affecting vertebrates. Proc. Royal Society B 277, 829–834 (2010)
7. Bracke, M., Edwards, S.A., Engel, B., Buist, W.G., Bo Algers, B.: Expert opinion as 'validation' of risk assessment applied to calf welfare. Acta Veterinaria Scandinavica 50(29) (2008), doi:10.1186/1751-0147-50-29
8. Kurvers, R.H., Prins, H.H., van Wieren, S.E., van Oers, K., Nolet, B.A., Ronald, C., Ydenberg, R.C.: The effect of personality on social foraging: shy barnacle geese scrounge more. Proc. Royal Society B 277, 601–608 (2010)

9. Griffiths, D.: A Pragmatic Approach to Spearmans Rank Correlation Coefficient. Teaching Statistics 2, 10–13 (1980)
10. Helgason, C.M., Jobe, T.H.: Perception based reasoning and fuzzy cardinality provide direct measures of causality sensitive to initial conditions in the individual patient. International Journal of Computational Cognition 1, 79–104 (2003)
11. Helgason, C.M., Watkins, F.A., Jobe, T.H.: Measurable differences between sequential and parallel diagnostic decision processes for determining stroke subtype: A representation of interacting pathologies. Thromb. Haemost. 88, 210–212 (2002)
12. Henrysson, S.: Gathering, analyzing, and using data on test items. In: Thorndike, R.L. (ed.) Educational Measurement, pp. 130–159. American Council on Education, Washington D.C (1971)
13. Kendler, K.S., Parnas, J.: Philosophical Issues in Psychiatry: Explanation, Phenomenology, and Nosology. Johns Hopkins University Press (2008)
14. Kendall, M.G.: Rank correlation methods, 4th edn. Charles Griffin & Co., London (1970)
15. Moller, A.P., Alatalo, R.V.: Good-genes effects in sexual selection. Proc. Royal Society Lond. B 266, 85–91 (1999)
16. Myers, J.L., Well, A.W.: Research Design and Statistical Analysis, 2nd edn. Lawrence Erlbaum (2003)
17. Noether, G.: Why Kendall Tau? Teaching Statistics 3(2), 41–43 (1981)
18. Rodgers, J.L., Nicewander, W.A.: Thirteen Ways to Look at the Correlation Coefficient. The American Statistician 42(1), 59–66 (1988)
19. Szmidt, E., Baldwin, J.: Intuitionistic Fuzzy Set Functions, Mass Assignment Theory, Possibility Theory and Histograms. In: 2006 IEEE World Congress on Computational Intelligence, pp. 237–243 (2006)
20. Szmidt, E., Kacprzyk, J.: Remarks on some applications of intuitionistic fuzzy sets in decision making. Notes on IFS 2(3), 22–31 (1996c)
21. Szmidt, E., Kacprzyk, J.: On measuring distances between intuitionistic fuzzy sets. Notes on IFS 3(4), 1–13 (1997)
22. Szmidt, E., Kacprzyk, J.: Group Decision Making under Intuitionistic Fuzzy Preference Relations. In: IPMU 1998, pp. 172–178 (1998)
23. Szmidt, E., Kacprzyk, J.: Distances between intuitionistic fuzzy sets. Fuzzy Sets and Systems 114(3), 505–518 (2000)
24. Szmidt, E., Kacprzyk, J.: On Measures on Consensus Under Intuitionistic Fuzzy Relations. In: IPMU 2000, pp. 1454–1461 (2000)
25. Szmidt, E., Kacprzyk, J.: Entropy for intuitionistic fuzzy sets. Fuzzy Sets and Systems 118(3), 467–477 (2001)
26. Szmidt, E., Kacprzyk, J.: Analysis of Consensus under Intuitionistic Fuzzy Preferences. In: Proc. Int. Conf. in Fuzzy Logic and Technology, pp. 79–82. De Montfort Univ., Leicester (2001)
27. Szmidt, E., Kacprzyk, J.: Analysis of Agreement in a Group of Experts via Distances Between Intuitionistic Fuzzy Preferences. In: Proc. 9th Int. Conf. IPMU 2002, pp. 1859–1865 (2002a)
28. Szmidt, E., Kacprzyk, J.: An Intuitionistic Fuzzy Set Based Approach to Intelligent Data Analysis (an application to medical diagnosis). In: Abraham, A., Jain, L., Kacprzyk, J. (eds.) Recent Advances in Intelligent Paradigms and Applications, pp. 57–70. Springer (2002b)
29. Szmidt, E., Kacprzyk, J.: An Intuitionistic Fuzzy Set Based Approach to Intelligent Data Analysis (an application to medical diagnosis). In: Abraham, A., Jain, L., Kacprzyk, J. (eds.) Recent Advances in Intelligent Paradigms and Applications, pp. 57–70. Springer (2002c)
30. Szmidt, E., Kacprzyk, J.: Distances Between Intuitionistic Fuzzy Sets: Straightforward Approaches may not work. In: IEEE IS 2006, pp. 716–721 (2006)

31. Szmidt, E., Kacprzyk, J.: An Application of Intuitionistic Fuzzy Set Similarity Measures to a Multi-criteria Decision Making Problem. In: Rutkowski, L., Tadeusiewicz, R., Zadeh, L.A., Żurada, J.M. (eds.) ICAISC 2006. LNCS (LNAI), vol. 4029, pp. 314–323. Springer, Heidelberg (2006)
32. Szmidt, E., Kacprzyk, J.: Some Problems with Entropy Measures for the Atanassov Intuitionistic Fuzzy Sets. In: Masulli, F., Mitra, S., Pasi, G. (eds.) WILF 2007. LNCS (LNAI), vol. 4578, pp. 291–297. Springer, Heidelberg (2007)
33. Szmidt, E., Kacprzyk, J.: A New Similarity Measure for Intuitionistic Fuzzy Sets: Straightforward Approaches not work. In: 2007 IEEE Conf. on Fuzzy Systems, pp. 481–486 (2007a)
34. Szmidt, E., Kacprzyk, J.: A new approach to ranking alternatives expressed via intuitionistic fuzzy sets. In: Ruan, D., et al. (eds.) Computational Intelligence in Decision and Control, pp. 265–270. World Scientific (2008)
35. Szmidt, E., Kacprzyk, J.: Amount of Information and Its Reliability in the Ranking of Atanassov's Intuitionistic Fuzzy Alternatives. In: Rakus-Andersson, E., Yager, R.R., Ichalkaranje, N., Jain, L.C. (eds.) Recent Advances in Decision Making. SCI, vol. 222, pp. 7–19. Springer, Heidelberg (2009)
36. Szmidt, E., Kacprzyk, J.: Ranking of Intuitionistic Fuzzy Alternatives in a Multi-criteria Decision Making Problem. In: Proceedings of the Conference: NAFIPS 2009, Cincinnati, USA, June 14-17, pp. 978–971. IEEE (2009) ISBN: 978-1-4244-4577-6
37. Szmidt, E., Kacprzyk, J.: Correlation of Intuitionistic Fuzzy Sets. In: Hüllermeier, E., Kruse, R., Hoffmann, F. (eds.) IPMU 2010. LNCS (LNAI), vol. 6178, pp. 169–177. Springer, Heidelberg (2010)
38. Szmidt, E., Kacprzyk, J.: The Spearman rank correlation coefficient between intuitionistic fuzzy sets. In: 2010 IEEE Int. Conf. on Intelligent Systems (IEEE IS 2010), London, pp. 276–280 (2010)
39. Szmidt, E., Kacprzyk, J.: The Kendall Rank Correlation between Intuitionistic Fuzzy Sets. In: Proc. World Conference on Soft Computing, San Francisco, CA, USA, May 23-26 (2011)
40. Szmidt, E., Kukier, M.: Classification of Imbalanced and Overlapping Classes using Intuitionistic Fuzzy Sets. In: IEEE IS 2006, London, pp. 722–727 (2006)
41. Szmidt, E., Kukier, M.: A New Approach to Classification of Imbalanced Classes via Atanassov's Intuitionistic Fuzzy Sets. In: Wang, H.-F. (ed.) Intelligent Data Analysis: Developing New Methodologies Through Pattern Discovery and Recovery, pp. 85–101. Idea Group (2008)
42. Zadeh, L.A.: Fuzzy sets. Information and Control 8, 338–353 (1965)
43. Zadeh, L.A.: Toward a perception-based theory of probabilistic reasoning with imprecise probabilities. Journal of Statistical Planning and Inference 105, 233–264 (2002)
44. Zadeh, L.A.: Generalized theory of uncertainty (GTU) principal concepts and ideas. Computational Statistics and Data Analysis 51, 15–46 (2006)
45. Zeng, W., Li, H.: Correlation coefficient of intuitionistic fuzzy sets. Journal of Industrial Engineering International 3(5), 33–40 (2007)
46. `http://archive.ics.uci.edu/ml/datasets/Iris`

Bideterminant and Generalized Kronecker-Capelli Theorem for Fuzzy Relation Equations

Irina Perfilieva* and Jiri Kupka

Centre of Excellence IT4Innovations, Division OU
Institute for Research and Applications of Fuzzy Modeling, University of Ostrava,
30. dubna 22, 701 03 Ostrava, Czech Republic
{Irina.Perfilieva,Jiri.Kupka}@osu.cz

Abstract. The aim of this contribution is to elaborate generalized notions of determinant and rank (of a matrix) and to show that the theory of fuzzy relation equations can be investigated with the help of them. We recall the notion of bideterminant of a matrix and investigate its properties in a semilinear space. We introduce three different notions of a rank of a matrix and compare them. Finally, we investigate solvability of a system of fuzzy relation equations in terms of discriminant ranks of its matrices (generalized Kronecker-Capelli theorem).

Keywords: semiring, semilinear space, residuated lattice, bideterminant, rank.

1 Introduction

Linear spaces are widely used in mathematics due to their clear and relatively simple structure. Many sophisticated problems can be solved after the so called linearization (projection on a certain linear space) is applied to an original formulation. The main tool in the theory of linear spaces is the theory of solvability of systems of linear equations. Besides its theoretical meaning, it is successfully used in many applications that are processed using special numerical methods.

In [4], a linear-like behavior of fuzzy systems has been described. It has been shown that the well known Compositional Rule of Inference (CRI) is modeled by a homomorphism between semilinear spaces of fuzzy sets. Therefore, many problems that involve the CRI in its formulation can be formulated in the language of semilinear spaces. Very often they lead to solvability of related systems of fuzzy relation equations which are analogues of systems of linear equations.

The aim of this contribution is to elaborate generalized notions of determinant and rank (of a matrix) and to show that, analogously to the classical case, the theory of fuzzy relation equations can be studied with the help of them. The notion of bideterminant was originally introduced by Kuntzman in [10]. Some combinatorial properties of bideterminants were investigated by M. Minoux, see

* Corresponding author.

R.R. Yager et al. (Eds.): *Soft Computing: State of the Art Theory*, STUDFUZZ 291, pp. 55–70.
DOI: 10.1007/978-3-642-34922-5_5

[11] and few subsequent papers. In our contribution, we list the most important properties of bideterminants that are similar to the classical ones.

Three generalized notions of a matrix rank will be introduced. All of them are equivalent if the underlying space is linear. However, in a semilinear space they are different. We analyze relationships among these notions in a vectorial semilinear space (Section 5). In this respect, we continue research initiated in [1], where the notion of a factor rank (one of those mentioned above) was introduced.

In the classical case, a solvability criterion (compatibility criterion) for a system of linear equations is given by the Kronecker-Capelli theorem. It compares ranks of two matrices: the matrix of coefficients and its extension by the right-hand side vector. If ranks are equal, then the system is solvable and vice-versa. In this contribution, we analyze the applicability of the Kronecker-Capelli theorem to systems of equations in a semilinear space. For two matrix ranks, namely, for discriminant and factor ranks, we will prove the first part (necessary condition) of the Kronecker-Capelli theorem (Section 6). Then we will show that the sufficient condition of the Kronecker-Capelli theorem is not valid for any type of ranks mentioned above. Last but not least, we will analyze the problem of solvability of systems of fuzzy relation equations in dual semilinear spaces (Section 7).

2 Basic Algebraic Constructions

In this section, we introduce semi-structures which naturally arise when we analyze solvability of systems of fuzzy relation equations. They are: a commutative semiring and a semimodule over a commutative semiring (also called a semilinear space) [4,6,13]. The latter can be considered as a weak variant of a vector (linear) space. Below, we recall necessary definitions.

Definition 1. *A semiring* $R = (R, +, \cdot, 0, 1)$ *is an algebra where*

1. $(R, +, 0)$ *is a commutative monoid,*
2. $(R, \cdot, 1)$ *is a monoid,*
3. *for all* $\alpha, \beta, \gamma \in R$, $\quad \alpha \cdot (\beta + \gamma) = \alpha \cdot \beta + \alpha \cdot \gamma, \quad (\beta + \gamma) \cdot \alpha = \beta \cdot \alpha + \gamma \cdot \alpha,$
4. *for all* $\alpha \in R$, $\quad 0 \cdot \alpha = \alpha \cdot 0 = 0.$

A semiring is commutative if $(R, \cdot, 1)$ *is a commutative monoid. A semiring is idempotent if* $+$ *is an idempotent operation.*

Example 1. (i) The set of non-negative integer numbers $(\mathbb{Z}_+, +, \cdot, 0, 1)$ with operations of addition and multiplication is the background example of a commutative semiring.

(ii) A reduct $[0,1]_{Ł} = ([0,1], \vee, \otimes, 0, 1)$ of Łukasiewicz algebra with operations

$$x \vee y = \max\{x, y\}, \quad x \otimes y = \max\{0, x + y - 1\}$$

is a commutative semiring.

(iii) The min-plus algebra $(\mathbb{R}_+, \wedge, +, +\infty, 0)$ over the set $\mathbb{R}_+$ of non-negative real numbers extended by $+\infty$, where $x \wedge y = \min(x, y)$, is a commutative semiring.

(iv) Let $L = (L, \vee, \wedge, *, \rightarrow, 0, 1)$ be an integral, residuated, commutative l-monoid (a *residuated lattice* shortly), and $L_\vee = (L, \vee, *, 0, 1)$ be its reduct. Then $L_\vee$ is a commutative $\vee$-semiring.

Definition 2. *Let $R = (R, +, \cdot, 0, 1)$ be a commutative semiring and $V = (V, +, \bar{0})$ a commutative monoid. We say that V is a (left) semilinear space over R if an external (left) multiplication $\lambda : \bar{x} \mapsto \lambda\bar{x}$ where $\lambda \in R$ and $\bar{x} \in V$ is defined. Moreover, the following properties are fulfilled for all $\bar{x}, \bar{y} \in V$ and $\lambda, \mu \in R$:*

1. $\lambda(\bar{x} + \bar{y}) = \lambda\bar{x} + \lambda\bar{y}$,
2. $(\lambda + \mu)\bar{x} = \lambda\bar{x} + \mu\bar{x}$,
3. $(\lambda \cdot \mu)\bar{x} = \lambda(\mu\bar{x})$,
4. $1\bar{x} = \bar{x}$,
5. $\lambda\bar{0} = \bar{0}$.

Example 2. Let $R = (R, +, \cdot, 0, 1)$ be a generic commutative semiring.

(i) Denote R^n $(n \geq 1)$ the set of n-dimensional vectors over R, i.e.

$$R^n = \{\bar{x} = (x_1, \ldots, x_n)^T \mid x_1 \in R, \ldots, x_n \in R\},$$

where $(x_1, \ldots, x_n)^T$ denotes the transpose of $(x_1, \ldots, x_n)$. Let $\bar{0} = (0, \ldots, 0)^T$ and

$$\bar{x} + \bar{y} = (x_1, \ldots, x_n)^T + (y_1, \ldots, y_n)^T = (x_1 + y_1, \ldots, x_n + y_n)^T.$$

For any $\lambda \in R$, let us define the external multiplication $\lambda\bar{x}$ by

$$\lambda\bar{x} = \lambda(x_1, \ldots, x_n)^T = (\lambda \cdot x_1, \ldots, \lambda \cdot x_n)^T.$$

Then $R^n = (R^n, +, \bar{0})$ is a semilinear space over R (it is sometimes called a *vectorial semilinear space* over R).

(ii) Denote $R^{n\times m}$, $n, m \geq 1$, a set of $n \times m$ matrices over R. Similarly to the case *(i)* above, define the operation $+$ and the external multiplication by an element from R. Let O be the zero matrix. Then $R^{n\times m} = (R^{n\times m}, +, O)$ is a semilinear space over R. The space $R^{n\times n}$ of all square $n \times n$ matrices over R will be denoted by $M_n(R)$.

Let us define the operation of multiplication of two matrices $A \in R^{n\times m}$ and $B \in R^{m\times k}$, $n, m, k \geq 1$. The resulting (product) matrix $C = AB$ where $C \in R^{n\times k}$ will be defined by the same rule as in the classical linear algebra:

$$C = \begin{pmatrix} \sum_{j=1}^m a_{1j} \cdot b_{j1} & \cdots & \sum_{j=1}^m a_{1j} \cdot b_{jk} \\ \vdots & \ddots & \vdots \\ \sum_{j=1}^m a_{nj} \cdot b_{j1} & \cdots & \sum_{j=1}^m a_{nj} \cdot b_{jk} \end{pmatrix}.$$

If in the definition of multiplication we let $k = 1$, so that $B \in R^{m\times 1}$ is a (column-) vector, then the product $C = AB$ of matrix $A \in R^{n\times m}$ and vector B is a vector from $R^{n\times 1}$.

3 Linear Dependence

In this section, we will recall the notion of bideterminant which has been originally introduced by Kuntzman in [10]. Some combinatorial properties of bideterminants were investigated by M. Minoux, see [11] and some subsequent papers.

Definition 3. *Assume that* $R = (R, +, \cdot, 0, 1)$ *is a commutative semiring and* R^n *is a vectorial semilinear space over* R.

- *Vector*
$$\bar{y} = \alpha_1\bar{x}_1 + \cdots + \alpha_k\bar{x}_k$$
is called a linear combination *of vectors* $\bar{x}_1, \ldots, \bar{x}_k \in R^n$ $(k \geq 1)$ *with coefficients* $\alpha_1, \ldots, \alpha_k \in R$.
- *Vectors* $\bar{x}_1, \ldots, \bar{x}_k \in R^n$ $(k \geq 1)$ *are* linearly dependent *if at least one vector among* $\bar{x}_1, \ldots, \bar{x}_k$ *can be represented as a linear combination of others.*
- *Vectors* $\bar{x}_1, \ldots, \bar{x}_k \in R^n$ $(k \geq 1)$ *are* linearly independent *if they are not linearly dependent.*

Example 3. Let $[0,1]_{\text{Ł}} = ([0,1], \vee, \otimes, 0, 1)$ be a semiring from Example 1, case *(ii)*, and $[0,1]^3_{\text{Ł}} = ([0,1]^3, \vee, \bar{0})$ be a vectorial semilinear space over L. Consider vectors $\bar{x}_1, \bar{x}_2, \bar{x}_3, \bar{x}_4 \in [0,1]^3$

$$\bar{x}_1 = \begin{pmatrix} 0.9 \\ 0.6 \\ 0.6 \end{pmatrix}, \quad \bar{x}_2 = \begin{pmatrix} 0.8 \\ 1 \\ 0.8 \end{pmatrix}, \quad \bar{x}_3 = \begin{pmatrix} 0.6 \\ 1 \\ 0.8 \end{pmatrix}, \quad \bar{x}_4 = \begin{pmatrix} 0.4 \\ 0.5 \\ 0.3 \end{pmatrix}.$$

Then $\bar{x}_4$ is a linear combination of $\bar{x}_1, \bar{x}_2$ and $\bar{x}_3$ - namely,

$$\bar{x}_4 = 0.5\bar{x}_1 \vee 0.5\bar{x}_2 \vee 0.3\bar{x}_3.$$

Let us remark that there are other notions of a linear dependence in the literature, see, e.g., [2], [7].

4 Bideterminant

In this section, we will recall the notion of bideterminant which has been originally introduced by Kuntzman in [10]. Some combinatorial properties of bideterminants were investigated by M. Minoux, see [11] and some subsequent papers. We assume that throughout this section, $R = (R, +, \cdot, 0, 1)$ is a commutative semiring.

Definition 4. *Let* A *be a* $n \times n$ *matrix from* $M_n(R)$. *Denote* P *(respectively,* Q*) the set of even (respectively, odd) permutations of the set* $\{1, 2, \ldots, n\}$. *A* bideterminant $\det(A)$ *of* A *is an ordered pair*

$$\det(A) = (\det_1(A), \det_2(A)),$$

such that $\det_1(A), \det_2(A) \in R$, *and*

$$\det_1(A) = \sum_{\sigma \in P} a_{1,\sigma(1)} \cdot a_{2,\sigma(2)} \cdot \dots \cdot a_{n,\sigma(n)} \tag{1}$$

and

$$\det_2(A) = \sum_{\sigma \in Q} a_{1,\sigma(1)} \cdot a_{2,\sigma(2)} \cdot \dots \cdot a_{n,\sigma(n)} \tag{2}$$

Obviously, $\det(A)$ is an element of R^2 and moreover, it is an element of the semilinear space R^2. We say that $\det(A)$ is *zero* (notation $\det(A) \equiv 0$) if $\det_1(A) = \det_2(A)$. Otherwise $\det(A)$ is *nonzero* ($\det(A) \not\equiv 0$).

Example 4. Let $[0,1]_{\text{Ł}} = ([0,1], \vee, \otimes, 0, 1)$ be the semiring from Example 1, case *(ii)*. Consider the following matrices $A, B \in [0,1]^{3\times 3}$:

$$A = \begin{pmatrix} 0.9 & 0.8 & 0.6 \\ 0.6 & 1 & 1 \\ 0.6 & 0.8 & 0.8 \end{pmatrix} \quad \text{and} \quad B = \begin{pmatrix} 0.9 & 0.4 & 0.6 \\ 0.6 & 0.6 & 1 \\ 0.6 & 0.3 & 0.8 \end{pmatrix}$$

Then (apply the rule of Sarrus and compute separately positive and negative products) $det(A)$ and $det(B)$ are as follows:

$$\begin{aligned} \det(A) = (&.9 \otimes 1 \otimes .8 \vee .8 \otimes 1 \otimes .6 \vee .6 \otimes .6 \otimes .8, \\ &.6 \otimes 1 \otimes .6 \vee .8 \otimes .6 \otimes .8 \vee .9 \otimes 1 \otimes .8) = (0.7, 0.7), \\ \det(B) = (&.9 \otimes .6 \otimes .8 \vee .4 \otimes 1 \otimes .6 \vee .6 \otimes .6 \otimes .3 \\ &.6 \otimes .6 \otimes .6 \vee .4 \otimes .6 \otimes .8 \vee .9 \otimes 1 \otimes .3) = (0.3, 0.2). \end{aligned}$$

By the agreement, $\det(A) \equiv 0$ while $\det(B) \not\equiv 0$.

Definition 5. *Let A be a $m \times n$ matrix over R, B a square $s \times s$ submatrix of A where $1 \le s \le \min(m,n)$. Bideterminant* $\det(B)$ *is called a (s-order)* minor *of A.*

It is not difficult to show that many properties of a classical determinant are valid for a bideterminant too. Below, we will list some of them in **(P1)** - **(P8)**.

(P1) If $E \in M_n(R)$ is the unit matrix, i.e. $a_{i,j} = 1$ if $i = j$, and $a_{i,j} = 0$ otherwise, then $\det(E) = (1, 0)$.

(P2) Let $A \in M_n(R)$ be a matrix such that for some $k \in \{1, 2, \dots, n\}$ and every $j = 1, 2, \dots, n$,

$$a_{k,j} = b_{k,j} + c_{k,j}. \tag{3}$$

Then

$$\det(A) = \det \begin{pmatrix} a_{1,1} & \dots & a_{1,n} \\ \vdots & \ddots & \vdots \\ a_{k-1,1} & \dots & a_{k-1,n} \\ b_{k,1} & \dots & b_{k,n} \\ a_{k+1,1} & \dots & a_{k+1,n} \\ \vdots & \ddots & \vdots \\ a_{n,1} & \dots & a_{n,n} \end{pmatrix} + \det \begin{pmatrix} a_{1,1} & \dots & a_{1,n} \\ \vdots & \ddots & \vdots \\ a_{k-1,1} & \dots & a_{k-1,n} \\ c_{k,1} & \dots & c_{k,n} \\ a_{k+1,1} & \dots & a_{k+1,n} \\ \vdots & \ddots & \vdots \\ a_{n,1} & \dots & a_{n,n} \end{pmatrix}$$

where the operation "+" is considered over elements from R^2.

(P3) Let $\lambda \in R$, and $A \in M_n(R)$ where for some $k \in \{1, 2, \dots, n\}$ and every $j = 1, 2, \dots, n$, we have $a_{k,j} = \lambda \cdot b_{k,j}$. Then

$$\det(A) = \lambda \det \begin{pmatrix} a_{1,1} & \dots & a_{1,n} \\ \vdots & \ddots & \vdots \\ a_{k-1,1} & \dots & a_{k-1,n} \\ b_{k,1} & \dots & b_{k,n} \\ a_{k+1,1} & \dots & a_{k+1,n} \\ \vdots & \ddots & \vdots \\ a_{n,1} & \dots & a_{n,n} \end{pmatrix}$$

where $\lambda \det(\cdot)$ denotes the external multiplication over the element from R^2.

(P4) Let $A \in M_n(R)$ be a matrix where for some k and for some l such that $k \neq l$, $a_{k,j} = a_{l,j}$, $j = 1, \dots, n$. Then $\det(A) \equiv 0$.

(P5) If A^T denotes the transpose of A then $\det(A) = \det(A^T)$.

(P6) Let $A, \tilde{A} \in M_n(R)$ and $\tilde{A}$ is obtained from A by a transposition of two rows. If $\det(A) = (\det_1(A), \det_2(A))$, then $\det(\tilde{A}) = (\det_2(A), \det_1(A))$. In particular, if $\det(A) \equiv 0$ then $\det(\tilde{A}) \equiv 0$.

(P7) Let $A \in M_n(R)$ where for at least one $k \in \{1, 2, \dots, n\}$ and every $j = 1, 2, \dots, n$, $a_{k,j} = 0$. Then $\det(A) = (0, 0)$.

(P8) Let $A \in M_n(R)$. By $A'_{i,j}$ we denote the matrix which is obtained from A by removing the i-th row and the j-th column. The following analog of the known row expansion is valid (see also [6]). Below, it is given for the i-th row, $i = 1, 2, \dots, n$:

$$\det(A) = \sum_{\{j \le n \mid i+j \text{ is even}\}} a_{i,j}(\det_1(A'_{i,j}), \det_2(A'_{i,j})) + \sum_{\{j \le n \mid i+j \text{ is odd}\}} a_{i,j}(\det_2(A'_{i,j}), \det_1(A'_{i,j})),$$

where $a_{i,j}(\det(\cdot), \det(\cdot))$ and "+" denote respective operations over elements of the semilinear space R^2.

Remark 1. The above listed properties **(P1)** - **(P8)** are formulated for rows of matrix. It is easy to see (and it follows from the property **(P5)**) that similar properties are valid for columns of matrix as well.

5 Ranks of a Matrix

In this section, three generalized notions of a rank of matrix will be introduced. All of them are equivalent if the underlying space is linear. However, in a semilinear space they are different. Below, we will analyze a relationship between various ranks in a vectorial semilinear space over $R = (R, +, \cdot, 0, 1)$.

Let A be an $n \times m$ matrix over R. We will associate two vectorial semilinear spaces R^n and R^m with A. R^m is a space which contains rows of A, and R^n is a space which contains columns of A. Rows of A will be denoted by $\bar{a}_1, \bar{a}_2, \dots, \bar{a}_n$, while columns of A will be denoted by $\bar{a}^1, \bar{a}^2, \dots, \bar{a}^m$. Moreover, we will write $A(\bar{a}_1, \bar{a}_2, \dots, \bar{a}_n)$ if A is a matrix with the respective rows and similarly, we will write $A(\bar{a}^1, \bar{a}^2, \dots, \bar{a}^m)$ if A is a matrix with the respective columns.

Definition 6. *Let A be an $n \times m$ matrix over R, i.e. $A \in R^{n \times m}$.*

- *A* rank $r(A)$ *of A (we will call it a* discriminant rank *to distinguish from other ones), is a maximal number k of rows $\bar{a}_{i_1}, \dots, \bar{a}_{i_k}$ (columns $\bar{a}^{j_1}, \dots, \bar{a}^{j_k}$) such that there exists a nonzero k-order minor of the $k \times m$ matrix $A(\bar{a}_{i_1}, \dots, \bar{a}_{i_k})$ ($n \times k$ matrix $A(\bar{a}^{j_1}, \dots, \bar{a}^{j_k})$). It is obvious that $r(A) \leq \min(n, m)$.*
- *A* column rank $r_c(A)$ *of A is the least number of linearly independent column vectors of A that are generators* [1] *of the set $\{\bar{a}^1, \bar{a}^2, \dots, \bar{a}^m\}$.*
- *A* factor rank *(see also [1] for a similar definition) $r_f(A)$ is the least positive integer k, $k \leq \min(m, n)$, such that there exist matrices $B \in R^{n \times k}$, $C \in R^{k \times m}$, that satisfy $A = BC$.*

Example 5. Let $[0,1]_{\text{Ł}} = ([0,1], \vee, \otimes, 0, 1)$ be the semiring from Example 1, case *(ii)*, and matrices $A, B \in [0,1]_{\text{Ł}}^{3\times 3}$ be from Example 4 where we have shown that $\det(A) \equiv 0$ and $\det(B) \not\equiv 0$. Therefore, their discriminant ranks are estimated as $r(B) = 3$ and $r(A) < 3$. It can be easily verified that $r(A) = 2$, because the the following 2×2 submatrix of A

$$A' = \begin{pmatrix} 0.9 & 0.8 \\ 0.6 & 1 \end{pmatrix}$$

has the nonzero bideterminant $\det(A') = (0.9, 0.4)$.

Example 6. Let $(\mathbb{R}_+, \wedge, +, +\infty, 0)$ be the semiring from Example 1, case *(iii)*. Then the bideterminant of the matrix

$$A = \begin{pmatrix} 5 & 3 \\ 10 & 6 \end{pmatrix}$$

is nonzero (namely, $\det(A) = (11, 13)$). Therefore, the discriminant rank $r(A) = 2$.

---

[1] Linearly independent vectors are generators of a set of vectors if any element of this set can be represented as a linear combination of generators.

5.1 Column and Factor Ranks

In this section, we analyze a relationship between column and factor ranks. We will prove that the latter is smaller than the former (cf. [1]).

Proposition 1. *Let R be a commutative semiring and $A \in R^{n\times m}$. Then $r_f(A) \leq r_c(A)$.*

Proof. Let A be an $n \times m$ matrix over R and $r_c(A) = k$. If $k = m$ then the conclusion follows immediately from the obvious representation $A = AE$ where E is the $m \times m$ unit matrix over R. Assume that $k < m$ and that the first k columns of A are linearly independent and generate the whole set of columns $\{\bar{a}^1, \bar{a}^2, \dots, \bar{a}^m\}$. This means that, for every $l = k+1, \dots, m$, there exist coefficients $\beta_{l1}, \dots, \beta_{lk} \in R$ such that $\bar{a}^l = \beta_{l1}\bar{a}^1 + \cdots \beta_{lk}\bar{a}^k$. Denote $\bar{\beta}^l = (\beta_{l1}, \dots, \beta_{lk})^T$ and let $\bar{e}^i \in R^k$ be the i-th unit column vector. Then it is a technical exercise to verify that $A = BC$ where $B_{n\times k} = B(\bar{a}^1, \bar{a}^2, \dots, \bar{a}^k)$ and $C_{k\times m} = C(\bar{e}^1, \dots, \bar{e}^k, \bar{\beta}^{k+1}, \dots, \bar{\beta}^m)$. The latter representation proves that $r_f(A) \leq k$.

Let us remark that the above given proof is simpler than the proof of a similar assertion in [1]. The following corollary (cf. [22]) easily follows from Proposition 1.

Corollary 1. *Let $A \in R^{n\times m}$ and $r_f(A) = m$. Then column vectors of A are linearly independent.*

Proof. Assume that $r_f(A) = m$, but the column vectors $\bar{a}^1, \bar{a}^2, \dots, \bar{a}^m$ of A are linearly dependent. Then at least one vector, say $\bar{a}^m$, can be represented as a linear combination of others:

$$\bar{a}^m = \beta_1\bar{a}^1 + \cdots \beta_{m-1}\bar{a}^{m-1}.$$

Therefore, $A = BC$ where $B_{n\times(m-1)} = B(\bar{a}^1, \bar{a}^2, \dots, \bar{a}^{m-1})$ and $C_{(m-1)\times m} = C(\bar{e}^1, \dots, \bar{e}^{m-1}, \bar{\beta})$ where $\bar{e}^i \in R^{m-1}$ is the i-th unit column vector, $i = 1, \dots, m-1$, and $\bar{\beta} = (\beta_1, \dots, \beta_{m-1})^T$. It follows that $r_f(A) \leq m-1$, which contradicts the assumption.

Remark 2. There are matrices A for which $r_f(A) < r_c(A)$ (see [1]).

5.2 Column and Discriminant Ranks

Below, a relationship between column and discriminant ranks is analysed. We will see that the column rank is greater than or equal to the discriminant rank.

The following statement is a direct consequence of the notion of linear dependence (see Section 3) and basic properties of bideterminant.

Proposition 2. *If row-vectors $\bar{a}_1, \dots, \bar{a}_n \in R^n$ (column-vectors $\bar{a}^1, \dots, \bar{a}^m \in R^n$) of a square matrix $A \in R^{n\times n}$ are linearly dependent then*

$$\det(A(\bar{a}_1, \dots, \bar{a}_n)) \equiv 0 \quad (\det(A(\bar{a}^1, \dots, \bar{a}^n)) \equiv 0).$$

Proof. The proof easily follows from the properties **(P2)** and **(P3)** of a bideterminant.

It is worth to remark that the reverse implication is not true. The counterexample can be found in [6].

Corollary 2. *If row-vectors* $\bar{a}_1, \ldots, \bar{a}_k \in R^m$, $k \leq \min(n, m)$, *(column-vectors* $\bar{a}^1, \ldots, \bar{a}^k \in R^n$*) of* $A \in R^{n \times m}$ *are linearly dependent then*

$$r(A(\bar{a}_1, \ldots, \bar{a}_k)) < k \quad (r(A(\bar{a}^1, \ldots, \bar{a}^k)) < k).$$

Corollary 3. *Let* $A \in R^{n \times m}$, *then*

$$r(A) \leq r_c(A).$$

Proof. If $r_c(A) = m$, then the conclusion follows from the trivial restriction $r(A) \leq \min(n, m)$. Assume that $r_c(A) = k$ and $k < m$. Then any s-order minor of A, where $s \geq k + 1$, is equal to zero. Therefore, by Definition 6, $r(A) < k + 1$ and $r(A) \leq r_c(A)$.

6 Kronecker-Capelli Theorem in a Semilinear Space

Let $R = (R, +, \cdot, 0, 1)$ be a commutative semiring, $m, n \geq 1$, and R^m, R^n vectorial semilinear spaces over R. Let moreover, $A = [a_{ij}]$ be a $n \times m$ matrix and $\bar{b} = (b_1, \ldots, b_n)^T$ vector over R. The following system of equations

$$\begin{aligned} a_{11} \cdot x_1 + \cdots + a_{1m} \cdot x_m &= b_1, \\ \cdots\cdots\cdots\cdots\cdots\cdots & \\ a_{n1} \cdot x_1 + \cdots + a_{nm} \cdot x_m &= b_n, \end{aligned} \tag{4}$$

is considered with respect to the unknown vector $\bar{x} = (x_1 \ldots, x_m)^T \in R^m$. The short denotation of (4) is as follows:

$$A\bar{x} = \bar{b},$$

where $A\bar{x}$ is the product of matrix A and vector $\bar{x}$. The matrix A and vector $\bar{b}$ will be referred to as the *matrix of coefficients* and the *right-hand side* vector of (4).

The following proposition easily follows from definitions above.

Proposition 3. *The system* $A\bar{x} = \bar{b}$ *is solvable if and only if the vector* $\bar{b}$ *can be represented as a linear combination of vector-columns* $\bar{a}^1, \ldots, \bar{a}^m$ *of the matrix of coefficients* A.

Thus, solvability of (4) depends on the relationship between A and $\bar{b}$. In the foregoing text, we have generalized notions of *determinant* and *rank* of matrix. In the text below, we are going to prove that solvability of (4) implies coincidence between the rank of A and the rank of its extension by $\bar{b}$. In linear algebra, this result is known as the Kronecker-Capelli theorem.

Thus, the Kronecker-Capelli theorem gives the criterion of solvability of a system of linear equations. More precisely, it compares ranks of two matrices: the matrix of coefficients and its extension by the right-hand side vector. If they are equal then the system is solvable and vice versa.

In this section, we will see that in a semilinear space, the Kronecker-Capelli theorem is valid only in its first part, i.e. only the necessary condition of solvability is true. Moreover, we will see that this necessary condition is true only for two matrix ranks, namely, for discriminant and factor ranks. The sufficient condition of solvability in the statement of the Kronecker-Capelli theorem is not true for all three introduced above ranks.

The statement below will be used in the proof of the Kronecker-Capelli theorem in a semilinear space.

Proposition 4. *Let A be an $n \times m$ matrix with columns $\bar{a}^1, \dots, \bar{a}^m \in R^n$. Let a column-vector $\bar{a}^{m+1} \in R^n$ be a linear combination of $\bar{a}^1, \dots, \bar{a}^m$. Then discriminant ranks of matrices $A = A(\bar{a}^1, \dots, \bar{a}^m)$ and $A' = A(\bar{a}^1, \dots, \bar{a}^m, \bar{a}^{m+1})$ coincide.*

6.1 Necessary Condition of the Kronecker-Capelli Theorem

Under the assumption that the system (4) is solvable we will examine the relationship between ranks of A and $A\bar{b}$ for three introduced above ranks.

Theorem 1. *If the system (4) is solvable then discriminant ranks of A and $A\bar{b}$ are equal, i.e. $r(A) = r(A\bar{b})$.*

Proof. Assume that the system (4) is solvable. Then vector $\bar{b}$ is a linear combination of the column vectors of A. Therefore, the conclusion of the theorem follows from Proposition 4.

Below, we will show that the claim of Theorem 1 is not valid for the column rank. The following example gives the justification.

Example 7. Let $B = (B_4, \vee, \wedge, ', 0, 1)$ be a boolean algebra on the partially ordered set $B_4 = \{0, a, b, 1\}$, see Fig. 1. The reduct $B_s = (B_4, \vee, \wedge, 0, 1)$ is a commutative semiring.

Let us consider the following system of equations over B_s:

$$\begin{aligned} (a \wedge x_1) \vee (b \wedge x_2) &= 1, \\ 1 \wedge x_3 &= 0, \end{aligned}$$

where the matrix of coefficients and the right-hand side vector are as follows:

$$A = \begin{pmatrix} a & b & 0 \\ 0 & 0 & 1 \end{pmatrix}, \quad \bar{b} = \begin{pmatrix} 1 \\ 0 \end{pmatrix}.$$

The system is solvable, and $x_1 = 1, x_2 = 1, x_3 = 0$ is its solution. However, $r_c(A) \neq r_c(A\bar{b})$ because $r_c(A) = 3$, while $r_c(A\bar{b}) = 2$.

Finally, let us remark that the claim of Theorem 1 is valid for the factor rank. This fact has been shown in [22], and we will not repeat the arguments.

6.2 Sufficient Condition of the Kronecker-Capelli Theorem

Let us study the converse statement to that in Theorem 1: under the assumption that ranks of A and $A\bar{b}$ are equal deduce that the system (4) is solvable. In the below given examples, we will show that for three introduced above ranks this converse statement is not true.

In the Example 8, the discriminant rank is used.

Example 8. Let R be a semiring from Example 1, case *(iii), i.e.* $R = (\mathbb{R}_+, \wedge, +, +\infty, 0)$. *Then the following system of equations is an instance of (4):*

$$(5 + x_1) \wedge (3 + x_2) = 7, \tag{5}$$

$$(10 + x_1) \wedge (6 + x_2) = 20, \tag{6}$$

where the matrix of coefficients and the right-hand side vector are as follows:

$$A = \begin{pmatrix} 5 & 3 \\ 10 & 6 \end{pmatrix}, \quad \bar{b} = \begin{pmatrix} 7 \\ 20 \end{pmatrix}.$$

Then $det(A) = (11, 13) \not\equiv 0$, *see Example 6. It follows that* $r(A) = 2$ *and* $r(A\bar{b}) = 2$ *as well. Thus,* $r(A) = r(A\bar{b})$. *However, the system of equations (5)-(6) is not solvable. In order to prove this, let us assume the opposite. Then by (5), either* $x_1 = 2$, $x_2 \geq 4$, *and (6) is transformed to*

$$12 \wedge (6 + x_2) = 20,$$

or $x_1 \geq 2$, $x_2 = 4$, *and (6) is transformed to*

$$(10 + x_1) \wedge 10 = 20.$$

None of the latter equations is solvable.

In the Example 9, the column rank is used.

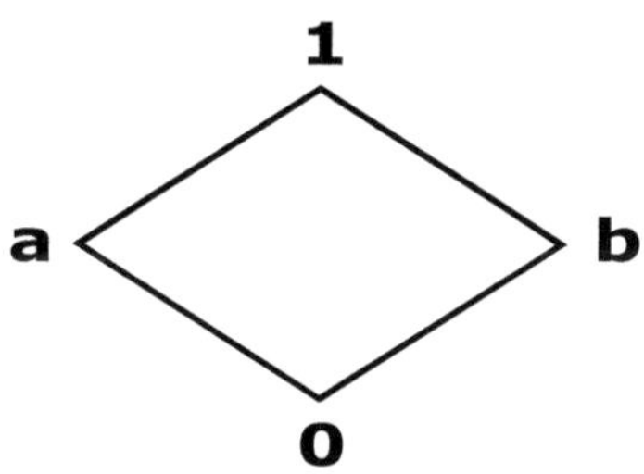

Fig. 1. Partially ordered set $B_4 = \{0, a, b, 1\}$

Example 9. Let $[0,1]_{Ł} = ([0,1], \vee, \otimes, 0, 1)$ be a semiring from Example 1, case *(ii)*, and $A\bar{x} = \bar{b}$ a system of equations over $[0,1]_{Ł}$ where

$$A = \begin{pmatrix} \frac{1}{2} & \frac{1}{2} \\ \frac{1}{2} & \frac{1}{4} \end{pmatrix}, \ \bar{b} = \begin{pmatrix} 1 \\ 1 \end{pmatrix}.$$

Because for any $x_1, x_2 \in L$,

$$\frac{1}{2} \otimes x_1 \vee \frac{1}{2} \otimes x_2 \le \frac{1}{2}, \qquad \frac{1}{2} \otimes x_1 \vee \frac{1}{4} \otimes x_2 \le \frac{1}{2},$$

the system is not solvable. It can be easily checked that $\bar{a}^1, \bar{a}^2$ are linearly independent, and thus, $r_c(A) = 2$. Analogously, $\bar{b}, \bar{a}_2$ are linearly independent too. However, due to $\bar{a}_1 = 1/2\bar{b}$, we have that $r_c(A\bar{b}) = 2$.

In the Example 10, the factor rank is used.

Example 10. Let $[0,1]_{Ł} = ([0,1], \vee, \otimes, 0, 1)$ be a semiring from Example 1, case *(ii)*, and $A\bar{x} = \bar{b}$ a system of equations over $[0,1]_{Ł}$ where

$$A = \begin{pmatrix} \frac{1}{2} & \frac{1}{4} \\ \frac{1}{2} & \frac{1}{4} \end{pmatrix}, \ \bar{b} = \begin{pmatrix} \frac{3}{4} \\ \frac{3}{4} \end{pmatrix}.$$

By the similar argumentation as in the Example 9, the above given system is not solvable (indeed, for any $x_1, x_2 \in L$, $\frac{1}{2} \otimes x_1 \vee \frac{1}{4} \otimes x_2 \le \frac{1}{2}$, and thus, it cannot be equal to $\frac{3}{4}$). On the other side, A (respectively, $A\bar{b}$) can be expressed as a product of

$$B = \begin{pmatrix} \frac{3}{4} \\ \frac{3}{4} \end{pmatrix} \text{ and } C = \begin{pmatrix} \frac{3}{4} & \frac{1}{2} \end{pmatrix} \left(\textit{respectively, } C = \begin{pmatrix} \frac{3}{4} & \frac{1}{2} & 1 \end{pmatrix}\right),$$

Because B is a 2×1 matrix, then $r_f(A) = r_f(A\bar{b}) = 1$.

Thus, the converse statement to that in Theorem 1 is not true for all three above considered ranks.

7 Solvability of Systems of Fuzzy Relation Equations

Two below given systems (8) and (9) of *fuzzy relation equations* (FRE) are important instances of the generic system of equations (4). Both systems were extensively investigated in the literature, see e.g., [5,3,8,9,12,17,14,16,18,19,20,21] for various results about their solvability. In this section, we will show how the generalized Kronecker-Capelli theorem can be used in the analysis of solvability of (8) and (9).

Let throughout this section, $L = (L, \vee, \wedge, *, \to, 0, 1)$ be a residuated lattice extended by the operation of negation:

$$\neg a = a \to 0. \tag{7}$$

Let $L_\vee = (L, \vee, *, 0, 1)$ be a commutative $\vee$-semiring from Example 1, case *(iv)*, and $L_\vee^n = (L^n, \vee, \bar{0})$ be a corresponding vectorial semilinear space over $L_\vee$ (see Example 2), case *(i)*. Below, we introduce another vectorial semilinear space over $L_\vee$ which can be considered as a dual with respect to the former.

Example 11. Let the semiring $L_\vee = (L, \vee, *, 0, 1)$ be as above. Define $L^n_\wedge = (L^n, \wedge, \bar{1})$, $n \geq 1$, as a dual vectorial semilinear space over $L_\vee$, where L^n is a set of n-dimensional vectors with components from L, and $\bar{1} = (1, \dots, 1)^T \in L^n$. Operations (addition and external multiplication) are as follows:

$$(x_1, \dots, x_n)^T \wedge (y_1, \dots, y_n)^T = (x_1 \wedge y_1, \dots, x_n \wedge y_n)^T,$$

and for any $\lambda \in L$,

$$\lambda \diagdown (x_1, \dots, x_n)^T = (\lambda \to x_1, \dots, \lambda \to x_n)^T.$$

Then it can be shown that $L^n_\wedge$ is a $\wedge$-semilinear space over $L_\vee$.

Let $m, n \geq 1$, and $n \times m$ matrix $A = [a_{ij}]$ and vector $\bar{b} = (b_1, \dots, b_n)^T$ have components in L. The first system of FRE

$$\begin{aligned} a_{11} * x_1 \vee \cdots \vee a_{1m} * x_m &= b_1, \\ \cdots\cdots\cdots\cdots\cdots\cdots & \\ a_{n1} * x_1 \vee \cdots \vee a_{nm} * x_m &= b_n, \end{aligned} \tag{8}$$

is written in the language of the vectorial semilinear space $L^n_\vee = (L^n, \vee, \bar{0})$ over $L_\vee$. It is known as a system of FRE with the sup-$*$ composition.

The second system of FRE

$$\begin{aligned} (a_{11} \to x_1) \wedge \cdots \wedge (a_{1m} \to x_m) &= \neg b_1, \\ \cdots\cdots\cdots\cdots\cdots\cdots & \\ (a_{n1} \to x_1) \wedge \cdots \wedge (a_{nm} \to x_m) &= \neg b_n, \end{aligned} \tag{9}$$

is written in the language of the dual vectorial semilinear space $L^n_\wedge = (L^n, \wedge, \bar{1})$ over $L_\vee$. It is known as a system of FRE with the inf-$\to$ composition.

7.1 Verification of Solvability

In this section, we will see how the generalized Kronecker-Capelli theorem can be used in proving that the system (8) is not solvable. For this purpose, we will use a contrapositive formulation of Theorem 1 with the discriminant rank of both matrices. This is because a computation of this type of rank can be performed algorithmically. The contrapositive assertion claims that if the rank of a matrix of coefficients A is not equal to the rank of its extension $A\bar{b}$, then the system (8) is not solvable.

At first, let us discuss a relationship between systems (8) and (9). We claim that they are not equivalent, but can be transformed into two systems that are *equivalent*. The latter means that every solution of one system is a solution of the other one and vice versa. The details are given below and are proven in [15]. Corollary 4 characterizes relationship between (8) and (9) in more details.

We will first transform (8) by applying $\neg$ to both sides of every equation:

$$\begin{aligned} \neg(a_{11} * x_1 \vee \cdots \vee a_{1m} * x_m) &= \neg b_1, \\ \cdots\cdots\cdots\cdots\cdots\cdots\cdots\cdots & \\ \neg(a_{n1} * x_1 \vee \cdots \vee a_{nm} * x_m) &= \neg b_n. \end{aligned} \tag{10}$$

Because in general, $\neg$ is not involutive, systems (8) and (10) are not equivalent. However, all solutions of (8) are solutions of (10). Moreover, if the latter has no solutions, so does the former.

At second, we will transform (9) by substituting negation $\neg\bar{x}$ for $\bar{x}$ where $\neg\bar{x} = (\neg x_1, \ldots, \neg x_n)^T$:

$$\begin{aligned} (a_{11} \rightarrow \neg x_1) \wedge \cdots \wedge (a_{1m} \rightarrow \neg x_m) &= \neg b_1, \\ \cdots\cdots\cdots\cdots\cdots\cdots\cdots\cdots & \\ (a_{n1} \rightarrow \neg x_1) \wedge \cdots \wedge (a_{nm} \rightarrow \neg x_m) &= \neg b_n. \end{aligned} \tag{11}$$

Proposition 5 ([15]). *Systems (10) and (11) are equivalent.*

Corollary 4. *If $\bar{x}^0 = (x_1^0 \ldots, x_m^0) \in L^m$ is a solution of (8) then $\neg\bar{x}^0 = (\neg x_1^0, \ldots, \neg x_m^0)$ is a solution of (9).*

Remark 3. Without going into specific details, let us remark that if an underlying residuated lattice L is an MV-algebra (L is divisible and keeps the law of double negation) then systems (8) and (9) are equivalent. In other words, vector $\bar{x}^0 = (x_1^0 \ldots, x_m^0) \in L^m$ is a solution of (8) if and only if vector $\neg\bar{x}^0 = (\neg x_1^0 \ldots, \neg x_m^0)$ is a solution of (9).

In the rest of this section, we will demonstrate, how the generalized Kronecker-Capelli theorem can be used in proving that both systems (8) and (9) are not solvable. We will choose the semiring $L = ([0,1], \vee, \otimes, 0, 1)$ as in Example 1, case *(ii)*. Let us remark that this semiring is a reduct of Łukasiewicz algebra, which is an example of an MV-algebra. In accordance with Remark 3, both systems (8) and (9) are equivalent, so that it is sufficient to analyze the solvability of (8).

Let the system (8) be determined by the matrix of coefficients A and the right-hand side vector $\bar{b}$ where

$$A = \begin{pmatrix} 0.9 & 0.8 & 0.6 \\ 0.6 & 1 & 1 \\ 0.6 & 0.8 & 0.8 \end{pmatrix}, \quad \bar{b} = \begin{pmatrix} 0.4 \\ 0.6 \\ 0.3 \end{pmatrix}.$$

It has been shown in Example 4 that $\det(A) = (0.7, 0.7)$ and $\det(B) = (0.3, 0.2)$ where B is obtained from A after we replace the second column of A by $\bar{b}$, i.e. $B = B(\bar{a}^1, \bar{b}, \bar{a}^3)$. Moreover, in Example 5, we showed that discriminant ranks of A and B are equal to $r(A) = 2$ and $r(B) = 3$. It is easy to see that $r(B) = r(A\bar{b}) = 3$. Therefore, $r(A) \neq r(A\bar{b})$, and by Theorem 1, system (8) determined by the matrix of coefficients A and the right-hand side vector $\bar{b}$, is not solvable. Because in the considered case, systems (8) and (9) are equivalent, then the system (9) is not solvable as well.

8 Conclusion

In this contribution, we recalled the notion of a semilinear (vector) space as a commutative semimodule over a commutative semiring. The main studied problem was solvability of a system of linear-like equations in a semilinear space. We investigated applicability of classical tools which use the notions of determinant and rank of a matrix of coefficients. With this purpose we investigated a bideterminant and a rank of a matrix in a semilinear space. We proved the necessary condition of the generalized Kronecker-Capelli theorem and showed that the converse statement does not hold in semilinear spaces. Finally, we demonstrated how the generalized Kronecker-Capelli theorem is used in verification that system of linear-like equations has no solution.

Acknowledgment. This work was supported by the European Regional Development Fund in the IT4Innovations Centre of Excellence project (CZ.1.05/ 1.1.00/02.0070).

References

1. Beasley, L.B., Pullman, N.J.: Semiring Rank vs. Column Rank. Linear Algebra Appl. 101, 33–48 (1988)
2. Cuninghame-Green, R.A., Butkovic, P.: Bases in max-algebra. Linear Algebra and its Applications 389, 107–120 (2004)
3. De Baets, B.: Analytical solution methods for fuzzy relation equations. In: Dubois, D., Prade, H. (eds.) The Handbooks of Fuzzy Sets Series, vol. 1, pp. 291–340. Kluwer, Dordrecht (2000)
4. Di Nola, A., Lettieri, A., Perfilieva, I., Novák, V.: Algebraic analysis of fuzzy systems. Fuzzy Sets and Systems 158, 1–22 (2007)
5. Di Nola, A., Sessa, S., Pedrycz, W., Sanchez, E.: Fuzzy Relation Equations and Their Applications to Knowledge Engineering. Kluwer, Boston (1989)
6. Golan, J.S.: Semirings and their Applications. Kluwer Academic Pulishers, Dordrecht (1999)
7. Gondran, M., Minoux, M.: Graphs, Dioids and Semirings. Springer, New York (2008)
8. Gottwald, S.: On the existence of solutions of systems of fuzzy equations. Fuzzy Sets and Systems 12, 301–302 (1984)
9. Klawonn, F.: Fuzzy Points, Fuzzy Relations and Fuzzy Functions. In: Novák, V., Perfilieva, I. (eds.) Discovering the World with Fuzzy Logic, pp. 431–453. Springer, Berlin (2000)
10. Kuntzman, J.: Théorie des réseaux graphes, Dunod (Libraire), Paris (1972)
11. Minoux, M.: Bideterminants, arborescences and extension of the matrix- tree theorem to semirings. Discrete Math. 171, 191–200 (1997)
12. Pedrycz, W.: Inverse problem in fuzzy relational equations. Fuzzy Sets and Systems 36, 277–291 (1990)
13. Perfilieva, I.: Semilinear spaces – basic structures for fuzzy systems. In: Proc. Conf. IPMU 2006, Paris, France, pp. 1087–1094 (2006)
14. Perfilieva, I., Gottwald, S.: Solvability and approximate solvability of fuzzy relation equations. Int. J. of General Systems 32, 361–372 (2003)

15. Perfilieva, I., Kupka, J.: Can fuzzy relation equations be treated like linear equations? In: Proc. World Conference on Soft Computing, pp. 179–185. San Francisco State University, San Francisco (2011)
16. Perfilieva, I., Nosková, L.: System of fuzzy relation equations with inf $\rightarrow$ composition: complete set of solutions. Fuzzy Sets and Systems 159, 2256–2271 (2008)
17. Perfilieva, I., Tonis, A.: Compatibility of systems of fuzzy relation equations. Int. J. of General Systems 29, 511–528 (2000)
18. Sanchez, E.: Resolution of composite fuzzy relation equations. Information and Control 30, 38–48 (1976)
19. Sessa, S.: Finite fuzzy relation equations with unique solution in complete Brouwerian lattices. Fuzzy Sets and Systems 29, 103–113 (1989)
20. Wang, X.-P.: Infinite fuzzy relational equations on a complete Brouwerian lattice. Fuzzy Sets and Systems 138, 657–666 (2003)
21. Wang, X.-P., Xiong, Q.-Q.: Some properties of sup-min fuzzy relational equations on infinite domains. Fuzzy Sets and Systems 151, 393–402 (2005)
22. Zhao, S., Wang, X.-P.: Invertible matrices and semilinear spaces over commutative semirings. Information Sciences 180, 5115–5124 (2010)

The Z-Number Enigma: A Study through an Experiment

Romi Banerjee and Sankar K. Pal

Centre for Soft Computing Research
Indian Statistical Institute
Kolkata 700 108, India
rm.banerjee@gmail.com, sankar@isical.ac.in

Abstract. The Z-number, proposed by Zadeh in the year 2011, is a new fuzzy-theoretic approach to the Computing With Words (CWW) paradigm. It aspires to capture the uncertainty of information conveyed by a sentence, and serve as a model for the precisiation and linguistic summarization of a natural language statement. The Z-number thereby, lends a new dimension to CWW – uniting CWW with Natural Language Processing (NLP). This article is an illumination upon our exploration of the Z-number approach to CWW. Here, we enlist the probable contributions of the Z-number to CWW, present our algorithm for CWW using the Z-number, and describe a simulation of the technique with respect to a real-life example of CWW. In the course of the simulation, we extend the interpretation of the set-theoretic intersection operator to evaluate the intersection of perceptions and discover some of the challenges underlying the implementation of the Z-number in the area of CWW.

Keywords: Computing With Words (CWW), cognition, fuzzy sets, linguistics, machine learning, text summarization, natural language processing, perceptions, soft computing, natural computing.

1 Introduction

Words encode perceptions and are inherently imprecise; not all events are expressible through precise numbers and symbols. Cognitive text or speech comprehension begins with the identification of the meaning of the constituent words, as per their usage, to arrive at the meaning of sentences; and a union of these sentence-perceptions leads to the comprehension of the complete text or speech sample.

The Computing With Words (CWW) paradigm, coined by Zadeh in 1996, draws inspiration from the remarkable perception-based decision-making ability of the human brain; the perceptions being encoded in the words and phrases used to describe events. The Intelligent Systems Revolution symbolizes the generation of machines that possess high levels of Machine IQ [1], [2], and undeniably, an implementation of the CWW paradigm would be a step in that direction.

Realization of the paradigm requires the machine to comprehend word-perceptions as well as a human being. Consequently, not only does the machine need to learn words – both existing and new, it also needs to apply them to form semantically, and

R.R. Yager et al. (Eds.): *Soft Computing: State of the Art Theory*, STUDFUZZ 291, pp. 71–88.
DOI: 10.1007/978-3-642-34922-5_6

ideally syntactically, correct natural language statements. Meanings of words, thus, need to be translated into some symbolic form. CWW is deeply engrained in the concepts of fuzzy logic [3], fuzzy linguistics [4], test score semantics [5] and PNL [6], and is the precursor to "Computing With Perceptions (CTP) [7]".

The Z-number [8], proposed by Zadeh, is a very new approach to the CWW paradigm. Besides unifying the concepts of fuzzy logic, fuzzy linguistics, test score semantics and PNL, the Z-number incorporates a measure of the reliability of the information in the sentence. Consequently, the Z-number could be sought as a medium of extension of the basic element of computation of CWW from word-perceptions to sentence-perceptions. We envision that this would lead to the development of a model of the natural process of comprehension in human beings.

In this article, we present a comprehensive study of the Z-number approach to CWW from an NLP perspective [9]. We begin with a recapitulation on sentences and its types (Section 2), followed by an overview of the basics of the Z-number (Section 3.1), and then move on to predict some of the contributions of the Z-number methodology to CWW (Section 3.2) and formulate a basic algorithm for the Z-number based CWW (Section 3.3).

The article, thereafter, focuses on our experiment where the Z-number approach is applied to a real-life example of CWW (an attempt at the simulation of the natural CWW in a machine). The experiment begins with an algorithm underlying the intuitive process of book selection in human beings (Section 4.1), which is then extended to include the Z-number components to simulate the book selection process in a machine (Section 4.2.B). The experiment calls for the extension of the basic intersection operator to allow the intersection of perceptions (Section 4.2.A) and leads to the discernment of some of the inherent challenges underlying the Z-numbers in CWW (Section 4.2.C).

2 A Recapitulation on Sentences [10]

A system that is to compute with words, essentially deals with computing with the words in sentences. In acknowledgement to the diversity of human speech and the different ways the same event can be described, this section prepares the base for the different kinds of sentences that such a system needs to anticipate and comprehend. Each type of sentence has its own unique set of rules and the component words and phrases need to be processed accordingly.

In the article, we use the words 'statement' and 'sentence' to mean the same – "Sentences are collections of words that make complete sense. The sense is not complete, unless something is being said about something".

Sentences are classified into a number of categories:

i. Based on the number of independent clauses –
   a. Simple sentences: Sentences with a single independent clause.
      E.g.: I love to read.
   b. Compound sentences: Sentences with two or more independent clauses.
      E.g.: He came back tired for he had been working all day.

c. Complex sentences: Sentences with a single independent clause and more than one dependent clause.
   E.g.: They went for a movie after they had completed their homework.

ii. Based on the nature of the sentences –
   a. Declarative sentences: Sentences that are used to make a simple statement. A declarative sentence ends with a full stop.
      E.g.: I love to read.
   b. Interrogative sentences: Sentence that are used to make queries. An interrogative sentence ends with a question mark.
      E.g.: Do you love to read?
   c. Exclamatory sentences: Sentences that are used to emphasize a fact or convey an emotion. An exclamatory sentence ends with an exclamation mark.
      E.g.: What a wonderful read!
   d. Imperative sentences: Sentences that are used to command or request, with the pronoun 'you' implied.
      E.g.: Please get me something to read.
   e. Conditional sentences: Sentences that indicate dependencies between conditions, a "cause and effect" relationship.
      E.g.: If I had a billion dollars, I would buy a castle of books.

Though not in common use, types of sentences also include –

i. Rhetorical Question: A question that is posed for persuasive effect without the expectation of an answer.
   E.g.: When will people learn the consequences of turning every bit of open space into cement monstrosities?
ii. Paradox: A paradox is a logical statement or group of statements that lead to a contradiction or a situation which (if true) defies logic or reason.
   E.g. : One thing that I know is that I know nothing (Socrates' Paradox).

3 The Z-Number [8]

Actions rely on decisions which again depend on the information provided. Thus, greater the reliability of the information better is the decision made. The Z-number approach to CWW includes a measure of the reliability of the information in a statement, along with other parameters that result in the linguistic summarization of the sentence. The Z-number could thus form the basis of CWW-based discourse-oriented systems.

3.1 An Overview of the Z-Number

Given a natural language statement, Y, the Z-number of Y is defined as a 2-tuple $Z = \langle A, B\rangle$, where A, a linguistic value, implies the restriction (constraint) on the values of X, a real valued linguistic variable – interpreted as the subject of Y, and B is a measure of the reliability (certainty) of A. Typically, A and B are expressed as words or clauses, and are both fuzzy numbers.

Examples :

i. Y_1 = Dinner's usually served at 9:00 pm.
Therefore, X = time of dinner service, and Z = <by 9:00pm, usually>

ii. Y_2 = This book has been a wonderful read.
Therefore, X = quality of book, and Z = <wonderful, certainly>

Understandably, A is context-dependent and explicitly mentioned while B is based on the perception of certainty presented by the statement. The value of B could be explicitly quoted in the statement (as in example i) or it could be implicit (as in example ii).

The ordered 3-tuple $\langle X, A, B\rangle$ is referred to as a Z-valuation. A Z-valuation is equivalent to an assignment statement X is $\langle A, B\rangle$. As for example,

i. The Z-valuation of Y_1: <time of dinner service , by 9:00 pm, usually>
Implication: [time of dinner service] is (by 9:00 pm, usually)

ii. The Z-valuation of Y_2: <quality of book, wonderful, certainly>
Implication: [quality of book] is (wonderful, certainly).

A collection of Z-valuations is referred to as Z-information. The Z-information provides the impetus to a decision-making process.

Preliminary rules of computation using the Z-numbers:

i. For the purpose of computation, the values of A and B need to be precisiated through association with membership functions, μ_A, μ_B respectively.

ii. X and A together define a random event in R, and the probability of this event, p, may be expressed as:

$$p = \int_R \mu_A(u) p_X(u) du \quad (1)$$

where, u is a real valued generic value of X and p_X is the underlying (hidden) probability density of X.

iii. The Z-valuation $\langle X, A, B\rangle$ is viewed as a generalized constraint on X, and is defined by:

Probability (X is A) is B,

or, $p = \int_R \mu_A(u) p_X(u) du$ is B (2)

iv. Equation 2 is mathematically equivalent to the expression:

$$\mu_B\left(\int_R \mu_A(u) p_X(u) du\right) \quad (3)$$

Subject to,

$$\int_R p_X(u)du = 1 \tag{4}$$

v. Computation using the Z-numbers is based on the Principle of Extension. As for example, considering a problem statement of the form:
"It is likely that the desert is good. What is the probability that it is not?"
Let, X = desert, A = good, B = likely, C = not good, D = degree of certainty;
μ_A, μ_B, μ_C, μ_D are the membership functions associated with A, B, C and D respectively;
p_X is the underlying (hidden) probability density of X;
u is a real valued generic value of X.
Therefore, we have, (X is A) is B, and
We need to evaluate: (X is C) is ?D.
Thus, using the Principle of Extension and Equations 2, 3 and 4:

$$\frac{<X, A, B>}{<X, C, ?D>} = \frac{\mu_B(\int_R \mu_A(u) p_X(u)du)}{(\int_R \mu_C(u) p_X(u)du) is ?D}$$

$$\text{or, } \mu_D(w) = \sup_{p_X} \mu_B(\int_R \mu_A(u) p_X(u)du) \tag{5}$$

Subject to,

$$w = (\int_R \mu_C(u) p_X(u)du) \text{ and}$$
$$\int_R p_X(u)du = 1 \tag{6}$$

3.2 Probable Contributions of the Z-Number to CWW

The definition of the Z-number subtly identifies the following as the contributions of the concept to the arena of CWW:

i. The Z-number serves as a model for the precisiation of natural language statements – unifying CWW and NLP.
ii. The Z-number summarizes the perception of a single simple sentence. If complex or compound sentences be deconstructed to their simple sentence components, the Z-number of each of the individual sentences can be evaluated to receive the Z-information for the entire sentence.
iii. The Z-information summarizes the perception of a group of statements.
iv. The parameters of Z-numbers help in the identification of the context of discourse of the sentence under observation – thus allowing sentences to be grouped into context-sensitive granules.
v. By virtue of points i though iv, the Z-number is visibly in agreement with the intuitive process of comprehension and reasoning in human beings.
vi. The Z-number could be used to extract 'knowledge' from a sentence.
vii. The parameters of the Z-numbers are context-independent.
viii. Translation from the Z-numbers to simple sentences is straightforward.

3.3 Algorithm for CWW Following the Z-Number Technique

Drawing inspiration from the possible areas of contribution of the Z-number to CWW, we aspire to design CWW-based systems that utilize the Z-number technique. It is thus that we have developed the following algorithm (Algorithm 1) for CWW based on the Z-number methodology. The algorithm takes natural language sentences as input and results in the natural language response to them as well.

Algorithm 1.

Input: Natural language sentence (I)

Output: Natural language response (O) to I

Assumptions:

i. The system is capable of distinguishing between relevant and irrelevant sentences

ii. The system comprehends the total perception of a complex or a compound sentence (Y) by –
   a. Extraction of the simple sentence components of Y
   b. Comprehension of the meaning of each of these simple sentence components
   c. Combining these component perceptions with respect to the connectives in Y

[Assumption ii. follows the natural process of cognition in human beings]

Steps:

1. If I is irrelevant
   Then
   Goto step 10
   Else
   Goto step 2
2. If I is a simple sentence
   Then
   Goto step 3
   Else
   i. Extract the simple sentence component set (I') of I
   ii. Repeat steps 3 through 4 for each sentence in I'
   iii. Goto step 5.
3. Extract the values of X, A and B in the sentence to evaluate the Z-valuation (Z_I)
4. Convert Z_I into equivalent logical expression (Z_E)
5. Combine all Z_E to the logical expression (E) guided by the connectives in I
6. Convert E to the equivalent mathematical expression (M)
7. Evaluate M to receive a set of Z-valuations (Z_O) in response
8. Translate Z_O into simple sentences (S)
9. If step 8 results in more than one simple sentence
   Then

If some or all the sentences in S can be compiled into a single sentence
Then

a. Assimilate all compatible simple sentences into a single complex or a compound sentence (S')

b. If S' does not include all the sentences in S
Then

b.1. $S'' = S - S'$

b.2. $O = S' \cup S''$

Else

$O = S'$

c. Goto step 10

Else

a. $O = S$

b. Goto step 10

Else

i. $O = S$

ii. Goto step 10

10. Stop

4 Experiment

In this section, we describe an experiment where we strive to simulate the natural CWW by human beings. The machine tries to model the intuitive process of reasoning involved in the selection of a book at a bookstore.

Assuming that a human reader has a genre and content preference, the process of book selection intuitively follows the algorithm (Algorithm 2) outlined in Section 4.1.

Algorithm 2 is then extended to include elements of Z-number based CWW, as is described in Algorithm 1, so as to simulate the book selection process in a machine.

4.1 The Intuitive Algorithm Underlying the Process of 'Selection of a Book'

Algorithm 2.

Input:

i. The summary set (S) of a set of n ($n \geq 1$) books (B), where $S = \{S_1, S_2, \ldots, S_n\}$, $B = \{B_1, B_2, \ldots, B_n\}$ and S_i is the summary of B_i – extracted from the book jacket

ii. Event set (E) of the reader's expectation of the literary contents of the book to be chosen

iii. Lower threshold (T) indicating the minimum number of events, out of E, that need to exist in the S_i of the selected B_i

iv. Additional selection criteria (C) with respect to the external characteristics of the book required.

Output: Book selected

Steps:

1. For every $B_i \in B$
   Repeat steps 2 though 4
2. Read S_i
3. Extract event set E_S in S_i
4. If $| E \cap E_S | \geq T$
   Then
   Assign B_i a membership of selection (> 0)
   Else
   Membership of selection of $B_i = 0$
5. Sort B_i in the descending order of memberships of selection
6. If every book is assigned a unique membership of selection
   Then
   Select B_i with the highest membership as the book to be purchased
   Else
   i. Select all B_i with the highest membership
   ii. Identify the B_i that satisfies most of C
   iii. This B_i is selected as the book to be purchased
7. Stop

Note:

i. Each event in the event set (E) is implied by 'keywords' – the basis for the natural CWW by the reader.
ii. Besides the summary of the book, the following are natural contributing factors to the selection process and are elements of C :
   a. The book being read or present in the possession of the reader
   b. Cost of the book
   c. Condition of the book – brittleness, ink marks, crumpled pages, pages missing
   d. Presence and quality of pictures
   e. Font type and font size
   f. Rarity of the book
iii. In this experiment, the selection process is entirely platonic – devoid of any emotions occurring in real-life situations that lead to contemplation of an increase in budget or to the selection of a book quite different from that planned.

4.2 Simulation of the Process of 'Selection of a Book' Using the Z-Number Technique

Intersection of Perceptions Conveyed by the Event-set (E) of the Reader's Expectations and the Event-set (E_s) Expressed in the Summary of the Book under Observation.

An intersection of the perceptions conveyed by E and E_s should practically imply the extraction of the common meaning conveyed by these perceptions. It is thus that we define the perception-intersection operator $(\cap_p)$ as follows:

Let,

E and E_s consist of Z-valuations of the sentences expressing the reader's expectations and the summary of the book, respectively; and

$$E' = \langle X_1, A_1, B_1\rangle \text{ and } E_S' = \langle X_2, A_2, B_2\rangle \text{ where } E' \in E \text{ and } E_S' \in E_S$$

Then,

$$(E' \cap_p E_S') = \langle X_1, A_1, B_2\rangle \text{ if } (A_1 \cap A_2) \neq \varnothing \tag{7}$$

X_1 and A_1 respectively represent a requirement and the corresponding expected value, while B_2 reflects the level of certainty with which the requirement is satisfied by the book under consideration.

The perception-intersection operator defined above could come of use in scenarios where it is imperative to verify the certainty with which a current situation satisfies a given requirement.

The Algorithm for the Process of 'Selection of a Book' Using the Z-number Technique

Combining elements of Algorithms 1 and 2, the algorithm for the book selection process using the Z-number Technique is outlined as follows:

Algorithm 3.

Input:

i. The summary set (S) of a set of n ($n \geq 1$) books (B), where $S = \{S_1, S_2, \ldots, S_n\}$, $B = \{B_1, B_2, \ldots, B_n\}$ and S_i is the summary of B_i – extracted from the book jacket
ii. Event set (E) – as Z-valuations – of the reader's expectation of the literary contents of the book to be chosen
iii. Lower threshold (T) indicating the minimum number of events, out of E, that need to exist in the S_i of the selected B_i
iv. Additional selection criteria (C) with respect to the external characteristics of the book required.

Output: Book selected

Assumptions:

i. A sentence is considered relevant if it contains at least one keyword from E
ii. Any word (W) in the summary that is a synonym of a keyword (W') in E, is treated as W'
iii. The system comprehends the total perception of a complex or a compound sentence (Y) by –
    a. Extraction of the simple sentence components of Y
    b. Comprehension of the meaning of each of these simple sentence components

c. Combining these component perceptions with respect to the connectives in Y

iv. The memberships of selection are assigned on the basis of the principle of extension, explained in Section 3.1

v. The reader does not read the summary of a book he/she has read or possesses.

Steps:

1. For every B_i in B
   Repeat steps 2 though 10
2. Initialize $E_S = \varnothing$ [E_S = Event set for the current B_i]
3. For every sentence (I) in S_i
   Repeat steps 4 through 7
4. If I is irrelevant
   Then
      i. Discard I
      ii. Goto next I
   Else
      Goto step 5
5. If I is a simple sentence
   Then
      Goto step 6
   Else
      i. Extract the simple sentence component set (I') of I
      ii. Repeat steps 6 through 7 for each sentence in I'
      iii. Goto step 8.
6. Extract the values of X, A and B in the sentence to evaluate the Z-valuation (Z_I)
7. Combine all Z_I to the logical expression (E') guided by the connectives in I
8. $E_S = E_S \cup E'$
9. If $| E \cap_p E_S | \geq T$
   Then
      i. Convert ($E \cap_p E_S$) to the logical expression ($\tilde{E}$)
      ii. Convert E and $\tilde{E}$ to the mathematical expressions M and $\tilde{M}$ respectively.
      iii. Evaluate the membership of selection of B_i by applying the principle of extension on M and $\tilde{M}$
   Else
      Membership of selection of $B_i = 0$
10. Sort B_i in the descending order of memberships of selection
11. If every book is assigned a unique membership of selection
    Then

Select B_i with the highest membership as the book to be purchased

Else

i. Select all B_i with the highest membership

ii. Identify the B_i that satisfies most of C

iii. This B_i is selected as the book to be purchased

12. Stop

Simulation of Algorithm 3

Assumptions:

i. The machine (*Mc*) is capable of annotating the words in a given text sample into the correct parts of speech and is capable of identifying the type of the sentence under consideration.

ii. *Mc* has read thirty works of fiction in the 'Mystery' genre.

iii. *Mc*'s vocabulary consists of one hundred and sixty five keywords (X). Each of these words is assigned a probability of occurrence (p_x), based on the number of books the words are found in.

The words in *Mc*'s vocabulary are: abduct, accomplice, advocate, agent, alibi, allegation, ammunition, anonymous, arms, assistant, awkward, baffle, blood, blunder, bury, case, catch, chief, chilling consequence, clue, cold-blooded, conspiracy, constable, convict, corpse, crime, criminologist, crooked, curious, danger, death, deceive, deduce, desperate, detective, discover, doctor, drug, duplicate, eavesdrop, enemy, evidence, evil, exhume, fake, fatal, fear, figure-out, find-out, fingerprint, follow, forbidden, forget, foul play, gang, gore, graveyard, gray cells, guilty, headquarters, hidden, hoax, homicide, how, illegal, illegitimate, illicit, impersonate, ingenuity, innocence, inquest, inspector, instinct, intrigue, investigate, jewels, judge, juvenile, kidnap, kill, lawyer, letters, locate, loot, macabre, Marple, mask, missing, mistake, motive, murder, mystery, nab, notorious, overhear, peculiar, plan, plot, plunder, Poirot, poison, police, post mortem, practical joke, prison, problem, proof, prosecution, psychology, puzzle, quarrel, question, racket, ransack, ransom, realize, red-handed, remember, remorse, remorseless, replicate, revenge, robber, sabotage, scandal, scheme, Scotland Yard, secret, sentence, shocking, shoot, sinister, soldier, solve, spy, stab, stolen, strange, suicide, superintendent, surprise, suspect, suspicious, symbols, terror, thief, tragic, trap, trial, trouble, underground, unknown, vengeance, verdict, victim, vile, violence, warning, weapons, what, when, where, who, whom, why, witness.

Mc is aware of synonyms like: (kill, murder), (problem, puzzle), (deduce, find-out, figure-out), (anonymous, unknown), (suspicious, curious), (verdict, sentence), (plot, scheme) etc.

Mc is aware of the polysemous/homonymous nature and different forms of certain words like: (puzzle (noun, verb), puzzled, puzzling), (judge (noun, verb)), (mystery, mysterious), (murder (noun, verb), murderer) etc.

iv. The words in X are subdivided into categories like: mystery_category, detective_name, events, verdicts, and so on where each such generic value is mapped to a real number – following the definition of the Z-number.

v. The words are further clustered into semantic nets or groups of words that occur together or are semantically linked e.g. Murder_net <murder, motive, quarrel, post mortem, police, exhume, fatal, clue, revenge> etc.

Inputs:

i. *Mc* wants to buy a new book with the requirements as follows:

Table 1. Requirement Set of M

Requirements – Event Set (*E*)	
Natural Language Statements	**Z-valuations**
The book needs to be a Mystery story	Z_1 = <Book Genre, mystery, certainly> A = mystery, u_1 = book genre
Preferably a murder mystery solved by Miss Marple	Z_{21} = <Mystery category, murder, ideally> A = murder, u_2 = mystery category Z_{22} = <Detective, <Marple, ideally> A = Marple, u_3 = detective
A mystery involving a grand robbery and solved by Marple or Poirot would be okay	Z_{31} = <Mystery category, robbery, possibly> A = robbery, u_2 = mystery category Z_{32} = <Detective, Marple, probably> A = Marple, u_3 = detective Z_{33} = <Detective, Poirot, probably> A = Poirot, u_3 = detective
Requirements – Others (*C*)	
Price $\leq$ Rs. 300/-	
New – no crumpled pages or ink marks	
M should not have already read the book	

Note:

a. Parameters A and B are quintessentially maintained and processed as fuzzy set types specific to the system.

b. [11] professes the Interval-Type2 Fuzzy Set being the most appropriate model of word perceptions.

Thus, E can be summarized by the logical expression:

$$[Z_1 \wedge ((Z_{21} \wedge Z_{22}) \vee (Z_{31} \wedge (Z_{32} \vee Z_{33})))] \tag{8}$$

And following Equations 2, 3 and 4, as discussed in Section 3.1, Equation 8 may be rewritten as:

$$M = \mu_{certainly}(\int_R \mu_{mystery}(u_1)p_X(u_1)du_1) \wedge [\{\mu_{ideally}(\int_R \mu_{murder}(u_2)p_X(u_2)du_2) \wedge \mu_{ideally}(\int_R \mu_{Marple}(u_3)p_X(u_3)du_3)\} \vee [\mu_{possibly}(\int_R \mu_{robbery}(u_2)p_X(u_2)du_2) \wedge \{\mu_{probably}(\int_R \mu_{Marple}(u_3)p_X(u_3)du_3) \vee \mu_{probably}(\int_R \mu_{Poirot}(u_3)p_X(u_3)du_3)\}]] \quad (9)$$

where,

$$\int_R p_X(u_1)du_1 = 1, \int_R p_X(u_2)du_2 = 1, \int_R p_X(u_3)du_3 = 1$$

ii. $T = 2$

iii. The summaries of some new books in the store are as follows:

a. **Summary:** Lymstock is a town with more than its share of secrets – a town where even a sudden outbreak of anonymous hate-mail causes only a minor stir. But all that changes when one of its recipients, Mrs. Symmington, commits suicide. Her final note said, "I can't go on." Only Miss Marple questions the coroner's verdict of suicide. Was this work of a poison-pen? Or of a poisoner? – The Moving Finger (Agatha Christie)

   Other Properties: Price = Rs. 150/-; New book; Not read

b. **Summary:** "The curious case of the Maiden Eggesford Horror". When the doctor advises Bertie to live the quiet life, he and Jeeves head for the pure air and peace of Maiden Eggesford. However, they hadn't reckoned on Bertie's irrepressible but decidedly scheming Aunt Dahlia, around whom an imbroglio of impressive proportions develops involving The Cat Which Kept Popping Up When Least Expected. As Berties observes, whatever aunts are, they are not gentlemen. – Aunts Aren't Gentlemen (P. G. Wodehouse)

   Other Properties: Price = Rs. 250/-; New book; Not read

c. **Summary:** Gerry Wade had proved himself to be a champion sleeper; so the other house guests decided to play a practical joke on him. Eight alarm clocks were set to go off, starting at 6:30 a.m. But when morning arrived, one clock was missing and the prank had backfired with tragic consequences. Gerry never woke up. Was he murdered? – The Seven Dials Mystery (Agatha Christie)

   Other Properties: Price = Rs. 150/-; New book; Not read

Execution:

Table 2. Summarization of Book 1

Relevant sentences in the summary	Simple sentence components	Z-valuations of the simple sentence components
Lymstock is a town with more than its share of secrets – a town where even a sudden outbreak of anonymous hate-mail causes only a minor stir.	1. Lymstock is a town with secrets. 2. There is a sudden outbreak of anonymous hate-mail.	Z_{11} = <Location, Lymstock, supposedly> Z_{12} = <Location property, has secrets, supposedly> Z_{13} = <Event, anonymous letters, probably> [hate-mail = letters]
But all that changes when one of its recipients, Mrs. Symmington, commits suicide.	1. Recipient Mrs. Symmington commits suicide.	Z_2 = <Letter event, recipient commits suicide, probably>
Only Miss Marple questions the coroner's verdict of suicide.	1. Coroner's verdict is suicide. 2. Miss Marple questions verdict.	Z_{31} = <Coroner verdict, suicide, probably> Z_{32} = <Verdict event. Marple questions, probably>
Was this work of a poison-pen?	- Is a simple sentence -	Z_4 = <Suspect, poison-pen, expectedly>
Or of a poisoner?	- Is a simple sentence -	Z_5 = <Suspect, murderer, expectedly> [poisoner = murderer]

Therefore,

$$E_{S1} = (Z_{11} \wedge Z_{12} \wedge Z_{13}) \wedge Z_2 \wedge (Z_{31} \wedge Z_{32}) \wedge (Z_4 \vee Z_5);$$

$|E \cap_p E_{S1}| = 3 > T$ [By virtue of Z_{31} and Z_5 in the summary of book1; and the fact that quite a large number of words in the summary of book1 fall under the vocabulary of Mc, the book certainly pertains to the genre 'Mystery'];

$(E \cap_p E_{S1})$ = [<Book genre, mystery, certainly> ∧ <Detective, Marple, probably> ∧ <Mystery category, murder, expectedly>] (Using Equation 7) and the corresponding mathematical expression (using parameters of Equation 9) is:

$$\begin{aligned} \tilde{M} = {} & \mu_{certainly}\Big(\int_R \mu_{mystery}(u_1)\, p_X(u_1)\, du_1\Big) \wedge \\ & \Big(\mu_{probably}\Big(\int_R \mu_{Marple}(u_3)\, p_X(u_3)\, du_3\Big) \wedge \mu_{expectedly}\Big(\int_R \mu_{murder}(u_2)\, p_X(u_2)\, du_2\Big)\Big) \end{aligned} \tag{10}$$

The membership of selection for book1 is evaluated on the basis of the application of the principle of extension to Equations 9 and 10; and considering the degree of overlap, the membership of selection should ideally approximate 1.

Table 3. Summarization of Book 2

Relevant sentences in the summary	Simple sentence components	Z-valuations of the simple sentence components
The curious case of the Maiden Eggesford Horror.	- Is a simple sentence -	Z_1 = <Case, Maiden Eggesford Horror, supposedly>
However, they hadn't reckoned on Bertie's irrepressible but decidedly scheming Aunt Dahlia, around whom an imbroglio of impressive proportions develops involving The Cat Which Kept Popping Up When Least Expected.	1. Aunt Dahlia is decidedly scheming	Z_{21} = <Character, Aunt Dahlia, certainly> Z_{22} = <Character nature, scheming, decisively>
As Bertie observes, whatever aunts are, they are not gentlemen.	1. Bertie observes that aunts aren't gentlemen.	Z_{31} = <Character, Bertie, certainly> Z_{32} = <Character action, observation, decisively>

Therefore,

$$E_{S2} = (Z_1 \wedge Z_{21} \wedge Z_{22} \wedge Z_{31} \wedge Z_{32}) \; ;$$

$| E \cap_p E_{S2} | = 1 < T$ [By virtue of the fact that some words in the summary of book2 fall under the vocabulary of *Mc*, the book 'probably' pertains to the genre 'Mystery'].

Thus, book2 is assigned a membership of selection = 0.

Table 4. Summarization of Book 3

Relevant sentences in the summary	Simple sentence components	Z-valuation of the simple sentence components
Gerry wade had proved himself to be a champion sleeper; so the other house guests decided to play a practical joke on him.	1. The house guests played a practical joke on Gerry.	Z_1 = <Event, practical joke, supposedly>
But when morning arrived, one clock was missing and the prank had backfired with tragic consequences	1. One clock was missing. 2. Prank had backfired. 3. Consequences were tragic.	Z_{21} = <Event, clock missing, supposedly> Z_{22} = <Event, practical joke, supposedly>[prank = practical joke] Z_{23} = <Practical joke event, backfire, supposedly> Z_{24} = <Practical joke event, tragic consequence, supposedly>
Was he murdered?	- Is a simple sentence -	Z_3 = <Event, murder, expectedly>

Therefore,

$E_{S3} = (Z_1 \wedge Z_{21} \wedge Z_{22} \wedge Z_{23} \wedge Z_{24} \wedge Z_3)$;

$| E \cap_p E_{S3} | = 2 = T$ [By virtue of Z_3 in the summary of book3; and the fact that quite a large number of words in the summary of book1 fall under the vocabulary of *Mc*, the book certainly pertains to the genre 'Mystery'];

$(E \cap_p E_{S3})$ = [<Book genre, mystery, certainly> ∧ <Mystery category, murder, expectedly>] (Using Equation 7), and the corresponding mathematical expression (using parameters of Equation 9) is:

$$\tilde{M} = \mu_{certainly}(\int_R \mu_{mystery}(u_1) p_X(u_1) du_1) \wedge \mu_{expectedly}(\int_R \mu_{murder}(u_2) p_X(u_2) du_2)) \qquad (11)$$

The membership of selection for book3 is evaluated on the basis of the application of the principle of extension to Equations 9 and 11, and considering the degree of overlap, the membership of selection should lie in the range (0, 1) and should be less than that of Book1.

Thus, on the basis of the interpretations of Equations 10 and 11, *Mc* evidently selects book1 – "The Moving Finger" by Agatha Christie. This decision by *Mc* coincides with the judgment a human being would make, given the scenario.

D. Observations

The experiment leads to the following illuminating observations:

i. The Z-number methodology blends in seamlessly with the intuitive algorithm followed for the example studied.

ii. CWW based on the Z-number calls for:

   a. A powerful Part-Of-Speech (POS) tagger – to identify the parts of speech of the words in the sentences – an insight into the probable values of *X*, *A* and *B* indicate noun phrases, adjectives and adverbs respectively.

   b. Resolution of the dependencies between the nouns and the corresponding pronouns used – such that the sentences on a common topic might be identified and processed accordingly.

   c. Methods to reduce complex or compound sentences to their simple sentence counterparts and vive versa.

   d. A robust model for the representation of the perceptions of the values of *A* and *B*.

   e. Compilation of a comprehensive context-sensitive text corpus – such that the probable generic values for the parameter *X* can be listed and mapped to real numbers, and the probable values of the parameter *A* be identified and assigned membership values and probability distributions as required.

   f. An algorithm to identify keywords and relevant sentences.

   g. An algorithm to extract implicitly specified values of the parameter *B*.

   A possible solution could be the use of default values with respect to the type of the sentence, as is shown in Table 5. (All examples in this article use this technique.)

Table 5. Suggested default values for B, with respect to the type of sentence

Sentence Type	Remarks	Default Value for *B*
Declarative		Probably or Supposedly
Exclamatory	Conveys explicit emotion	Certainly
Imperative	Expect a definite course of action	Definitely or Decisively
Interrogative	Expects an answer	Expectedly

Note: In a natural conversation scenario – the behavioural parameters (general attitude) of the participants in a discourse influence the value of *B*.

h. An algorithm to identify and process polysemous and homonymous words, capitonyms and synonyms.
i. Methods to identify new words, map the meanings of these words to synonyms existing in the system vocabulary and to learn to use them as well.
j. Schemes to identify and process new sub-contexts under the context of discourse.
k. Schemes to translate Z-valuations to logical expressions, mathematical expressions and natural language statements.
l. Well defined operators and rules of computation – as is guided by the context and the word-perception models assumed.

Each of these aforementioned points (from ii.a though ii.l) represent the basic challenges underlying the implementation of the Z-numbers for CWW.

iii. The system should respond within the average human response time (150 – 300 ms).

5 Conclusion

This article is an elucidation on our study of the Z-number approach to CWW. The Z-number not only aspires to provide a framework for the precisiation of the meaning of a natural language statement, but includes parameters that permit extraction of the information and the uncertainty of the information conveyed by the sentence. The Z-numbers predictably have a radical role to play in the realm of CWW and NLP. The probable contributions of the Z-number to CWW have been clearly outlined in the article.

Besides the strengths, we present here algorithms that illustrate the seamless merge of intuition with Z-number elements to simulate a real-life CWW scenario. These algorithms illuminate the need for intensive research on a fair number of issues that need to be resolved prior to the actual implementation of the Z-numbers. Some of the prime areas that need delving into are models of word perceptions, algorithms for the extraction of the Z-number parameters from natural language statements, well-defined rules of computation, and algorithjms to convert Z-numbers to logical/mathematical expressions and natural language statements. The article also defines an operator for the intersection of perceptions.

The Z-number is indeed an intriguing field of study in the arena of CWW and we are in the process of excavating the answers to the challenges highlighted in this article.

Acknowledgement. This project is being carried out under the guidance of Professor Sankar K. Pal. The work was supervised by Prof. Pal as a J.C. Bose Fellow of the Government of India.

References

[1] Turing, A.M.: Computing machinery and intelligence. Mind 49, 433–460 (1950)

[2] Zadeh, L.A.: Intelligent systems revolution: Is it Real (1998), http://www-bisc.cs.berkeley.edu/zadeh/papers

[3] Zadeh, L.A.: Fuzzy logic = Computing with words. IEEE Trans. Fuzzy Systems 4(2), 103–111 (1996)

[4] Zadeh, L.A.: Outline of a new approach to the analysis of complex systems and decision processes. IEEE Trans. Syst. Man Cyberns. 3(1), 28–44 (1973)

[5] Zadeh, L.A.: Test-score semantics for natural languages and meaning representation via PRUF. Empirical Semantics, 281–349 (1982)

[6] Zadeh, L.A.: A new direction in AI: Toward a computational theory of perceptions. AI Magazine 22(1), 73–84 (2001)

[7] Zadeh, L.A.: Precisiated Natural Language (PNL). AI Magazine 25(3), 74–91 (1994)

[8] Zadeh, L.A.: A note on Z–Numbers. Information Sciences 181, 2923–2932 (2011)

[9] Kacprzyk, J., Zadrozny, S.: Computing with words is an implementable paradigm: Fuzzy queries, linguistic data summaries and natural language generation. IEEE Trans. Fuzzy Systems 18(3), 461–472 (2010)

[10] Nesfield, J.C.: English Grammar Series: Book IV: Idioms. grammar and synthesis. Macmillan, London (1956)

[11] Mendel, J.M., Wu, D.: Perceptual computing – Aiding people in making subjective judgments. IEEE Press, N.J (2010)

On Rule Learning Methods: A Comparative Analysis of Classic and Fuzzy Approaches

Marcos E. Cintra[1], Maria C. Monard[1], and Heloisa A. Camargo[2,*]

[1] University of Sao Paulo (USP) – Mathematics and Computer Science Institute
CEP 13566 - 590, Sao Carlos, SP
mecintra@gmail.com, mcmonard@icmc.usp.br
[2] Computer Science Department – Federal University of Sao Carlos
Rod. Washington Luis, Km 235, PO Box 676, 13565-905, Sao Carlos, SP, Brazil
heloisa@dc.ufscar.br

Abstract. Classification is an important task widely researched by the machine learning and fuzzy communities. In this paper, we present and compare methods from both communities, in order to support the selection of a suitable method, according to two conflicting objectives: accuracy × interpretability. Two groups of rule-based methods are analysed: decision tree-based and genetic-based approaches. For the tree-based approaches, C4.5, PART and FuzzyDT, a fuzzy version of the C4.5 algorithm, are used. For the genetic-based approaches, MPLCS, a method from the machine learning community to generate rule-based models, SLAVE and FCA-Based, both fuzzy-based, are analysed. Since accuracy and interpretability are usually conflicting objectives, in this paper, we briefly present these methods and then discuss the models generated by them. Comparisons take into account the error rates and syntactic complexity of the produced models. Ten benchmark datasets are used in the experiments with a 10 fold cross-validation strategy. Results show that FCA-Based and MPLCS are able to obtain good accuracy and interpretability.

Keywords: Fuzzy Systems, Decision Trees, Genetic Algorithms.

1 Introduction

Classification is an important task widely researched by the machine learning and fuzzy communities. Classic and fuzzy algorithms for supervised machine learning are concerned with the development of methods that extract patterns from data in order to make intelligent decisions based on these patterns. The interpretability is an important issue when classification methods are proposed. Interpretability of classification models, in spite of its subjectivity, can be defined as the quality of how easily the model, as a whole, can be understood and abstracted by its users. Thus, an approach that induces interpretable models must be concerned with the total number of rules and the amount of conditions

* This work was supported by the Brazilian State Council FAPESP.

R.R. Yager et al. (Eds.): *Soft Computing: State of the Art Theory*, STUDFUZZ 291, pp. 89–104.
DOI: 10.1007/978-3-642-34922-5_7

of these rules, *i.e.*, the syntactic complexity of the model. In general, highly accurate models tend to have high syntactic complexity, whereas, models with low syntactic complexity tend to be less accurate.

In this sense, decision trees (DT) [1] are powerful as they produce models with low syntactic complexity which are quite intuitive and whose structures can be interpreted as rules. The induction process of DTs is usually fast and the induced models are competitive, accuracy wise, with the ones generated by other interpretable machine learning methods. Another desirable quality of DTs is their embedded feature selection process that allows it to use only the most relevant features in the model, which are selected according to certain measures, improving the generated model interpretability.

Some of the most well-known and relevant DT based algorithms are ID3, CART, and C4.5 [1, 2]. These algorithms generate a tree structure through recursively dividing the feature space until this decision space is completely partitioned into a set of non-overlapping subspaces. Specifically, C4.5 uses the information gain and entropy measures when deciding on the importance of the features [2]. In order to optimize their estimated error rates, DTs usually use a pruning process. Pruning also simplifies the whole models, which consequently become more interpretable. PART [3] is an example of a DT-based approach for rule generation. This method repeatedly generates various DTs extracting the best rule of each DT at a time to construct the rule set of a classifier.

Fuzzy rule based classification systems are based on the fuzzy set and fuzzy logic theories proposed by Loft A. Zadeh. Two advantageous characteristics of fuzzy systems regarding interpretability are: i) the system uses semantically meaningful fuzzy sets to define attributes; ii) fuzzy rules are built by linguistic variables and linguistic terms, such as "temperature is high" or "speed is low", adding interpretability to the induced model.

The knowledge base and inference mechanism are the two basic components of a fuzzy classification system. The knowledge base is formed by the Fuzzy Data Base (FDB) and the Fuzzy Rule Base (FRB). The fuzzy data base contains the definitions of the features (also named attributes or variables) in terms of fuzzy sets, while the fuzzy rule base contains a set of rules defining the given problem. The inference mechanism derives the conclusions (or outputs) of the system based on the knowledge base and on the inputs to the system.

In the literature, it is possible to find several fuzzy approaches for the induction of fuzzy classifiers, amongst them, fuzzy rule-based systems [4, 5], genetic fuzzy rule-based systems [6–8], fuzzy DTs [9, 10], and evolutionary (other than genetic) fuzzy rule-based systems [11–13]. Regarding the genetic rule-based fuzzy systems, their advantages include: i) Genetic Algorithms (GAs) perform a global search and do not get stuck in local maxima; ii) it is possible to address the interpretability $\times$ accuracy problem during the search process by means of multi-objective fitness functions; iii) it is possible to adjust rules and fuzzy sets during the genetic process in order to improve the model performance and interpretability [14, 15].

Although the genetic generation of fuzzy systems might be one of the most researched topic in the fuzzy community, GAs usually have a high computational cost due to their global search and, for some approaches, also, due to the process required to form the genetic search space. A well-known genetic fuzzy approach is SLAVE – Structural Learning Algorithm on Vague Environment [16]. SLAVE uses the iterative approach to learn fuzzy rules, performing an embedded feature selection process as well as a rule selection post process.

An alternative to the high cost of GAs is the DT-based approach. For this purpose, we have recently proposed a fuzzy version of the classic C4.5 DT in [17]. Our approach is quite similar to the classic one and its more relevant characteristics are described in Section 2.

The aim of this paper is to experimentally compare different proposals based on DTs and GAs, from both, the machine learning and fuzzy communities. Since there is a large number of genetic fuzzy approaches proposed, we selected two of them, the well known approach named SLAVE, and another one proposed by us, FCA-Based. On the other hand, due to the special characteristics of low computational cost, competitive and highly interpretable induced models, we also include two classic DT-based approaches and a fuzzy DT in our experiments. Comparisons were performed taking into account the accuracy and syntactic complexity of the generated models. The goal of this research is to provide substantial information on these approaches, indicating their most relevant qualities and drawbacks.

The remainder of this paper is organized as follows. Section 2 describes the FCA-Based, SLAVE and MPLCS methods, which use the genetic paradigm. Section 3 describes and compares the classic and the fuzzy C4.5 DT approaches, as well as PART. Section 4 presents the experiments and results, followed by the conclusions and future work in Section 5.

2 Classification Methods Based on the Genetic Paradigm

GAs [18] are a part of the evolutionary algorithms, which are techniques inspired on the biological evolution. GAs have been applied in several areas. They usually require a randomly generated initial population of hypotheses. For rule-based classification, the initial population is usually formed by rules or rule sets. The hypothesis of GAs is that the fittest members of a population have better chances of producing offspring. This way, by generating several populations, it is possible to evolve solutions and reach satisfactory results. This population, whose members are called chromosomes or individuals, encodes candidate solutions to a given problem. The first population gives rise to the following population by means of genetic operators, such as selection, mutation, crossover, and elitism, among others. At each generation, the hypotheses in the current population are evaluated relative to a given fitness measure, with the fittest hypotheses selected probabilistically as seeds to produce the next generation, and so on. Usually, GAs stop by reaching a maximum number of generations or when a satisfactory fitness level is reached.

Chromosomes are usually represented by an array of elements. These arrays can contain indexes to a preselected list of solutions (called search space). By representing hypotheses using arrays of elements with fixed length, the crossover operator is simple to apply due to the alignment of the chromosomes forming the population. Variable length representations are also used, but they require higher computational effort for the use of genetic operators. The fitness function evaluates the quality of the represented solution and it is always directly connected to the type of problem to be tackled. For classification problems using rule bases, the usual fitness measures adopted are related to the accuracy and interpretability of the generated models.

Next, we present the SLAVE, FCA-Based, and MPLCS fuzzy classification systems, which are the GA-based learning methods to generate classification rules used in this work.

2.1 Fuzzy Classification Systems

The classification task can be roughly described as: given a set of objects $E = \{e_1, e_2, ..., e_n\}$, also named *examples, cases, or instances*, which are described by m features, assign a class c_i from a set of classes $C = \{c_1, c_2, ..., c_j\}$ to an object e_p, described by its feature values $e_p = (a_{p_1}, a_{p_2}, ..., a_{p_m})$.

Fuzzy classification systems are rule based fuzzy systems that granulate the domains of their features by means of fuzzy sets and partitions. The linguistic variables in the antecedent part of the rules represent features, and the consequent part represents a class. A typical fuzzy classification rule can be expressed by

R_k : **IF** X_1 is A_{1l_1} **AND** X_2 is A_{2l_2} **AND** ... **AND** X_m is A_{ml_m}
THEN $Class = c_i$

where R_k is the rule identifier, $X_1, ..., X_m$ are the features of the set of examples considered in the problem (represented by linguistic variables), $A_{1l_1}, ..., A_{ml_m}$ are the linguistic values used to represent the feature values, and $c_i \in C$ is the class. Notice that not all identifiers participate in a general classification rule. The inference mechanism compares the input example to each rule in the fuzzy rule base aiming at determining the class it belongs to.

The classic and general fuzzy reasoning methods are widely used. Given a set of fuzzy rules, *i.e.*, a FRB, and an input instance, the classic fuzzy reasoning method classifies this input instance using the class of the rule with maximum compatibility to the input instance, while the general fuzzy reasoning method calculates the sum of compatibility degrees for each class and uses the class with highest sum to classify the input instance.

2.2 SLAVE

SLAVE [16] is a genetic learning algorithm that uses the iterative approach to generate a FRB. In the iterative approach, chromosomes usually represent individual rules, and a single rule is selected at each iteration of the GA. The set of selected rules form the rule base of the model. SLAVE includes an embedded feature selection process. The preselection of attributes minimizes the problems caused by large search spaces, such as excessive execution time, while improving the interpretability of the generated models.

The main idea of SLAVE is to reduce the original problem of obtaining a complete set of rules to a simpler problem which consists in obtaining only one rule at a time. In this approach, each chromosome of the population represents a single rule, but only the best individual in each iteration is considered, the remaining chromosomes being discarded. In fact, in the iterative model, one execution of the GA provides a partial solution (a rule) to the learning problem.

Regarding the feature selection process adopted by SLAVE, it dynamically explores the set of possible variables in order to find the most useful rule and the most relevant variables for this rule. Thus, this feature selection process is implemented for each single rule, not for the whole set of rules. The basic schema of this process consists of modifying the rule representation in the search mechanism of SLAVE in order to allow the learning algorithm to search not only for the best rule, but also the best set of variables for each rule. SLAVE produces rules with different weights, which are used by the inference mechanism to improve the classification performance. SLAVE usually produces reasonably small rule sets.

2.3 FCA-Based

The FCA-Based method [8] forms the GA search space by using the theory of Formal Concept Analysis (FCA) [19]. FCA is a mathematical technique for extracting concepts and structures from data. It was introduced in the 1980s and is becoming increasingly popular due to its nice visual representation of data and relations found in data. The basic data structure in FCA is the formal context, which is a representation of the relations between objects and attributes. A formal context is usually represented in a table form where the columns represent the attributes and the rows represent the objects (objects are usually called instances or examples in classification). The most important difference between an attribute $\times$ value table and a formal context is that FCA only works with binary attributes. In order to handle continuous and multi-valued attributes, they must be transformed into binary attributes using a scaling process. The table representing the formal context contains 1 (true) in cell (i, j) if object i has attribute j, and 0 (false) otherwise. By extracting classification rules from data using the FCA theory to form the GA search space, the FCA-Based method is able to avoid the creation of a large number of useless rules, a task that has a high computational cost. After the rule extraction process, the FCA-Based method uses a GA to generate the fuzzy rule base.

The FCA-Based method uses an integer chromosome codification. The size of the chromosome is equal to the maximum number of rules considered acceptable for the final FRBs and it is initialized with the number of rules found in the rule base produced by the Wang & Mendel method [20] using the same FDB. This heuristic allows the definition of chromosomes with a reasonable number of rules.This approach requires a considerable extra computational cost compared to our approach. The integer codification uses the index of a rule in the search space generated by FCA in each of its genes. To allow the generation of rule bases with less rules than the maximum size of the chromosome, a ***-1*** value is used to indicate that a gene represents an inactive rule.

For the fitness calculation, and aiming at reducing the number of rules in the final FRB, the FCA-Based method uses the Correct Classification Rate (CCR) and the number of rules (NR) in the rule base represented by each chromosome during the search process. This evaluation process uses a self-adaptive algorithm that keeps and updates referential values of the ideal CCR and NR. After each generation, an update occurs if a better CCR is obtained with a number of rules equal or smaller than the best current NR. In the sequence, the NR is used in a penalization mechanism that decreases the fitness value of a chromosome when its NR is larger than the current reference NR.

In order to improve the interpretability of the final rule bases generated by the GA, FCA-Based has a simple post selection process that checks the ability of each individual rule to improve the classification power of the rule set. This process aims at removing as many rules as possible while keeping (or improving) the accuracy of the whole FRB.

2.4 MPLCS

MPLCS [21] stands for Memetic[1] Pittsburgh Learning Classifier System (MPLCS). The MPLCS method has many variants according to the adopted local search mechanism.

The version used in our experiments uses the local search, based on the rule set-wise operator. This local search has three main stages: i) an evaluation of the candidate rules; ii) the selection of the rules that will form the offspring rule set; iii) the generation of the final individual. In the first stage, all rules are evaluated with all the examples of the training set, producing a map of correct and incorrect classifications for each rule. The next stage uses this map to evaluate how much each candidate rule can contribute to improve the accuracy of the offspring rule set without re-evaluating the rule set.

2.5 Comparing the Models Generated by the SLAVE, FCA-Based, and MPLCS

One issue with the models generated by SLAVE is that they contain rules with sets of fuzzy label disjunctions in their antecedents. For example, one of the models generated for the Iris dataset, with three rules, is presented next.

1. If **X2** is {L0}, class is **Iris-setosa** (W 0.977)

---

[1] Memetics is a theory of mental content based on an analogy with the Darwinian evolution. Memes are similar to genes in GA, but represent ideas, believes, patterns of behaviour, which can reproduce.

2. If **X0** is {L0 L1} and **X2** is {L0 L1} and **X3** is {L0 L1}, class is **Iris-versicolor** (W 0.402)
3. If **X0** is {L1 L2} and **X2** is {L1 L2} and **X3** is {L0 L2}, class is **Iris-virginica** (W 0.719)

Although the model has only 3 rules, with a total of 7 conditions, 6 of these conditions contain disjunctions of fuzzy labels, impacting on the model interpretability. It is also possible to find the association of linguistic values that are not defined by neighbouring fuzzy sets, which makes each rule quite difficult do understand.

The models generated by MPLCS, similarly to SLAVE, contain conjunctions of disjunctions, and, for continuous attributes, the splitting points can be quite unnatural and difficult to be interpreted. As an example, the rule set for the Iris dataset, with 4 rules, is shown next.

1. If **sepalLength** is > 6.243 and **petalLength** is > 5.085, class is **Iris-virginica**
2. If **sepalLength** is < 6.340 and > 7.020 and **petalWidth** is > 1.627, class is **Iris-virginica**
3. If **petalLength** is < 1.983, class is **Iris-setosa**
4. Default rule: class is **Iris-versicolor**

The cutting points defined by the algorithm can be quite similar and close to each other. For instance, the splitting points for **sepalLength** in rules 1 and 2 discard values from a very close interval from 6.243 to 6.340, which makes the understanding of the whole model difficult.

The FCA-BASED models present only conjunctions of conditions in the antecedent of their rules, and, due to the fact that it is based on the fuzzy logic, the discretization of continuous attributes is done using highly interpretable linguist valued fuzzy sets. The rule set for the iris dataset is presented next.

1. If **sepalLenght** is **medium**, and **petalLenght** is **medium** and **petalWidth** is **medium**, **class** is **Iris-virginica**
2. If **petalLenght** is **large** and **petalWidth** is **medium**, **class** is **Iris-versicolor**
3. if **petalLength** is **small**, and **petalWidht** is **medium**, and **sepalWidth** is **small**, **class** is **Iris-setosa**
4. if **petalLength** is **medium**, and **petalWidht** is **small**, and **sepalWidth** is **large**, **class** is **Iris-virginica**

For experts and persons who are familiar with the domain, the fuzzy linguistic values **small**, **medium** and **large** are directly interpreted. Nevertheless, for those who are unfamiliar with the domain, it is necessary to check the FDB for the information regarding the number, type and distribution of the fuzzy sets defining each attribute, in order to interpret the rules and rule set. This process, although quite straightforward, requires extra effort. To reduce the time to understand the FDB, the information can be presented in graphs.

Next, we discuss DT based classification methods.

3 Classification Methods Based on Decision Trees

DTs are widely used in machine learning due to its simplicity of generation and powerful representation of knowledge. Fuzzy DTs have also been proposed in the literature. The classic C4.5 DT algorithm, PART (a DT-based approach for rule generation), and FuzzyDT, our fuzzy version of the C4.5 algorithm, are presented next, as well as a comparison of their generated models.

3.1 C4.5

DT algorithms generate a tree structure through recursively partitioning the feature space until the whole decision space is completely divided into a set of non-overlapping class subspaces (leaf nodes). They also perform an embedded selection of features during its partitioning process, so only relevant features are used in the tree, improving the time used to classify new examples as well as the interpretability of the model. C4.5 is one of the most well-know DT algorithms [2]. C4.5 uses the information gain and entropy measures when deciding on the importance of the features.

In order to avoid overfitting, a stopping criterion can be used to prevent some subsets of training examples from being subdivided. The pruning of a part of the DT structure helps preventing overfitting. Regarding the pruning process, C4.5 employs post-pruning, *i.e.*, the pruning takes place after the tree is completely induced assessing the error rates of the tree and its components directly on the set of training examples [1]. This assessment is related to the confidence level that the error obtained with the pruned tree, in relation to the error for the original tree, will represent the real error.

The default confidence level used by C4.5 is 25%. It is important to notice that the smaller the confidence limits, the higher the chances of pruning, while the higher the confidence limits, the smaller the chances of pruning. Thus, if we set the confidence limit to 100%, what we are saying is that we believe that the predicted error, obtained with the examples at hand, is equal to the real error and no pruning will be performed. This idea conflicts with the natural intution one might have that a 25% confidence limit will produce less pruning than an 80% confidence limit. This way, one should not associate the default 25% confidence limits of C4.5 with a 25% pruning of the tree.

3.2 PART

PART, as its name indicates, is an algorithm based on partial DTs [3]. A partial DT is an ordinary DT that contains branches to undefined subtrees. PART is a rule-induction procedure that adopts the separate-and-conquer strategy. In essence, it builds a rule, removes the instances it covers, and continues creating rules recursively for the remaining instances until none is left. In order to generate this rule, PART generates a DT and prunes all but one leaf (specifically the leaf with the largest coverage) and makes the branch of this leaf into a rule, discarding the rest of the tree. Its authors explain that using a pruned tree to obtain a rule, instead of building it incrementally by adding conjunctions of conditions one at a time, avoids the over-pruning problem of the basic separate-and-conquer rule learner.

In fact, using the separate-and-conquer methodology in conjunction with DTs adds flexibility and speed to the process. Since it is wasteful to build a full DT just to obtain a single rule, PART significantly accelerates the process described without sacrificing the above advantages by building a "partial" DT instead of a fully explored one. An integration of the construction and pruning operations is used in order to find a "stable" subtree that can be simplified no further. This way, once this subtree has been found, the tree induction ceases and a single rule is selected.

3.3 FuzzyDT

FuzzyDT [17] is a fuzzy implementation of the classic C4.5 algorithm. It uses the same measures of C4.5 (entropy and information gain) to decide on the importance of the features. The main difference between the classic and the fuzzy C4.5 is the fact that the fuzzy version discretizes continuous attributes using fuzzy sets before the induction of the tree.This way, the process can be seen as inducing a tree using only discrete features, since the continuous features are defined in terms of fuzzy sets and the training set is fuzzified before the DT induction.

3.4 Comparing the Models Generated by the C4.5, PART, and FuzzyDT

The model produced by C4.5, as by most of the DT algorithms, form a set of disjunct rules in which only one rule is fired to classify a new example. For FuzzyDT, on the other hand, the tree can be seen as a set of rules that are fired simultaneously. Since they are fuzzy rules, the degree of compatibility of each rule with a new example is calculated and used by the inference mechanism to classify this new example. This way, the inference of fuzzy DTs requires higher computational effort than the classic DTs. In spite of this additional cost, the compatibility information of the rule with the example guides the inference, which considers all attributes of a rule (branch), while classic DTs simply tests one attribute at a time, even if the input values are close to the test values.

PART generates a set of ordered rules. The inference process is quite straightforward: the first rule is checked, if it does not cover the example, the next rule is checked, until the example is classified or the last rule is reached.

Next we compare some important features and definitions of FuzzyDT, C4.5 and PART.

Evaluation of features – For the partitioning process, the three methods use the same measures, *i.e.*, entropy and information gain, to select the features to be used in the test nodes of their branches or rules;

Induction process – FuzzyDT and C4.5 repeatedly subdivide the feature space using the most informative features until a leaf node is reached or no features or examples remain. PART uses a similar approach to generate partial trees, but for each generated tree, only the branch of the tree that correctly classifies the largest number of examples is used. This process is repeated as many times as necessary.

Handling continuous features – PART and C4.5 split the domain according to the examples at hand by minimizing entropy and maximizing information gain. The drawback of this process is the discretization of continuous attributes, which might create unnatural divisions that reflect on a lower interpretability of the rules and rule set. Another issue with PART and C4.5 is that the number of divisions used to split continuous attributes, even if known a priori, cannot be informed to the algorithm. In fact, the splitting of continuous attributes is done dynamically by the algorithms, and might be distant from the patterns of the application domain or even from the representation used by an expert. FuzzyDT, on the other hand, can use the partitions (and thus, number of fuzzy sets) defined by an expert. Furthermore, even if this information is not available, fuzzy partitions can be automatically defined and are easily interpretable. The equalized universe method [22], which evenly splits the domain into a defined number of fuzzy sets, is a simple approach that prevents the creation of unnatural splitting points.

Reuse of features – for PART and C4.5, the same continuous feature can be used more than once in one single rule (for example, if the feature is temperature, a rule might present tests such as "temperature $\leq$ 95", "temperature $\leq$ 74", "temperature $\leq$ 10", and so on). This repetition of the same feature and subdivision of the domain degrades the interpretation of the rule. On the other hand, asFuzzyDT fuzzyfies ("discretizes") the attributes using fuzzy sets, a feature can be used only once in one rule, favoring the interpretability of the generated rules.

Inference – The C4.5 algorithm checks the root test and then the following triggered branch of the tree, to classify a new example. The process is intuitive and clear. Similarly, PART checks the ordered rule set in sequence. However, for continuous features, whenever the input values are located in the decision frontiers, misclassifications might occur due to the fact that the whole inference is done based on a single attribute at a time. For FuzzyDT, as stated before, the membership degree of the input example is calculated for each fuzzy set defining each attribute. These membership degrees are then used to calculate the confidence degree for each rule. Since all branches might be fired simultaneously, this confidence degree is used by the classification process, taking into consideration all the attributes included in each rule, instead of the approach used by the classic DTs of checking a single attribute at a time. This way, FuzzyDT gives more credibility to the final classification. Nevertheless, although many branches of the tree might not be fired by an example, if the DT is large, the FuzzyDT inference process will require a considerable computational effort when compared to PART and C4.5. The calculation can be reduced by defining a minimum threshold of membership degree to continue testing rules or not.

Next, we present the experiments and results.

4 Experiments

Our experiments were carried out using 10 datasets from the UCI Machine Learning Repository [23] and 10-fold cross-validation strategy. The KEEL framework [24] was used for the SLAVE, C4.5, PART, and MPLCS algorithms, all executed with default parameters, except for the number of fuzzy sets for SLAVE, which was set to 3. For FCA-BASED and FUZZYDT, we used our own implementations.

As previously stated, the motivation to compare such different approaches for the automatic generation of classifiers is to provide information and insight when selecting a classification method for a particular problem. For this purpose, we considered the performance of the models, in terms of error rates, and their interpretability, in terms of their syntactic complexity, which takes into consideration the average number of rules generated and the average number of conditions of these rules.

Table 1 summarizes the dataset characteristics giving the total number of examples (Examples); number of features (Features), including the number of continuous and discrete features in brackets; number of classes (Classes), and the majority error (ME), which is the error of the most naive algorithm, which always predicts the majority class of the dataset.

Table 1. General characteristics of the datasets

Dataset	Examples	Features	Classes	ME
Credit	653	15(6,9)	2	45.33
Cylinder	277	32(19,13)	2	35.74
Diabetes	769	8(8,0)	2	34.90
Glass	220	9(9,0)	7	65.46
Heart	270	13(13,0)	2	44.44
Ionosphere	351	34(34,0)	2	35.90
Iris	150	4(4, 0)	3	66.67
Segment	210	19(19,0)	7	85.72
Vehicle	846	18(18,0)	4	74.23
Wine	178	13(13,0)	3	59.74

Notice that all fuzzy methods used the same (FCA-BASED and FUZZYDT) or similar (SLAVE) fuzzy partitions, *i.e.*, the same number of fuzzy sets and their distribution, as well as the type of membership function.

Table 2 presents the mean error rates and standard deviation for the tested methods. The majority error (ME) is presented in the second column. The lasts lines present the average rank and the final rank for each approach. For the average rank and the final rank for each method, when computing the error rank for each dataset, if two or more error measures are equal, the fractional strategy for assigning rankings was used, *i.e.*, they receive the same ranking numbers, which is the mean of what they would have under ordinary rankings. The results for the DT-based approaches are presented in the first columns, and thus, the results for the GA-based approaches in the last columns. The smallest error rates are dark-gray shaded.

Table 2. Error rates and standard deviation for decision tree based approaches

Approaches		DT-based approaches						GA-based approaches					
		FuzzyDT		C4.5		PART		SLAVE		FCA-Based		MPLCS	
Datasets	ME	Error	SD	Error	SD	Error	SD	Error	SD	Error	SD	Error	SD
Credit	45.33	15.78	6.68	12.09	4.39	37.35	13.54	37.38	3.86	9.82	3.90	11.02	4.89
Cylinder	35.74	34.20	0.09	27.69	10.95	31.32	10.68	35.74	0.96	25.77	5.15	25.95	10.85
Diabetes	34.90	26.10	0.05	22.89	8.07	31.47	10.49	25.66	4.26	23.09	2.30	23.67	8.41
Glass	65.46	48.26	8.12	27.03	11.29	48.98	17.50	38.25	10.19	39.56	5.21	29.39	12.17
Heart	44.44	21.85	5.60	20.37	8.32	38.15	13.86	20.74	10.10	19.72	7.44	17.41	8.61
Iono	35.90	13.53	9.95	11.11	4.51	20.52	13.02	12.27	4.81	11.76	4.97	9.98	4.66
Iris	66.67	8.00	2.67	4.00	5.33	60.00	20.00	4.67	4.27	4.68	6.33	2.67	5.33
Segment	85.71	20.48	5.65	0.48	0.45	10.70	4.88	13.43	0.43	24.75	7.39	0.69	0.55
Vehicle	74.23	35.85	4.05	24.33	9.09	58.36	19.96	39.59	4.82	44.15	4.73	27.04	10.64
Wine	59.74	12.86	11.43	6.67	6.94	36.31	14.51	9.54	7.05	4.27	4.99	5.00	3.89
Final Rank		4.9		2.3		5.6		3.3		3.0		1.9	
Avg. Rank		5		2		6		4		3		1	

Considering only the DT-based approaches, C4.5 obtained the smallest error rates for all datasets. FuzzyDT presented smaller error rates than PART for 8 datasets. It should be observed that in most cases learning was very poor for PART, as well as for FuzzyDT using Cylinder, *i.e.*, the error rate for these models is similar to the most naive learning algorithm that always predicts the most frequent class in the dataset.

Considering only the GA-based approaches, MPLCS had the smallest error rates for 6 datasets while FCA-Based for 4 datasets. Moreover, the error rate of the model generated by SLAVE for Cylinder is the ME, thus, there was no learning.

Comparing all methods, *i.e.*, DT-based and GA-based, C4.5 obtained the smallest error rates for 4 datasets, while FCA-Based and MPLCS for three datasets each.

To test whether there was a statistically significant difference among the six algorithms we used the Friedman test [25] with the null-hypothesis that the performance of all algorithms, assessed in terms of the error rates, was comparable. The Friedman test found there is no statistically significant difference among the tested algorithms with a 95% confidence level.

As discussed previously, some methods tend to present good error rates, but low interpretability, or vice-versa. This way, to analyse the interpretability of the generated models, Table 3 presents the average number of rules and the Syntactic Complexity (SC) of the models generated by the six analysed algorithms, as well as the average rank and final rank of their SC. In this work, the SC is defined as the total number of conditions in each rule set. The dark-gray shaded cells highlight the smallest syntactic complexity values obtained in both approaches. Notice that although the rules produced by SLAVE and MPLCS present conjunctions of disjunctions, Table 3 does not consider the number of disjunctions in the rules of these models.

As one can observe, PART produced the models with the smallest syntactic complexity for 9 of the datasets and SLAVE for the remaining one. Observe that dataset Cylinder is the remaining one, which had the ME as error rate (Table 3).

Table 3. Average number of rules and syntactic complexity values

Approaches	DT-based approaches						GA-based approaches					
	FuzzyDT		C4.5		PART		SLAVE		FCA-Based		MPLCS	
Datasets	Rules	SC	Rules	SC	Rules	SC	Rules	SC	Rules	SC	Rules	SC
Credit	21.0	64.5	19.6	95.4	3.0	12.4	5.1	18.6	10.5	32.8	6.9	33.9
Cylinder	31.8	83.7	42.8	248.5	3.6	17.6	1.0	1.0	15.8	70.1	11.8	118.4
Diabetes	12.6	34.4	23.6	150.2	1.3	3.7	4.2	19.6	9.1	33.1	8.3	33.6
Glass	26.6	95.0	24.1	137.8	2.5	10.1	11.9	51.1	6.8	39.4	7.6	21.9
Heart	17.4	49.0	18.5	86.1	1.4	4.0	8.3	43.9	14.1	58.5	7.0	30.7
Iono	20.2	54.4	13.9	72.4	2.4	9.8	15.1	73.0	19.9	77.5	4.6	19.5
Iris	8.2	13.4	4.6	12.1	1.0	2.0	3.2	10.4	4.5	12.8	4.0	7.6
Segment	22.6	72.2	10.0	38.0	1.3	3.2	3.5	15.9	11.5	49.2	4.2	9.3
Vehicle	65.6	296.9	66.3	503.0	2.7	12.0	21.7	151.9	30.2	172.9	19.2	72.9
Wine	13.8	35.0	5.1	12.5	1.9	5.7	4.5	30.2	4.9	15.3	4.2	6.6
Final Rank	4.9		5.1		1.1		3.1		4.1		2.7	
Avg. Rank	5		6		1		3		4		2	

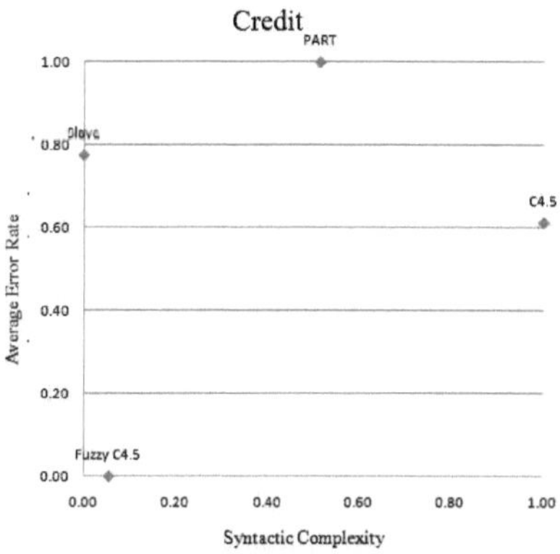

(a) Credit (error[9.8, 37.3], SC[11.2, 95.4])

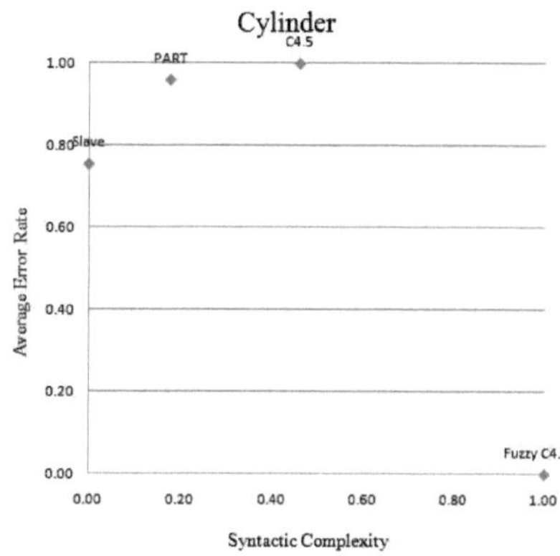

(b) Cylinder (error[25.8, 34.2], SC[1.0, 248.5])

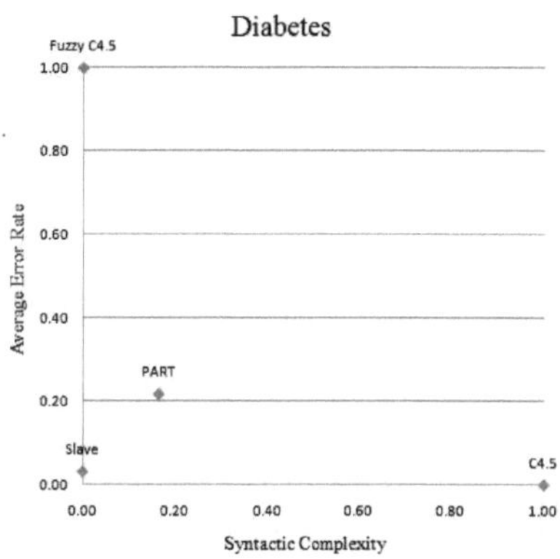

(c) Diabetes (error[22.9, 31.5], SC[3.7, 150.2])

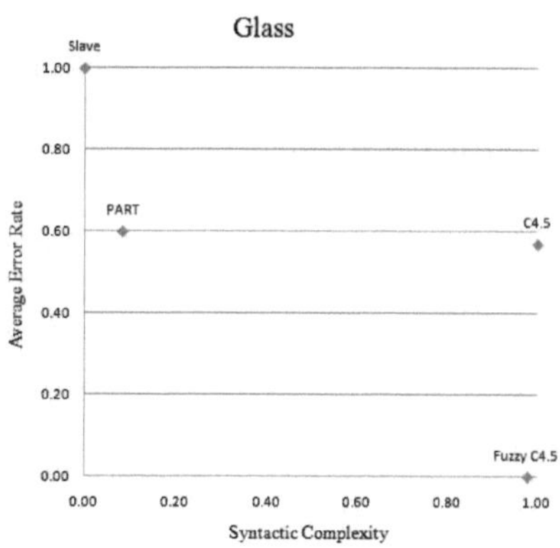

(d) Glass (error[27.0, 49.0], SC[10.1, 137.8])

Fig. 1. Error × SC for Credit, Cylinder, Diabetes, and Glass

Thus, the model generated is simply one rule assigning the majority class to any new instance. However, PART was ranked last in accuracy (Table 2). On the other hand, although C4.5 is ranked last regarding the SC, it is ranked second for accuracy. Furthermore, MPLCS is ranked second regarding the SC, and it is ranked first for accuracy.

In order to consider the performance of the methods both in terms of error rates and syntactic complexity, we used the normalized values of the error rate and SC to produce some graphs and visually analyse the results. To illustrate, Figure 4 presents the results for Credit, Cylinder, Diabetes, and glass, the first four datasets. Because the values are normalized, notice that the origins of the graphs do not represent null error and null syntactic complexity. Instead, the origins are defined by the smallest error rate and SC of the results of the tested methods for each dataset. Similarly, point (1,1) represents the maximum error and syntactic complexity obtained on the dataset. By using the normalized values, instead of the real ones, it is easier to choose the most appropriate algorithms for a specific dataset by focusing on the ones that are plotted closest to the origin of the graphs. The idea is to discard the methods whose values are plotted farthest from the origin and just compare and analyse those closest to the origin in order to obtain the best compromise between error rate and syntactic complexity.

For datasets Credit and Diabetes, the FCA-Based algorithm presents the smallest error rate and low SC. The second best would be MPLCS. However, it is important to notice that the rules produced by MPLCS contain the disadvantage of being formed by conjunctions of disjunctions, while FCA-Based produce quite clear and interpretable rules. For the Cylinder dataset, FCA-Based should be chosen, and MPLCS for the Glass dataset. C4.5, although having low error rates, had the worst SC for these datasets. PART and SLAVE had both good SC, but poor error rates.

Next, we present the final conclusions.

5 Conclusions

Classification is an important task in the machine learning and fuzzy communities. Many classification approaches have been proposed by both communities, some of them sharing similar cores. For instance, both communities have decision tree-based methods, genetic-based methods, methods based on artificial neural networks, among others. Aiming at comparing similar methods from both communities that produce interpretable models, two groups of rule-based methods are analysed in this work: decision tree-based and genetic-based approaches.

The decision tree-based group include C4.5, PART and FuzzyDT. The genetic-based group includes MPLCS, a method from the machine learning community to generate rule-based models, as well as SLAVE and FuzzyDT, both fuzzy-based. These methods were analysed according to their accuracy and syntactic complexity on ten benchmark datasets using a ten fold cross-validation strategy.

Results show that FCA-Based and MPLCS were able to obtain good accuracy and interpretability, while the other methods had good accuracy and poor syntactic complexity, or poor accuracy and good syntactic complexity. One important issue when comparing the models produced by FCA-Based and MPLCS is the fact that MPLCS, as well as SLAVE, produce rules with conjunctions of conditions which might contain sets of disjunctions. This characteristic makes MPLCS and SLAVE much less complex with respect to the SC than the ones produced by C4.5, PART, and FCA-Based, although the disjunctions impact on the readability of the rules, instead of improving it.

As future work, we intend to include other methods from both communities in the experiments and consider other important issues in our comparisons, such as the time taken to generate the models, and their ability to classify examples from datasets whose classes have different cost for misclassification, such as in medical domains. We also intend to use a larger set of datasets.

References

1. Quinlan, J.R.: C4.5: Programs for Machine Learning (Morgan Kaufmann Series in Machine Learning), 1st edn. Morgan Kaufmann (January 1993)
2. Quinlan, J.R.: Bagging, Boosting and C4.5. In: Proc. of the 13th Conf. on Artificial Intelligence, pp. 725–730 (1996)
3. Frank, E., Witten, I.H.: Generating accurate rule sets without global optimization. In: ICML 1998: Proceedings of the 15th Int. Conf. on Machine Learning, pp. 144–151. Morgan Kaufmann (1998)
4. Ishibuchi, H., Yamamoto, T.: Rule weight specification in fuzzy rule-based classification systems. IEEE Transactions on Fuzzy Systems 13, 428–435 (2005)
5. Nakashima, T., Schaefer, G., Yokota, Y., Ishibuchi, H.: A weighted fuzzy classifier and its application to image processing tasks. Fuzzy Sets and Systems 158, 284–294 (2007)
6. Mansoori, E., Zolghadri, M., Katebi, S.: Sgerd: A steady-state genetic algorithm for extracting fuzzy classification rules from data. IEEE Transactions on Fuzzy Systems 16(4), 1061–1071 (2008)
7. Cintra, M.E., Camargo, H.A.: Fuzzy rules generation using genetic algorithms with self-adaptive selection. In: IEEE International Conference on Information Reuse and Integration - IRI, pp. 261–266 (2007)
8. Cintra, M.E., Monard, M.C., Camargo, H.A., Martin, T.P.: A hybrid approach for the automatic generation of fuzzy systems using fuzzy formal concepts. In: 2012 IEEE International Conference on Fuzzy Systems, FUZZ-IEEE 2012 (accepted for publication, 2012)
9. Olaru, C., Wehenkel, L.: A complete fuzzy decision tree technique. Fuzzy Sets and Systems 138(2), 221–254 (2003)
10. Cintra, M.E., Monard, M.C., Camargo, H.A.: An evaluation of rule-based classification models induced by a fuzzy method and two classic learning algorithms. In: The Brazilian Symposium on Artificial Neural Network (SBRN), vol. 1, pp. 188–193 (2010)
11. Ahmadizar, F., Soltanpanah, H.: Reliability optimization of a series system with multiple-choice and budget constraints using an efficient ant colony approach. Expert Systems with Applications 38(4), 3640–3646 (2011)

12. Marinaki, M., Marinakis, Y., Stavroulakis, G.E.: Fuzzy control optimized by a multi-objective particle swarm optimization algorithm for vibration suppression of smart structures. Struct. and Multidisciplinary Optimization 43(1), 29–42 (2011)
13. Prakash, A., Deshmukh, S.: A multi-criteria customer allocation problem in supply chain environment: An artificial immune system with fuzzy logic controller based approach. Expert Systems with Applications 38(4), 3199–3208 (2011)
14. Angelov, P.P.: An evolutionary approach to fuzzy rule-based model synthesis using indices for rules. Fuzzy Sets and Systems 137, 325–338 (2003)
15. Chiou, Y., Lan, L.W.: Genetic fuzzy logic controller: an iterative evolution algorithm with new encoding method. Fuzzy Sets and Systems 152, 617–635 (2005)
16. Gonzalez, A., Perez, R.: Selection of relevant features in a fuzzy genetic learning algorithm. IEEE Transactions on Systems and Man and and Cybernetics and Part B: Cybernetics 31(3), 417–425 (2001)
17. Cintra, M.E., Camargo, H.A.: Feature Subset Selection for Fuzzy Classification Methods. In: Hüllermeier, E., Kruse, R., Hoffmann, F. (eds.) IPMU 2010. CCIS, vol. 80, pp. 318–327. Springer, Heidelberg (2010)
18. Mitchell, T.M.: Machine Learning. McGraw-Hill (1997)
19. Wille, R.: Restructuring lattice theory: An approach based on hierarchies of concepts. In: Rivals, I. (ed.) Ordered Sets, vol. 23, pp. 445–470 (1982)
20. Wang, X.Z., Wang, Y.D., Xu, X.F., Ling, W.D., Yeung, D.S.: A new approach to fuzzy rule generation: fuzzy extension matrix. Fuzzy Sets and Systems 123, 291–306 (2001)
21. Bacardit, J., Krasnogor, N.: Performance and efficiency of memetic pittsburgh learning classifier systems. Evolutionary Computation 17(3), 307–342 (2009)
22. Chen, M.S., Wang, S.W.: Fuzzy clustering analysis for optimizing membership functions. Fuzzy Sets and Systems 103, 239–254 (1999)
23. Frank, A., Asuncion, A.: UCI machine learning repository (2010)
24. Bull, L., Bernadó-Mansilla, E., Holmes, J.: Learning Classifier Systems in Data Mining. Springer (2008)
25. Demsar, J.: Statistical comparison of classifiers over multiple data sets. Journal of Machine Learning Research 7(1), 1–30 (2006)

Evolutionary Fuzzy Rules for Ordinal Binary Classification with Monotonicity Constraints

Christian Moewes and Rudolf Kruse

Faculty of Computer Science, Otto-von-Guericke University of Magdeburg, Germany
cmoewes@ovgu.de, kruse@iws.cs.uni-magdeburg.de

Abstract. We present an approach to learn fuzzy binary decision rules from ordinal temporal data where the task is to classify every instance at each point in time. We assume that one class is preferred to the other, e.g. the undesirable class must not be misclassified. Hence it is appealing to use the Variable Consistency Dominance-based Rough Set Approach (VC-DRSA) to exploit preference information about the problem. In this framework, the VC-DomLEM algorithm has been used to generate the minimal set of consistent rules. Every attribute is then fuzzified by first applying a crisp clustering to the rules' antecedent thresholds and second using the cluster centroids as indicator for the overlap of neighboring trapezoidal normal membership functions. The widths of the neighboring fuzzy sets are finally tuned by an evolutionary algorithm trying to minimize the specificity of the current fuzzy rule base.

1 Introduction

In many real-world problems, complex systems need to be described in a highly comprehensible way. The explaining descriptions are usually based on observations of several variables describing the state of the system. For many systems, it is quite common that the number of variables is large, say around 50. Thus, it can be quite difficult for human experts to build a model that describes the system in an adequate way. Usually, a trade-off has to be found to balance both the correctness and the complexity of the model.

Complex models are naturally regarded skeptically since a proof of correctness is hard to obtain. In our highly technologized information society, rapidly increasing quantities of data coming from complicated systems are stored without any possibility of being analyzed manually. Therefore, it is extremely desirable to have machines that learn from empirical data *and* guarantee both interpretable and correct models.

Here, we will restrict ourselves to machines that solve supervised classification problems based on numerical data. A Fuzzy Rule-based Classifier (FRBC) is such kind of machine. Nowadays, an FRBC is the state of the art in many real-world applications, e.g. automobile controllers (Schröder et al, 1997). It has nice properties that are demanded and appreciated by safety experts. That is, an FRBC is linguistically interpretable and its behavior is easy to approve by these experts. An open research problem is the question how such an FRBC can be found and tuned automatically. We want to

R.R. Yager et al. (Eds.): *Soft Computing: State of the Art Theory*, STUDFUZZ 291, pp. 105–112.
DOI: 10.1007/978-3-642-34922-5_8

develop machine learning tools that come up with interpretable fuzzy rule bases for such systems.

The online discrimination of vehicle crashes to deploy certain stages of a restraint system is such complex problem (Moewes, 2007; Moewes et al, 2008, 2010). Fuzzy binary decision rules shall be obtained from ordinal temporal data. Every instance is classified at each point in time. The so-called fire class is preferred to the no-fire class, e.g. we must not deploy in a no-fire case. To the best of our knowledge, no suitable (fuzzy) rule learner exists for this safety-related problem. An exhaustive discussion and recent recent approaches to tackle safety-related problems can be found in (Nusser, 2009).

In this paper, we show that the crash discrimination problem corresponds to ordinal binary classification with monotonicity constraints. Hence, the idea is to use a rule-based ordinal classifier that exploits monotonicity. This important property is implicitly handled by the Variable Consistency Dominance-based Rough Set Approach (VC-DRSA) (Greco et al, 2001). In this framework, the minimal set of consistent rules has been generated by the VC-DomLEM algorithm (Błaszczyński et al, 2011).

Another novelty in this paper is a proposal to fuzzify the crisp classifier to an FRBC. Every attribute is fuzzy partitioned in three steps. First, a crisp clustering is applied to the rules' antecedent thresholds. Second, the cluster centroids are used as indicator for the overlap of neighboring trapezoidal normal membership functions. Finally, the widths of the neighboring fuzzy sets are tuned by an evolutionary algorithm that minimizes the specificity of the current fuzzy rule base. The fuzzification step is basically performed to compress the original rule base for interpretation issues.

2 Fundamentals

2.1 Ordinal Binary Classification

Given an object x which can be described by attribute values $x = (x_1, \ldots, x_n)$, the aim in ordinal classification is to predict an unknown class label ω from an ordered set $\Omega = \{0, \ldots, k-1\}$. A meaningful order between classes is assumed which corresponds to a natural order between the labels $\omega \in \Omega$. When dealing with $|\Omega| > 2$, ordinal classification problems are found in many real-world applications, e.g. recommender systems. There, users can rate items on a finite scale, e.g. a school grade from 1 to 5. The ultimate goal is to predict the rates of a new user given the known rates. For $|\Omega| = 2$, we say that the ordinal classification problem is binary.

2.2 Rough Set Approach

The rough set approach is performed in two steps to extract knowledge from observations (Błaszczyński et al, 2011). First, all inconsistencies in the data are found by computation of lower and upper approximations of all observations. In the indiscernibility-based rough set approach (IRSA) (Pawlak, 1991), these sets correspond to decision classes. Data points do not obey any order and can be therefore compared by the indiscernibility relation.

The dominance-based rough set approach (DRSA) (Greco et al, 2001) assumes that the decision classes are ordered. Furthermore DRSA monotonically relates observations' attributes and their decision classes based on some dominance principle, e.g. "the higher the attribute value, the more possible the instance is positive". Naturally, DRSA is therefore suitable to handle ordinal classification problems with monotonicity constraints. So, certain decision rules based on indiscernibility might be inconsistent w.r.t. the dominance principle. In IRSA and DRSA, approximations are based on granules of knowledge and the lower approximation of a set is defined by including granules of knowledge in the approximated set.

To relax the definition of the lower approximation to allow somewhat adequately consistent rules, different probabilistic generalizations of rough sets have been suggested. The variable consistency dominance-based rough set approach (Greco et al, 2001) is one of them. It permits an extension of the lower approximation of a set by observations with might belong to the set up to a sufficient degree.

The second step of RSA extracts decision rules from the observations approximations given in the approximations. Certain (or consistent) rules are received from the lower approximations. Possible rules are obtained by the upper approximations. Approximate rules are found in the boundary, i.e. the difference of lower and upper approximation. The set of minimal decision rules can be found by sequential covering, as it is performed by the algorithm VC-DomLEM (Błaszczyński et al, 2011). This algorithm tries to maximize the accuracy on the training set by a minimal number of rules covering all consistent training observations.

3 Generation of Fuzzy Rules

We chose the VC-DomLEM algorithm (Błaszczyński et al, 2011) to obtain a minimal set of crisp rules of the form

$$R_i : \text{if } x_1 \geq \theta_{i,1} \text{ and } \dots \text{ and } x_p \geq \theta_{i,p} \text{ then class } = +1.$$

We considered only *certain* rules for the positive class. That is, we want to prevent false positive under all circumstances. Hence the obtained rule base only covers positive observations. The negative class is predicted when no positive rule fires. This reasoning will be elucidated more detailed based on a real-world application in Section 4.

In order to discretize the real value sets of the given attributes, we fuzzy partitioned the value sets based on the rule thresholds $\theta_{i,j}$. The fuzzification shall serve as a compression of rules. Furthermore it shall enable the user to obtain a more comprise linguistic description of the obtained decision rules.

The partitioning of every attribute x_j with $1 \leq j \leq p$ is performed by several heuristic steps. First, the clause thresholds θ_{ij} of all rules R_i with $1 \leq i \leq r$ are collected. Second, an arbitrary number of splits k_j is determined, e.g. by the user. Note that x_j is partitioned into $k_j + 1$ fuzzy sets. Then, we group all elements of $\{\theta_{ij} \mid 1 \leq i \leq r\}$, e.g. using k-means clustering (MacQueen, 1967). Finally, the cluster centroids $c_{1,j}, \dots, c_{k,j}$ represent the intersecting points of neighboring trapezoidal fuzzy sets $\mu_{i,j}, \mu_{i+1,j}$ such that

$$\mu_{i,j}(c_{i,j}) = \mu_{i+1,j}(c_{i,j}) = 0.5.$$

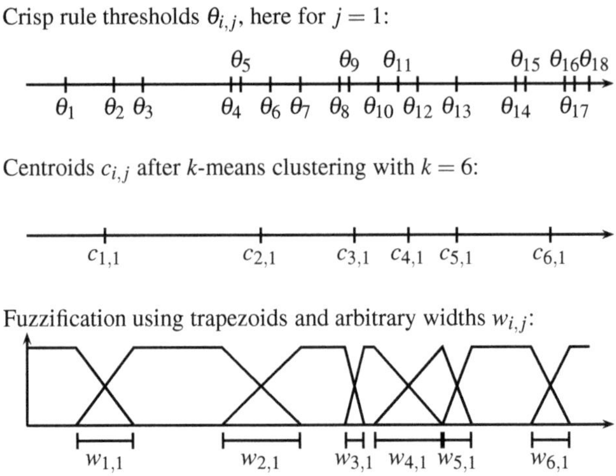

Fig. 1. Fuzzy partitioning of attributes: First, the crisp rule thresholds $\theta_{i,j}$ are clustered with k-means and $k = 6$. Then, the cluster centroids are used as intersecting points of neighboring fuzzy sets. The arbitrary widths $w_{i,j}$ of the overlapping fuzzy sets are found by an EA.

Algorithm 1. Evolutionary Algorithm

Input: training data, crisp rule base, centroids, k

```
t ← 0
top-k ← some integer
pop(t) ← create population of size μ
pop(t) ← repair pop(t)
while t < t_max {
    for each individuals A ∈ pop(t) {
        create fuzzy rules with crisp rules and fuzzy partition
        fitness A.F ← compute specitivity
    }
    rank all individuals according to their fitness values
    pop_1 ← select top-k individuals
    fill up pop_1 by shuffle-crossover of best individuals
    apply Gaussian mutation to all but the best A ∈ pop_1
    t ← t+1,    top-k ← top-k − 1
    pop(t) ← repair pop_1
}
return best individual from pop(t)
```

The complete process of fuzzification is shown exemplarily for x_1 in Fig. 1.

The only parameters left for tuning are the intersecting widths $w_{i,j}$ of the overlapping trapezoidal fuzzy sets. We simply suggest to apply a steady-state evolutionary algorithm (EA) to find good assignments of every width. The whole algorithm is shown in Alg. 1. Each candidate solution is represented by a vector of reals, i.e. all $w_{i,j}$'s. The widths are initialized by $w_{i,j} \in (0, \frac{\text{range}}{k}]$. The following procedure is called for evaluating the fitness of a candidate solution: First, we partition the attributes according to the fixed centroids and the variable widths. Second, we fuzzify every rule's clause using the fuzzy partition. Finally, the fitness is given by the *specificity*

$$\text{specificity} = \frac{\text{TN}}{\text{TN} + \text{FP}}$$

of the fuzzy rule-based classifier. That is, the ratio of true negatives divided by the sum of true and false negatives. The four possible outcomes of binary classification are shown in Tbl. 1.

Table 1. FP = false positive/alarm, FN = false negative

		given class	
		+1	-1
predicted	+1	TP	FP
outcome	-1	FN	TN

Note that since we already optimized the accuracy

$$\text{accuracy} = \frac{\text{TP} + \text{TN}}{\text{TP} + \text{FP} + \text{TN} + \text{FN}}$$

using the VC-DRSA, we may optimize the measure of specificity instead. It is possible to introduce different or more than this performance measure for evaluating a candidate solution, e.g. the *sensitivity*

$$\text{sensitivity} = \frac{\text{TP}}{\text{TP} + \text{FN}}.$$

However, due to the preference structure of the problem discussed in Section 4, it is sufficient to optimize the specificity in the second step.

4 Application

Nowadays Europeans travel three times as much as 20 years ago, mainly by car. Every year, around 40,000 people loose their lives on European roads. Many automobile manufacturers and suppliers therefore set themselves the objective of *zero accidents*. Indeed, looking at the number of fatalities recorded by the German Federal Highway Research Institute[1], we observe a strong tendency towards this vision of zero accidents. In 1970, there used to be 14 millions cars and more than 19,000 road fatalities in East and West Germany. More than 30 year later in 2006, the number of road fatalities decreased to 5,000 whereas the number of vehicles on German roads increased more than threefold to 46 millions. Until 1970 number of fatalities used to be directly proportional to the number of vehicles on German roads.

[1] Deutsche Bundesanstalt für Straßenwesen `http://www.bast.de`

Milestones of the automobile technology have played an important part in contributing to this contrasting trend. In 1972, a general speed limit had been introduced on (West) German roads. For years later, the requirement to wear a seat belt followed. Then in 1981, a big step towards the above described vision had been taken by introducing the airbag technology. Together with the seat-belt tensioner, it embodies the safest *restraint system* up to today. In the nineties, side airbags and nowadays knee airbags for the driver and front-seat passenger have been added to this system.

The development of these technologies would not have been possible without the design and implementation of standardized *crash tests*. Every millisecond today's cars must decide based on dozens of signals whether (and if so which) components of the restraint systems shall be deployed or not. For instance, it could be fatal if the airbag wrongly deployed during a first collision with a wild boar (a so-called *no-fire crash*) followed by a frontal collision with a tree. The airbag, for example, shall only deploy in severe crash situations, so-called *fire crashes*. Unfortunately in the first moments of a crash, the signals during a no-fire crash are very similar to those obtained by a fire crash.

Matters are complicated further by the crash test costs that usually come up to a car's price. Thus naturally, only few dozens of these prohibitive tests are possible during the development of every automobile series. In addition, dissimilar vehicles behave different in a standardized crash test. Neither all types of loads nor all possible technical equipments can be tested.

The decision logic developed by automobile supplies must therefore be accurately aligned to every car platform. Decision rules, e.g. based on fuzzy logic, are formulated by domain experts doing tedious detail work to meet most requirements of a car manufacturer and technical inspectors. The rule base to be generated must be both highly interpretable and very accurate regarding crash discrimination. It can take up to one month to cope with this task manually.

Although automobile suppliers started using technologies from the field of knowledge discovery in databases, the described problem to learn such decision logic is still open. Many attempts to induce fuzzy rules from observations (see e.g. (Nauck and Kruse, 1997; Ishibuchi et al, 2001; Wang and Lee, 2002)) fail since they generate either way too many rules composed of many clauses or too simple rules leading to a high error-proneness.

We try to find an algorithm which is capable to induce a decision logic that is accurate and interpretable. The problem to maintain high accuracy has been already solved using statistical learning methods, e.g. SVMs (Moewes, 2007; Moewes et al, 2008, 2010). Ambitious ideas (Moewes and Kruse, 2008) to induce interpretable and accurate fuzzy rules from SVMs failed (Moewes and Kruse, 2011).

Note that each algorithm must deal with the preference structure of this binary classification problem. That is, at each point in time which is labeled negatively (i.e. no-fire), a correct prediction is demanded. Every false positive may lead to severe injuries. On the contrary it is enough to classify one positive instance before a certain time has past. Recognizing a fire crash after this point in time could be fatal. In total, both decision classes (i.e. fire and no-fire) are preference-ordered: We prefer the fire class to the no-fire one.

Note that each signal can be seen as criterion, i.e. an attribute with preference-order domain. Clearly, the higher one of the signals, the more likely it is to fire at that point in time. This monotonicity relation facilitates a fire decision. We conclude that the task to solve is an ordinal classification problem with monotonicity constraints. VC-DRSA enables us to directly incorporate both the decision class preferences and the monotonicity constraint.

We would finally obtain a minimal set of certain rules that will only need a fraction of all possible attributes as fuzzy clauses. We expect the crisp VC-DomLEM classifier to outperform the FRBC on both accuracy and specificity. The discretization of all attributes, however, will increase the readability of the rules enormously.

5 Conclusions

We presented an approach to obtain fuzzy binary decision rules for ordinal binary classification problems with monotonicity constraints. It is based on VC-DRSA. A minimal set of crisp rules have been obtained by the VC-DomLEM algorithm. The rules have been fuzzified by a discretization procedure and an evolutionary algorithm. We expect our framework to outperform existing fuzzy classifiers in specificity and interpretability (i.e. number of rules, number of used clauses) on the crash discrimination problem. The advantages of our approach are the implicit feature reduction by RSA and the automatic generation of linguistically interpretable consistent rules.

We will investigate the generation of rules covering negative examples as well. Furthermore we plan to control the number of rules, e.g. by (fuzzy) confirmation measures. We expect temporal rules of a more general form to further exploit the problem description. The core idea will be to use DRSA for time preferences (so-called time dominance) (Greco et al, 2010). The integration of time series data mining techniques, e.g. motif discovery (Moewes and Kruse, 2009), might additionally boost the performance.

References

Błaszczyński, J., Słowiński, R., Szeląg, M.: Sequential covering rule induction algorithm for variable consistency rough set approaches. Information Sciences 181(5), 987–1002 (2011), doi:10.1016/j.ins.2010.10.030

Greco, S., Matarazzo, B., Słowiński, R., Stefanowski, J.: Variable Consistency Model of Dominance-Based Rough Sets Approach. In: Ziarko, W., Yao, Y. (eds.) RSCTC 2000. LNCS (LNAI), vol. 2005, pp. 170–181. Springer, Heidelberg (2001)

Greco, S., Matarazzo, B., Słowiński, R.: Dominance-based rough set approach to decision under uncertainty and time preference. Annals of Operations Research 176(1), 41–75 (2010), doi:10.1007/s10479-009-0566-8

Ishibuchi, H., Nakashima, T., Murata, T.: Three-objective genetics-based machine learning for linguistic rule extraction. Information Sciences 136(1-4), 109–133 (2001), doi:10.1016/S0020-0255(01)00144-X

MacQueen, J.: Some methods for classification and analysis of multivariate observations. In: Cam, L.M.L., Neyman, J. (eds.) Proceedings of the Fifth Berkeley Symposium on Mathematical Statistics and Probability, vol. 1, pp. 281–297. University of California Press, Berkeley (1967)

Moewes, C.: Application of support vector machines to discriminate vehicle crash events. Diploma thesis, School of Computer Science, University of Magdeburg, Universitätsplatz 2, 39106 Magdeburg, Germany (2007)

Moewes, C., Kruse, R.: Unification of fuzzy SVMs and rule extraction methods through imprecise domain knowledge. In: Verdegay, J.L., Magdalena, L., Ojeda-Aciego, M. (eds.) Proceedings of the International Conference on Information Processing and Management of Uncertainty in Knowledge-Based Systems (IPMU 2008), Torremolinos (Málaga), pp. 1527–1534 (2008)

Moewes, C., Kruse, R.: Zuordnen von linguistischen Ausdrücken zu Motiven in Zeitreihen (Matching of Labeled Terms to Time Series Motifs). Automatisierungstechnik 57(3), 146–154 (2009), doi:10.1524/auto.2009.0760

Moewes, C., Kruse, R.: On the usefulness of fuzzy SVMs and the extraction of fuzzy rules from SVMs. In: Galichet, S., Montero, J., Mauris, G. (eds.) Proceedings of the 7th Conference of the European Society for Fuzzy Logic and Technology (EUSFLAT 2011) and LFA-2011, Advances in Intelligent Systems Research, vol. 17, pp. 943–948. Atlantis Press, Amsterdam (2011), doi:10.2991/eusflat.2011.46

Moewes, C., Otte, C., Kruse, R.: Tackling Multiple-Instance Problems in Safety-Related Domains by Quasilinear SVM. In: Dubois, D., Lubiano, M.A., Prade, H., Ángeles Gil, M., Grzegorzewski, P., Hryniewicz, O. (eds.) Soft Methods for Handling Variability and Imprecision. AISC, vol. 48, pp. 409–416. Springer, Heidelberg (2008)

Moewes, C., Otte, C., Kruse, R.: Simple machine learning approaches to safety-related systems. In: De, R.K., Mandal, D.P., Ghosh, A. (eds.) Machine Interpretation of Patterns: Image Analysis and Data Mining. Statistical Science and Interdisciplinary Research, vol. 11, pp. 231–249. World Scientific Publishing Co. Inc., Hackensack (2010)

Nauck, D., Kruse, R.: A neuro-fuzzy method to learn fuzzy classification rules from data. Fuzzy Sets and Systems 89(3), 277–288 (1997), doi:10.1016/S0165-0114(97)00009-2

Nusser, S.: Robust learning in safety-related domains: Machine learning methods for solving safety-related application problems. PhD thesis, School of Computer Science, University of Magdeburg, Universitätsplatz 2, 39106 Magdeburg, Germany (2009)

Pawlak, Z.: Rough Sets: Theoretical Aspects of Reasoning about Data. Kluwer Academic Publishers, Norwell (1991)

Schröder, M., Petersen, R., Klawonn, F., Kruse, R.: Two paradigms of automotive fuzzy logic applications. In: Jamshidi, M., Titli, A., Zadeh, L., Boverie, S. (eds.) Applications of Fuzzy Logic: Towards High Machine Intelligence Quotient Systems. Environmental and Intelligent Manufacturing Systems Series, vol. 9, pp. 153–174. Prentice-Hall, Inc., Upper Saddle River (1997)

Wang, J., Lee, C.: Self-adaptive neuro-fuzzy inference systems for classification applications. IEEE Transactions on Fuzzy Systems 10(6), 790–802 (2002), doi:10.1109/TFUZZ.2002.805880

Social Network Analysis of Co-fired Fuzzy Rules

Jose M. Alonso[1], David P. Pancho[2], O. Cordón[2,3], A. Quirin[2], and L. Magdalena[2]

[1] Department of Electronics, University of Alcala
jose.alonso@depeca.uah.es
[2] European Centre for Soft Computing
{david.perez,oscar.cordon,arnaud.quirin,
luis.magdalena}@softcomputing.es
[3] DECSAI and CITIC-UGR
ocordon@decsai.ugr.es

Abstract. The popularity of modern online social networks has grown up so quickly in the last few years that, nowadays, social network analysis has become one of the hottest research lines in the world. It is important to highlight that social network analysis is not limited to the analysis of networks connecting people. Indeed, it is strongly connected with the classical methods widely recognized in the context of graph theory. Thus, social network analysis is applied to many different areas like for instance economics, bibliometrics, and so on. This contribution shows how it can also be successfully applied in the context of designing interpretable fuzzy systems. The key point consists of looking at the rule base of a fuzzy system as a fuzzy inference-gram (fingram), i.e., as a social network made of nodes representing fuzzy rules. In addition, nodes are connected through edges that represent the interaction between rules, at inference level, in terms of co-fired rules, i.e., rules fired at the same time by a given input vector. In short, fingram analysis consists of studying the interaction among nodes in the network for the purpose of understanding the structure and behavior of the fuzzy rule base under consideration. It is based on the basic principles of social network analysis which have been properly adapted to the design of fuzzy systems.

1 Introduction

Social networks [40] have existed since humans were aware of the great advantages derived from the fact of collaborating and living together in structured groups. Of course, this happened thousands of years ago. However, in the last few years the popularity of modern social networks has grown up very quickly because of the huge boom of new technologies for telecommunications. Nowadays, some websites like *facebook*[1], *twitter*[2] or *LinkedIn*[3] are widely known all around the world both for fun but also for professional purposes, with millions of users registered. Moreover, users of such social networks consider them as an essential part of their everyday life.

---

[1] A social utility for connecting with friends online at http://www.facebook.com

[2] A social utility for following people online at http://www.twitter.com

[3] A professional social network online at http://www.linkedin.com

R.R. Yager et al. (Eds.): *Soft Computing: State of the Art Theory*, STUDFUZZ 291, pp. 113–128.
DOI: 10.1007/978-3-642-34922-5_9

In consequence, social networks are attracting more and more attention from both industry and academia. Accordingly, lots of researchers have begun to work very actively on issues related to social networks [14] becoming a very flourishing field. There are studies in the context of all kinds of social sciences [25] such as bibliometrics [42], politics [11], medicine [13], economics [19], etc. There are also works dealing with industrial applications, for example supply chain management [27].

This paper introduces a new methodology for visualizing and analyzing fuzzy rule-based systems viewed as social networks. Hence, the main contribution consists in defining the so-called fuzzy inference-grams (*fingrams*).

Since the proposal of Zadeh and Mamdani's seminal ideas [28,43,44], interpretability [1] is widely recognized as one of the strongest points of fuzzy system identification methodologies. It represents the ability of fuzzy systems to model a real system in a human-friendly understandable way. To do so, the knowledge embedded into fuzzy systems is usually expressed in the form of linguistic variables and rules. Thus, the rule base of a fuzzy system becomes the main communication interface to users [31]. Moreover, a fuzzy rule base can be seen as a population made up of a set of individuals (fuzzy rules) which compete and collaborate among them with the aim of yielding both good generality-specificity and interpretability-accuracy trade-offs. In consequence, users can understand the system behavior through checking graphically existent relationships among rules. Fortunately, they can be easily analyzed by looking at the rule base as a fingram, i.e., as a social network made of nodes (representing fuzzy rules) and edges (representing the interaction among rules). Rule interaction is measured at inference level in terms of co-fired rules, i.e., rules fired at the same time by a given input vector.

The main goal of fingram analysis is the understanding of the structure and behavior of a fuzzy rule base under consideration. It is mainly based on the adaptation of given techniques for social network analysis to the design of fuzzy systems. As it will be thoroughly explained along the paper, the analysis of fingrams offers many possibilities: finding out the most significant rules, identifying potential inconsistencies among fuzzy rules, assessing the interpretability of fuzzy systems, etc.

The rest of the contribution is organized as follows. Section 2 starts with a brief overview on visual representation and analysis of fuzzy systems, then it presents some techniques for social network visualization and analysis, and it ends with the introduction of basic aspects related to interpretability assessment. Section 3 goes in detail with the generation and analysis of fingrams. It is important to notice that, as a first step, the general approach is particularized for the analysis of fuzzy rule-based classifiers (FRBCs), i.e., fuzzy rule-based systems for classification purposes. Section 4 summarizes the experiments carried out along with the achieved results. Finally, some conclusions and future work are sketched in Section 5.

2 Preliminaries

2.1 Visual Analysis of Fuzzy Rule-Based Systems

A complete analysis of visualization requirements for fuzzy systems is provided in [35]. It gives an overview on existing methodologies to yield 2D and 3D graphical representations of fuzzy systems. It comprises visualization of fuzzy data, fuzzy partitions, and fuzzy rules. Different alternatives are available depending on the requirements of the end-user. Moreover, requirements may change according to the visualization tasks to perform: interactive exploration; automatic computer-supported exploration; receiving feedback from users; and capturing users' profiles and adaptation.

The most relevant works to obtain visual representations of multi-dimensional fuzzy rules are those developed by Berthold et al. [6,15]. They make a mapping from a high-dimensional feature space onto a two-dimensional space which maintains the pair-wise distances between rules. The established mapping also displays an approximation of the rule spread and overlapping. As a result, it is possible to visualize and explore multi-dimensional fuzzy rule bases in a 2D graphical representation. Authors claim such representation yields a user friendly and interpretable exploratory analysis. However, the complexity of the analysis grows exponentially with the number of features and rules to be displayed. In consequence, in complex problems with many rules the interpretation of the resultant graph is not straightforward.

Unfortunately, there are not many papers tackling with visual analysis of the inference process of fuzzy systems, and most of them are limited to visual descriptions. Probably, this is due to the well-known linguistic expressivity of such systems that gives prominence to linguistic representations. However, when dealing with complex problems, even when the design is made carefully to maximize interpretability, the number of rules can become huge because of the curse of dimensionality characteristic of fuzzy rule-based systems. In those cases, looking for a plausible linguistic explanation of the inferred output, derived from the linguistic description of the fuzzy knowledge base, is not straightforward. Explaining the inferred output as an aggregation of all the involved rules is not easy when many rules are fired at the same time for a given input.

Some authors [22,23] have bet for searching visual explanations of the system output. Ishibuchi et al. established a set of design constraints with the aim of producing groups of rules with only two antecedent conditions that can be plotted in a two-dimensional (2D) space. They look for a visual representation able to explain the output of fuzzy rule-based classifiers to human users. Nevertheless, considering only two antecedents per rule is a strong limitation that may penalize the accuracy of the system.

2.2 Visual Analysis of Social Networks

Although there are several approaches for visualizing different kind of social networks, we will focus on co-citation social networks and the works published by Vargas-Quesada and Moya-Anegón [32,42] which strongly inspired our proposal. Indeed, the term *fingram* was coined by inspiration on the term *scientogram* firstly introduced by Vargas-Quesada and Moya-Anegón [32] as a novel tool for visualizing the structure of science [42].

Scientograms are visual science maps, i.e., visual representations of scientific domains in the form of social networks. They illustrate interactions among authors and papers through the basic notion of paper co-citation, representing the frequency with which documents are jointly cited by pairs. It is possible to group them by author, journal, or categories. Obviously, depending on the kind of regrouping, the information that can be extracted from the generated maps is different.

The standardized co-citation measure, firstly introduced by Salton and Bergmark, is computed by the next equation [38]:

$$MCN(ij) = \frac{Cc(ij)}{\sqrt{c(i) \cdot c(j)}} \tag{1}$$

where *Cc* means co-citation, *c* stands for citation, while *i* and *j* represent two different entities (authors, documents, journals, categories, institutions, countries, etc.).

In addition, network scaling (NS) is aimed to obtain simplified structures revealing the backbone, i.e., the underlying organization of the original network. NS is based on estimating the proximity between pairs of nodes by means of computing distances, similarities, correlations, and so on. Actually, NS is efficiently carried out by Pathfinder algorithm [8,12] that is essential to make feasible a good visual interpretation. Pathfinder is in charge of pruning the initial network while keeping only the most relevant links into the final Pathfinder networks (PFNETs). It is worthy to remark that the combination of entity co-citation and NS yields high quality, schematic network visualizations in several fields such as psychology (for representing the cognitive structure of a subject [39]), software development (for debugging of multi-agent systems [41]), or scientometrics (for analyzing large scientific domains [9]).

The next step is about the automatic visualization of PFNETs. For this porpuse, the spring embedder family of methods is the most widely used in the area of Information Science. Spring embedders assign coordinates to the nodes with the aim of producing aesthetical pleasant graphs. Vargas-Quesada and Moya-Anegón recommend the use of Kamada-Kawai's algorithm [26] which is one of the most extended methods to perform this task. Starting from a circular position of the nodes, it generates networks following aesthetic criteria: maximizing the use of available space, minimizing the number of crossed links, forcing the separation of nodes, building balanced maps, etc. Notice that, the combination of entities co-citation, PFNETs, and Kamada-Kawai makes the entities that share most sources with the rest, tend to be located toward the center.

Lastly, concerning the analysis of scientograms, according to [42] there are three main measures of centrality that yield useful information with the aim of identifying the most significant nodes of a PFNET: *Degree of Centrality* (regarding the number of direct links gathering in a node), *Centrality of Closeness* (measuring the distance among nodes), and *Centrality of Intermediation or Betweeness* (looking at nodes that act as link between other nodes contained in the shortest path).

2.3 Assessing Interpretability of Fuzzy Rule-Based Systems

Interpretability characterization and evaluation is a very subjective task which strongly depends on the skills and background (experience, preferences, knowledge, etc.) of the specific end-user who interprets the linguistic description of a fuzzy system with the aim of conceiving the significance of the system behavior.

Thus, assessing interpretability remains a trending and hot topic. Gacto et al. [17] have recently published a complete taxonomy about existent interpretability indexes. They identify four groups of indexes coming up from the combination of two different criteria, namely the nature of the index (complexity vs. semantic) and the considered elements (partitions vs. rule base) in the fuzzy system under study:

1. Complexity at partition level.
2. Complexity at rule base level.
3. Semantic-based interpretability at partition level.
4. Semantic-based interpretability at rule base level.

Most previous works [7,20] only deal with the readability of fuzzy systems. Therefore, most indexes correspond to groups (1) and (2). They usually make only basic analysis of complexity, i.e., they only count the number of elements (features, membership functions, rules, premises, etc.) included in the fuzzy system at partition level (group 1) and rule base level (group 2). Hence, they may be deemed as structural-based interpretability indexes.

On the other hand, group (3) contains works regarding structural properties of fuzzy partitions [34] such as distinguishability, coverage, and so on. They generally measure the degree of fulfillment of semantic constraints that should be overimposed during the design process. It is widely admitted that working with the so-called Strong Fuzzy Partitions (SFPs) [37] satisfies all semantic constraints required to have interpretable fuzzy partitions from the structural point of view.

Finally, only a few authors have begun recently to put emphasis on the importance of defining indexes in group (4). They advocate for extending the analysis of readability to evaluate the comprehensibility, i.e., the implicit and explicit semantics embedded in fuzzy systems [16,31]. There are also some papers dealing with the consistency of fuzzy rule bases and with the number of co-fired rules, i.e., rules simultaneously fired from a given input [4,10,30].

3 Proposal

This section thoroughly explains how to visualize and analyze FRBCs by means of fingrams. They represent a novel tool that arises from adopting a social network based approach inspired on the one proposed by Vargas-Quesada and Moya-Anegón for visualizing and analyzing the structure of science [42].

Fingrams are graphs which represent fuzzy rule bases as social networks. They contain nodes representing fuzzy rules and edges showing the interactions among them in terms of co-fired rules.

3.1 Fingram Generation, Scaling and Drawing

Given a fuzzy system containing N rules and an experimental dataset covering most possible situations, the $N \times N$ weight matrix M describes the interactions among the N rules in terms of frequency of co-firing.

$$M = \begin{pmatrix} 0 & m_{12} & \dots & m_{1N} \\ m_{21} & 0 & \dots & m_{2N} \\ \dots & \dots & \dots & \dots \\ m_{N1} & m_{N2} & \dots & 0 \end{pmatrix} \tag{2}$$

The co-firing measure (m_{ij}), inspired on the standardized co-citation measure (Eq. 1) proposed by Salton and Bergmark [38], is defined by the next equation:

$$m_{ij} = \begin{cases} \frac{SFR_{ij}}{\sqrt{FR_i \cdot FR_j}} & , i \neq j \\ 0 & , i = j \end{cases} \quad (3)$$

where SFR_{ij} means the number of data samples for which rules R_i and R_j are simultaneously fired, while FR_i and FR_j count respectively the total number of samples for which the same rules R_i and R_j are fired, without taking care if they are fired together or not. Notice that m_{ij} are normalized and M is symmetrical. Note also that the number of times a rule is fired is computed in an inferential way for all available data samples. Hence, it is extremely dependant on the goodness (quantity and quality) of the available experimental data.

An undirected graph is straightforwardly generated from the weight matrix M. This is made up by connecting N nodes using edges whose weights are directly taken from M. Thus, m_{ij} equals zero means that there is no link between nodes i and j.

Since the initial graph related to the matrix M is likely to be quite dense and difficult to analyze, it is worthy to apply a pruning mechanism before printing and exploring the generated fingram. To do so, a NS method like Pathfinder[4] [8,12] able to discover and keep only the most relevant links in M is very effective. It has already been successfully applied in the context of social networks. As result of running Pathfinder the initial graph representing M is translated into a pruned network called PFNET. This only keeps those links which do not violate the triangle inequality stating that the direct distance between two nodes must be lesser than or equal to the distance between them passing through any group of intermediate and connected nodes. Notice that, thanks to the properties of PFNETs, the pruned fingram preserves the underlying structure with all relevant information at global level in comparison to the original one.

Even though there are many different methods for the automatic visualization of social networks, the spring embedder family has become the most widely used in the area of Information Science. Spring embedders assign coordinates to the nodes in such a way that the final graph will be pleasing to the eye, and that the most important elements are located in the center of the representation. Among them, probably the most famous method is the one proposed by Kamada and Kawai [26]. Starting from a circular position of the nodes, it generates networks with aesthetic criteria such as the maximum use of available space, the minimum number of crossed links, the forced separation of nodes, the generation of balanced maps, etc. Notice that, the combination of rule co-firing, PFNETs, and Kamada-Kawai makes the most relevant rules, those exhibiting the highest interaction with the rest, tend to be located toward the center of the graphical representation.

The visual representation of the resultant graph is what we have called fingram. Furthermore, it can be enhanced with additional relevant information related to the specific problem under consideration. For instance, in the case of classification problems, the nodes represent fuzzy rules of FRBCs. More specifically, each rule is represented by a circular node whose size is proportional to the number of covered instances, and whose color corresponds to the class pointed out by the rule. Each node is labeled with the rule

[4] We have selected a recently published variant of Pathfinder algorithm (MST-PathFinder [36]) able to prune maps in cubic time.

identifier R_i but also with two very informative numbers, the percentage of instances in the dataset that are covered by the rule and the percentage of them matching with the rule output. Moreover, the number of border lines around a node indicates the number of linguistic propositions minus one in the rule description. In addition, each link among two nodes is characterized by an attached label that yields the related co-firing measure. The link thickness is proportional to its value. Furthermore, the link color is informative too. It is green for those rules pointing out the same class, and red in the case of rules pointing out different classes (potential inconsistencies).

Finally, it is important to highlight that our proposal is not affected by the well-known curse of dimensionality problem of fuzzy systems that implies that the number of fuzzy rules grows exponentially with the number of inputs. First, nodes represent directly rules instead of premises. Second, Pathfinder has been successfully applied to the analysis of large scientific domains representing thousands of co-cited entities [42]. In consequence, fingrams are able to display the interactions among thousands of rules in the form of highly interpretable graphs. Hence, even when the number of rules is huge the pruned fingram can be still comfortably viewed by any expert.

3.2 Fingram Exploratory Analysis and Interpretability Assessment

The expert analysis of fingrams can take profit of all tools already available for social network analysis. As a first approach, we advocate for the use of the so-called *Degree of Centrality* [42]. This means that we will point out the most significant rules, those corresponding to the nodes that concentrate the larger number of links in a fingram. Remind that thanks to the specific way scaling and drawing are done, the most salient links and nodes are likely to be placed in the center, and those less relevant in the periphery. Thus, those rules that correspond to nodes located in the periphery of a fingram, especially those connected with a low value (the weight of the associated link is small) to the remaining graph, are good candidates to be deleted. This could have an interesting collateral advantage since removing such rules is likely to increase interpretability while keeping almost the same accuracy. A basic simplification procedure may consist first on ranking rules according to their relevance and then finding out and removing those non-relevant ones, normally located at the periphery of the fingram.

Furthermore, the analysis of fingrams can report very useful information about the analysis and verification, at inference level, of the related fuzzy rule bases. For instance, one can directly analyze its global structure through exploring the number and location of apparent groups of rules, analyze the respective location of the rules coding for different classes, etc. As a result, it is easy to detect potential inconsistencies among fuzzy rules. They turn up when the co-fired rules yield different output classes. In addition, the higher the link weight (co-firing degree computed by Eq. 3) is, the larger the interaction among rules is, and the larger the degree of inconsistency results.

Notice that, even when a rule base is fully consistent at linguistic level, there may arise some possible inconsistencies at inference level because of the rule aggregation procedure made as part of the inference process. Such potential conflicts are difficult to detect mainly because they are partially hidden since they are typically produced by new unknown data samples that were not taken into account during the learning stage. For instance, it may happen that several rules are fired at the same time for a new given input vector and as result several outputs are activated with degrees higher than zero. When two different classes are activated with very similar degrees the situation can be labeled as an ambiguous case. Such situation is not desirable, no matter if the system is (or not) able to yield the right output class, because a slight modification in the input data may yield a wrong output. We can conclude that a FRBC producing many ambiguous cases is a non-reliable system and should be corrected.

With respect to interpretability, we assume that fuzzy partitions are interpretable and the matching among linguistic terms and fuzzy sets is supervised and approved by an expert. Notice that interpretable fuzzy partitions must represent prototypes that are meaningful for the end-user. Then, given a rule format along with an inference mechanism, the system interpretability can be evaluated looking only at rule level. Our assumption is the following: the larger the number of co-fired rules, the smaller the comprehensibility of the FRBC.

Fingrams give us all required information. Eq. 4 formalizes a novel interpretability index:

$$COFCI = \begin{cases} 1 - \sqrt{\frac{\sum_{i=1}^{N}\sum_{j=1}^{N}\left[(P_i+P_j)\cdot m_{ij}\right]}{MaxThr}}, & if \sum_{i=1}^{N}\sum_{j=1}^{N}\left[(P_i+P_j)\cdot m_{ij}\right] \leq MaxThr \\ 0, & otherwise \end{cases} \tag{4}$$

where $COFCI$ stands for Co-firing Based Comprehensibility Index. N is the total number of rules. P_i and P_j count the number of premises (antecedent conditions) in rules i and j, while m_{ij} is the measure of co-firing for the same rules i and j; it is computed by Eq. 3. In addition, $MaxThr$ is a threshold which represents a maximum value established to get a normalized measure in the interval [0,1]. It should be fixed by the designer of the FRBC, looking at the maximum number of rules that may be acceptable (by an end-user) for each specific problem according to its inherent complexity (number of inputs, output classes, available training data, etc.). According to our experimentations, we suggest setting $MaxThr$ greater or equal than one thousand times the multiplication of the number of classes (C) by the number of inputs (I) by the number of training samples (T):

$$MaxThr \geq 10^3 \cdot C \cdot I \cdot T \tag{5}$$

4 Experimental Analysis

This experimental study deals with an example of medical application where interpretability is of prime importance. Interpretability is a distinguishing capability of fuzzy systems which is really appreciated in most applications. Moreover, it becomes an essential requirement for those applications that involve an extensive interaction with humans. For instance, decision support systems in medicine [33] must be easily understandable, for both physicians and patients, with the aim of being widely accepted and successfully applicable.

We have chosen the well-known Wisconsin Breast Cancer Database (WBCD) [29] for illustrative porpuses. This dataset contains cases from a study that was conducted at the University of Wisconsin Hospitals, Madison, about patients who had undergone surgery for breast cancer. The task is to determine if the detected tumor is benign or malignant. Thus, the dataset contains 683 samples (we have removed the missing values), nine features (*Clump Thickness*, *Cell Size*, *Cell Shape*, *Marginal Adhesion*, *Epithelial Size*, *Bare Nuclei*, *Bland Chromatin*, *Normal Nucleoli*, and *Mitoses*) and one output class (*Benign / Malignant*). The whole dataset is freely available at the KEEL[5] machine-learning repository.

For simplicity, this analysis focuses only on FRBCs that were generated following the HILK (Highly Interpretable Linguistic Knowledge) fuzzy modeling methodology [3,5]. We have chosen HILK because it is especially thought for making easier the design process of interpretable FRBCs. To do so, it imposes several constraints (SFPs, global semantics, Mamdani rules [28], etc.) during the design phase. The rule base is made up of rules of form:

$$\textbf{If } \underbrace{X_a \text{ is } A_a^i}_{Proposition\ P_a} \textbf{ AND} \ldots \textbf{AND } \underbrace{X_z \text{ is } A_z^j}_{Proposition\ P_z} \textbf{ Then } Y \text{ is } C^n$$

where C^n is the selected output class; X_a is the name of the input variable a; and A_a^i represents the label i of such variable. Namely, A_a^i can be one of the elementary terms in the SFP or a composite term defined as a convex hull of adjacent elementary terms corresponding to OR and NOT combinations [21]. These kinds of rules are usually known as DNF rules. Notice that, the absence of an input in a rule means that it is not considered in the evaluation of such rule. This special kind of proposition is usually referred as *Don't care* [24] and it should be interpreted as ANY since it means that it is true no matter the selected linguistic term. Because several output classes can be activated since several fuzzy rules can be fired at the same time by the same input vector, the winner rule fuzzy reasoning mechanism is considered. Furthermore, the well-known minimum and maximum fuzzy operators are taken for conjunction and disjunction.

It is important to notice that HILK methodology is implemented as part of the free software tool GUAJE[6] [2]. Moreover, the new methodology for visualizing and exploring fuzzy rule bases proposed in this paper is also implemented in that tool. The drawing of the graphs themselves is done using another freeware tool named Graphviz[7] [18].

The rest of this section is devoted to show the utility of the new methodology proposed in this paper through some illustrative examples. As a starting point, the entire dataset has been randomly split into two subsets. The 75% of samples are considered as training set while the remaining 25% of samples compose the test set. Please notice that we do not apply cross-validation because, for the sake of clarity, we do not care about finding the best FRBC for the WBCD problem. We are aware that probably there are better rule bases for WBCD in the fuzzy literature, but our goal is to explain the new

[5] KEEL stands for Knowledge Extraction based on Evolutionary Learning. It is a free software tool available online at `http://sci2s.ugr.es/keel/`

[6] A free software tool for generating understandable and accurate fuzzy rule-based systems in a Java environment `http://www.softcomputing.es/guaje`

[7] A free software tool available online at `http://www.graphviz.org/`

methodology with a simple case instead of looking for the best solution for this specific problem.

Thus, we use GUAJE with the aim of building FRBCs automatically extracted from the available training data. Uniform SFPs with three triangular fuzzy sets are initially defined for each input. We are going to consider rules generated with the well-known Wang and Mendel (WM) and Fuzzy Decision Trees (FDT) algorithms both provided by GUAJE[8]. Hence, we generate two first set of rules corresponding to $FRBC_{WM}$ and $FRBC_{FDT}$. Moreover, we have simplified them with the simplification algorithm, also provided by GUAJE, in order to obtain two additional more compact FRBCs. Let's call them $FRBC_{WM-SIMP}$ and $FRBC_{FDT-SIMP}$. Two further simplifications guided by fingram analysis of $FRBC_{FDT-SIMP}$ have been carried out. They are named as $FRBC_{FDT-SIMP-F1}$ and $FRBC_{FDT-SIMP-F2}$.

Table 1 summarizes the main quality indicators characterizing those FRBCs previously generated. On the one hand, each column corresponds to one of the FRBCs under consideration. On the other hand, each row is related to one specific quality indicator.

Table 1. Quality evaluation of the generated FRBCs

	$FRBC_{WM}$	$FRBC_{WM-SIMP}$	$FRBC_{FDT}$	$FRBC_{FDT-SIMP}$	$FRBC_{FDT-SIMP-F1}$	$FRBC_{FDT-SIMP-F2}$
ACC_{TR}	0.998	0.998	0.975	0.975	0.943	0.939
ACC_{TS}	0.83	0.918	0.947	0.953	0.93	0.918
NR	195	23	35	9	3	2
TRL	1755	155	165	27	6	4
ARL	9	6.739	4.714	3	2	2
AFR_{TR}	6.043	2.977	2.625	1.488	1.133	1.093
AFR_{TS}	6.299	3.047	2.965	1.614	1.155	1.113
AFD_{TR}	0.555	0.797	0.766	0.865	0.859	0.878
AFD_{TS}	0.455	0.776	0.734	0.847	0.823	0.867
$COFCI$	0	0.675	0.510	0.880	0.960	0.969

Firstly, we take care of the achieved accuracy regarding both training (ACC_{TR}) and test (ACC_{TS}). Accuracy is computed as the percentage of samples properly classified. Secondly, we tackle with assessing interpretability. To do so, considering only one index is not enough as it was pointed out in Section 2.3. Therefore, we have considered several structural-based but also semantic-based interpretability indexes at rule base level. NR stands for number of rules. TRL means total rule length, that represents the total number of linguistic propositions into the whole rule base. ARL stands for average rule length, computed as TRL divided by NR. We have also reported the average number of fired rules with respect to both training (AFR_{TR}) and test (AFR_{TS}) sets. One rule is counted as fired by a given data sample only in the case in which it is activated with a confidence firing degree greater or equal than 0.1. In addition, we have computed the average confidence firing degree (AFD) regarding again training (AFD_{TR}) and test (AFD_{TS}) sets. AFD is measured as the firing degree of the winner rule for each

[8] The interested reader is referred to [2,3] for further details about algorithms provided by GUAJE.

data sample and then averaged for the whole dataset. Finally, $COFCI$ is the novel interpretability index proposed in this work. It is computed following Eq. 4 with $MaxThr$ equals 10^4.

Looking carefully to values shown in Table 1, we can draw some interesting conclusions. First, WM generates a lot of complete rules, i.e., each rule takes into account all inputs. In consequence, generated rules are quite specific and they are likely to be simplified. Moreover, rule base is affected by overfitting because $FRBC_{WM}$ exhibits very high ACC_{TR} while ACC_{TS} is not so good. Furthermore, it seems there is a lot of redundancy inside the rule base. Indeed, AFR_{TR} and AFR_{TS} achieve very high values while AFD_{TR} and AFD_{TS} remain quite low that implies a lot of overlapping among rules. Regarding all interpretability indicators (NR, TRL, ARL, AFR, AFD and $COFCI$), $FRBC_{WM}$ can be deemed as not interpretable at all. Such feeling is confirmed when observing the fingrams displayed in Fig. 1. Of course, the rule base is so complex that is not easy to make any useful interpretation neither focusing on the initial network (Fig. 1(a)) nor looking at the scaled one (Fig. 1(b)). Anyway, we can appreciate how the scaling process becomes very effective turning up a quite clear structure that was hidden.

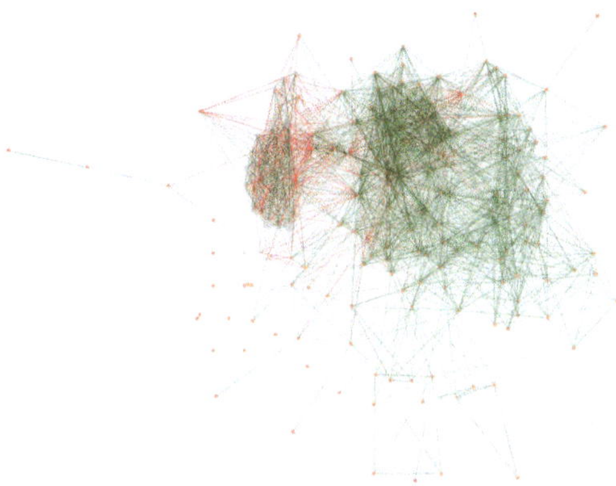

(a) Complete fingram

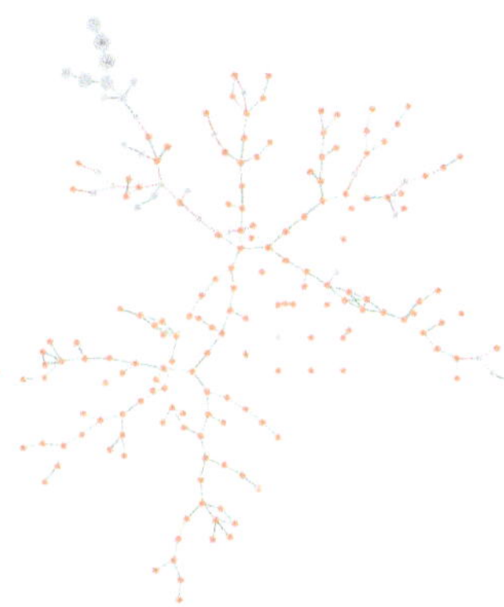

(b) Scaled fingram

Fig. 1. Fingrams related to $FRBC_{WM}$ before and after network scaling with Pathfinder

Second, FDT produces a smaller set of much more general incompletes rules minimizing the overfitting effect. Thus, $FRBC_{FDT}$ yields much closer values for both ACC_{TR} and ACC_{TS}. In comparison with $FRBC_{WM}$, ARL and AFR are significantly decreased while AFD is increased, so $COFCI$ increases accordingly. We can conclude that $FRBC_{FDT}$ yields a better interpretability-accuracy trade-off than $FRBC_{WM}$. Fingrams corresponding to $FRBC_{FDT}$ are depicted in Fig. 2. Obviously, they are much clearer than those ones previously presented in Fig. 1.

With respect to the effect of the initial simplification, not guided by fingrams, we appreciate an improvement in the generalization capabilities of the selected FRBCs. Of course, simplification is made considering only training data. It preserves ACC_{TR} while

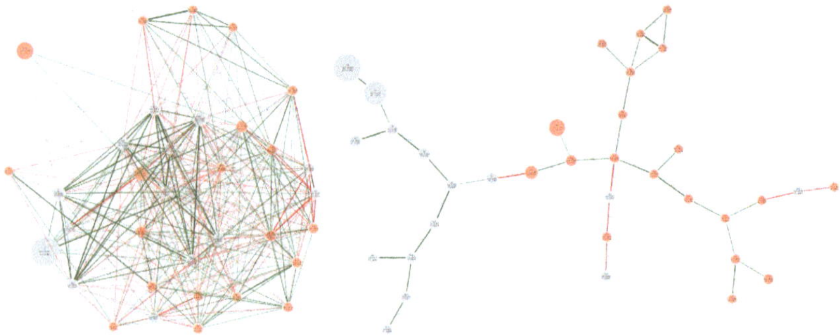

(a) Complete fingram (b) Scaled fingram

Fig. 2. Fingrams related to $FRBC_{FDT}$ before and after network scaling with Pathfinder

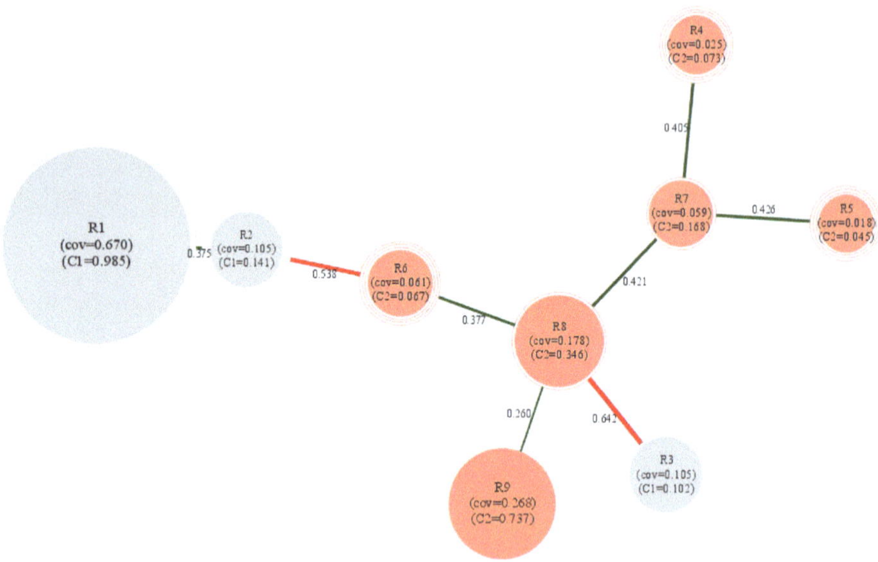

Fig. 3. Scaled fingram related to $FRBC_{FDT-SIMP}$

interpretability is strongly improved. As a side effect, ACC_{TS} is also increased. Furthermore, AFD grows up regarding both training and test. As a result, simplified FRBCs become much more trustworthy. Moreover, making a comparison between the two simplified FRBCs under study ($FRBC_{WM-SIMP}$ and $FRBC_{FDT-SIMP}$), it becomes obvious that $FRBC_{FDT-SIMP}$ yields the best interpretability-accuracy trade-off. $FRBC_{FDT-SIMP}$ is made up of only nine rules so its related fingram, plotted in Fig. 3, becomes very informative.

Each rule is represented by a circular node whose size is proportional to the number of covered instances, and whose color corresponds to the class pointed out by the rule. Each node is labeled with the rule identifier Ri but also with two very informative numbers, the percentage of instances in the dataset that are covered by the rule (cov) and the percentage of them matching with the rule output (Ci). In addition, the number of border lines around a node indicates the number of linguistic propositions minus one in the rule description. Each link between two nodes is characterized by an attached label that yields the related co-firing measure. The link thickness is proportional to its value. Furthermore, the link color is informative too; it is green for those rules pointing out the same class, while it is red in the case of rules pointing out different classes (potential inconsistencies).

From Fig. 3 we appreciate how most samples belonging to class C1 are handled by R1. On the other hand, rules R8 and R9 seem to be the most significant ones for class C2. If we keep only those three rules while removing the remaining, then we generate $FRBC_{FDT-SIMP-F1}$ whose quality indicators are detailed in Table 1. It is a very simple and highly interpretable FRBC, while its accuracy it is not strongly penalized with respect to $FRBC_{FDT-SIMP}$. Finally, looking carefully at rules R8 and R9 they may be merged into only one rule. In that case we obtain $FRBC_{FDT-SIMP-F2}$. Again, interpretability gets better while accuracy is only slightly reduced, as it was shown in Table 1.

5 Conclusions and Future Work

This paper has introduced a new methodology for exploratory analysis of fuzzy rule-based systems. In addition, we have proposed a novel interpretability index that takes into account the comprehensibility of fuzzy systems looking at the correspondence between their linguistic description and their inference process. It deals with semantic-based interpretability at rule base level and it is therefore aimed to cover the lack of such kind of indexes in the fuzzy literature.

We have shown the utility of our proposal in a simple but very illustrative classification problem where interpretability is highly appreciated because it copes with a medical diagnosis application. Achieved results are encouraging. The analysis of fingrams has helped us effectively in the hard task of searching for good interpretability-accuracy trade-offs. Anyway, in the future we will extensively validate the methodology and we will look for other co-firing metrics able to yield additional information about consistency, generality and/or specificity of rules.

Notice that, a software module for fingrams generation and analysis is available with the free software tool GUAJE. It can be freely downloaded at:

http://www.softcomputing.es/guaje

Acknowledgements. This work has been partly supported by the Spanish Government under program Juan de la Cierva JCI-2011-09839, the Spanish Ministry of Economy and Competitiveness under projects TIN2011-29824-C02-01, TIN2011-29824-C02-02, and TIN2009-07727, including EDRF funding, and the European Centre for Soft Computing (ECSC) located at Mieres (Asturias, Spain).

References

1. Alonso, J.M., Magdalena, L.: Editorial: Special issue on interpretable fuzzy systems. Information Sciences 181(20), 4331–4339 (2011)
2. Alonso, J.M., Magdalena, L.: Generating Understandable and Accurate Fuzzy Rule-Based Systems in a Java Environment. In: Petrosino, A. (ed.) WILF 2011. LNCS (LNAI), vol. 6857, pp. 212–219. Springer, Heidelberg (2011)
3. Alonso, J.M., Magdalena, L.: HILK++: an interpretability-guided fuzzy modeling methodology for learning readable and comprehensible fuzzy rule-based classifiers. Soft Computing 15(10), 1959–1980 (2011)
4. Alonso, J.M., Magdalena, L., Cordón, O.: Embedding HILK in a three-objective evolutionary algorithm with the aim of modeling highly interpretable fuzzy rule-based classifiers. In: 4th International Workshop on Genetic and Evolutionary Fuzzy Systems (GEFS), pp. 15–20 (2010)
5. Alonso, J.M., Magdalena, L., Guillaume, S.: HILK: A new methodology for designing highly interpretable linguistic knowledge bases using the fuzzy logic formalism. International Journal of Intelligent Systems 23(7), 761–794 (2008)
6. Berthold, M.R., Hall, L.O.: Visualizing fuzzy points in parallel coordinates. IEEE Transactions on Fuzzy Systems 11(3), 369–374 (2003)
7. Casillas, J., Cordón, O., Herrera, F., Magdalena, L.: Interpretability issues in fuzzy modeling. STUDFUZZ, vol. 128. Springer, Heidelberg (2003)
8. Chen, C.: Generalised similarity analysis and pathfinder network scaling. Interacting with Computers 10, 107–128 (1998)
9. Chen, C.: Information visualization: Beyond the horizon. Springer (2004)
10. Cheong, F., Lai, R.: Constraining the optimization of a fuzzy logic controller using an enhanced genetic algorithm. IEEE Transactions on SMC. Part B, Cybernetics 30(1), 31–46 (2000)
11. Cranmer, S.J., Desmarais, B.A.: Inferential network analysis with exponential random graph models. Political Analysis 19(1), 66–86 (2011)
12. Dearholt, D., Schvaneveldt, R.: Properties of pathfinder networks. In: Schvaneveldt, R. (ed.) Pathfinder Associative Networks: Studies in Knowledge Organization, pp. 1–30. Ablex Publishing Corporation (1990)
13. Dunn, A.G., Westbrook, J.I.: Interpreting social network metrics in healthcare organisations: A review and guide to validating small networks. Social Science & Medicine 72(7), 1064–1068 (2011)
14. Freeman, L.C.: The development of social network analysis: A study in the sociology of science. Empirical Press (2004)
15. Gabriel, T.R., Thiel, K., Berthold, M.R.: Rule visualization based on multi-dimensional scaling. In: FUZZ IEEE, pp. 66–71 (2006)
16. Gacto, M.J., Alcalá, R., Herrera, F.: Integration of an index to preserve the semantic interpretability in the multi-objective evolutionary rule selection and tuning of linguistic fuzzy systems. IEEE Transactions on Fuzzy Systems 18(3), 515–531 (2010)
17. Gacto, M.J., Alcalá, R., Herrera, F.: Interpretability of linguistic fuzzy rule-based systems: An overview of interpretability measures. Information Sciences 181, 4340–4360 (2011)
18. Gansner, E.R., North, S.C.: An open graph visualization system and its applications to software engineering. Software - Practice and Experience 30(11), 1203–1233 (1999)
19. Granovetter, M.: The impact of social structure on economic outcomes. The Journal of Economic Perspectives 19(1), 33–50 (2005)

20. Guillaume, S.: Designing fuzzy inference systems from data: An interpretability-oriented review. IEEE Transactions on Fuzzy Systems 9(3), 426–443 (2001)
21. Guillaume, S., Magdalena, L.: An or and not implementation that improves linguistic rule interpretability. In: 11th International Fuzzy Systems Association World Congress, Beijing, China, pp. 88–92 (2005)
22. Ishibuchi, H., Kaisho, Y., Nojima, Y.: Designing fuzzy rule-based classifiers that can visually explain their classification results to human users. In: 3rd International Workshop on Genetic and Evolutionary Fuzzy Systems (GEFS), pp. 5–10 (2008)
23. Ishibuchi, H., Kaisho, Y., Nojima, Y.: A visual explanation system for explaining fuzzy reasoning results by fuzzy rule-based classifiers. In: NAFIPS, pp. 1–6 (2008)
24. Ishibuchi, H., Murata, T., Türksen, I.B.: Single-objective and two-objective genetic algorithms for selecting linguistic rules for pattern classification problems. Fuzzy Sets and Systems 89, 135–150 (1997)
25. Kadushin, C.: Who benefits from network analysis: Ethics of social network research. Social Networks 27(2), 139–153 (2005)
26. Kamada, T., Kawai, S.: An algorithm for drawing general undirected graphs. Information Processing Letters 31(1), 7–15 (1989)
27. Kim, Y., Choi, T.Y., Yan, T., Dooley, K.: Structural investigation of supply networks: A social network analysis approach. Journal of Operations Management 29(3), 194–211 (2011)
28. Mamdani, E.H.: Application of fuzzy logic to approximate reasoning using linguistic systems. IEEE Transactions on Computers 26(12), 1182–1191 (1977)
29. Mangasarian, O.L., Street, W.N., Wolberg, W.H.: Breast cancer diagnosis and prognosis via linear programming. Operations Research 43(4), 570–577 (1995)
30. Marquez, A., Marquez, F., Peregrin, A.: A multi-objective evolutionary algorithm with an interpretability improvement mechanism for linguistic fuzzy systems with adaptive defuzzification. In: FUZZ IEEE, pp. 277–283 (2010)
31. Mencar, C., Castiello, C., Cannone, R., Fanelli, A.M.: Interpretability assessment of fuzzy knowledge bases: A cointension based approach. International Journal of Approximate Reasoning 52(4), 501–518 (2011)
32. Moya-Anegón, F., Vargas-Quesada, B., Herrero-Solana, V., Chinchilla-Rodríguez, Z., Corera-Álvarez, E., Muñoz-Fernández, F.: A new technique for building maps of large scientific domains based on the cocitation of classes and categories. Scientometrics 61(1), 129–145 (2004)
33. Nauck, D., Kruse, R.: Obtaining interpretable fuzzy classification rules from medical data. Artificial Intelligence in Medicine 16, 149–169 (1999)
34. de Oliveira, J.V.: Semantic constraints for membership function optimization. IEEE Transactions on SMC. Part A, Systems and Humans 29(1), 128–138 (1999)
35. Pham, B., Brown, R.: Analysis of visualization requirements for fuzzy systems. In: International Conference on Computer Graphics and Interactive Techniques (GRAPHITE), pp. 181–187 (2003)
36. Quirin, A., Cordón, O., Santamaría, J., Vargas-Quesada, B., Moya-Anegón, F.: A new variant of the pathfinder algorithm to generate large visual science maps in cubic time. Information Processing and Management 44, 1611–1623 (2008)
37. Ruspini, E.H.: A new approach to clustering. Information and Control 15(1), 22–32 (1969)
38. Salton, G., Bergmark, D.: A citation study of computer science literature. IEEE Transactions on Professional Communication 22, 146–158 (1979)
39. Schvaneveldt, R.W., Durso, F.T., Dearholt, D.W.: Network structures in proximity data. The Psychology of Learning and Motivation: Advances in Research and Theory (24), 249–284 (1989)

40. Scott, J.: Social Network Analysis: A Handbook, 2nd edn. Sage Publications (2000)
41. Serrano, E., Quirin, A., Botia, J., Cordón, O.: Debugging complex software systems by means of pathfinder networks. Information Sciences 180(5), 561–583 (2010)
42. Vargas-Quesada, B., Moya-Anegón, F.: Visualizing the structure of science. Springer (2007)
43. Zadeh, L.A.: Fuzzy sets. Information and Control 8, 338–353 (1965)
44. Zadeh, L.A.: Outline of a new approach to the analysis of complex systems and decision processes. IEEE Transactions on SMC 3, 28–44 (1973)

Learning Techniques in Presence of Uncertainty

Parisa D. Hossein Zadeh and Marek Z. Reformat

Dept. of Electrical and Computer Engineering
University of Alberta
Edmonton, Alberta, Canada
{Dehlehho,Marek.Reformat}@ualberta.ca

Abstract. In the last few years we have witnessed increased popularity of agent systems. This popularity is the result of agents' ability to work effectively and perform complex tasks in a wide range of applications. In this paper, we highlight the importance of learning mechanisms that are essential for behavioural adaptation of agents in complex environments. We provide a high-level introduction and overview of different types of learning approaches proposed in recent years. We also argue the necessity of dynamic learning processes for handling uncertainty, and propose an uncertainty-oriented architecture of agents together with a specialized knowledge base.

1 Introduction

Agent systems are becoming increasingly popular due to wide range of applications in which they can be deployed [1]. An agent is an autonomous system that acts in a dynamic environment towards achieving its goals. In some applications, a team of agents may work together towards realization of their goals. It is common that a human supervisor provides initial knowledge to the agent. However, the built-in primary knowledge may not suffice to allow an agent to operate in a highly dynamic environment. Agents' behaviour should not be limited to actions defined and supplied by a human. Agents should be able to adapt their behaviour via a continuous learning process [2]. Such agents, referred to as "software agents", use machine learning techniques to adapt to user's demands and dynamic environments [3]. In [3], authors analyze imitation learning as the foundation behind human infants' learning ability. Their research is based on extensive studies of psychologists observing developmental progress of human infants. Typically, software agents have limited processing capability, hence employed learning mechanisms should have low computational complexity.

Learning mechanisms are essential factors enabling agents to operate in complex environments and to achieve human-like behaviour. Software agents' learning processes should provide agents with abilities to perform two important tasks. First, agents have to know how to act upon receiving new information in terms of storing the new concepts in their memory, forming links to the already known concepts, and consciously updating the information. Second, agents should be able to select appropriate actions from their repository of behavioural patterns. In this context, if the agents' old behaviours do not provide an acceptable outcome new action patterns

R.R. Yager et al. (Eds.): *Soft Computing: State of the Art Theory*, STUDFUZZ 291, pp. 129–143.
DOI: 10.1007/978-3-642-34922-5_10

have to be learned so that agents can perform their tasks correctly. This means that appropriate actions have to be learned and carried out on new situations. The ability to adapt to the changes in an environment is a necessary feature of intelligent agents.

In this paper, we provide a brief survey of some state-of-the-art learning mechanisms. Furthermore, we focus on uncertainty that is a fundamental and unavoidable feature of any environment. We emphasize the fact that uncertainty is present in discovering and analyzing information, and agents' abilities to learn should accommodate methods and techniques capable of dealing with imprecision, ambiguity, lack of full information, and limited trust in information sources. We argue that agents' knowledge bases and architectures should be suitable for storing and reasoning about uncertainty.

2 Learning Mechanisms

The basic motivation for studying learning models of agent systems stems from the strong need for an efficient learning mechanism capable of performing in complex and uncertain environments. This model would be much less complicated if agents were dealing with certain, precise and complete knowledge.

A number of different learning approaches have been developed over the last few years. These approaches target different aspects of a learning process, and use variety of learning approaches and knowledge representation schemas. This section starts with a description of two learning tools: conceptual and behavioural. The former provides an agent with facts and items related to its domain knowledge, while the latter leads to a better selection of actions to be preformed. The case-base reasoning and human involvement are discussed as important components of a learning process. The adaptive mechanisms for behavioural rules are presented next, in which the rules to be fired are identified based on pre-conditions activated by an agent's perception. We also discuss elements of reinforcement learning that are used to enhance inference of a system that is built based on truth maintenance principles. An interesting approach of learning mechanisms that compares new knowledge with the one already known to an agent is presented next. Fuzzy clustering process is described, which is used to pre-process data and prepares input to a fuzzy controller. The concept of human involvement is presented as an example of interactive learning mechanisms. Different levels of human participation are described and evaluated.

2.1 Conceptual and Behavioural Learning

Two learning mechanisms - conceptual and behavioural learning – can be used to address the adaptability of agents in dynamic environments. Architectures and functionality of two cognitive software agents, namely, CMattie (conscious Mattie) and IDA (intelligent distribution agent) are investigated in [4]. A conscious software agent is defined as a system that senses the environment through its cognitive characteristics: decision making, reasoning, knowledge perception and processing. This enables the agent to cope with unusual situations. The processes of these two agents are implemented by small pieces of codes, called codelets. The agents' architectures are composed of two main sections. In the first, a slipnet contains the agent's domain knowledge that initially consists of limited numbers of built-in concepts. In the second, a behaviour net holds a set of actions and their links to each other.

Based on [4], a first step in any learning mechanism is to identify newly encountered situations by an agent. For this purpose, authors embed a function in an agent's perception module, which is triggered by observing words or phrases that have never been experienced by an agent. Next, a conceptual learning is applied as a learning mechanism that is founded on case-based memory and case-based reasoning. In conceptual learning, an agent views the newly encountered situations in terms of its past experiences. Thus, relevant functions are retrieved for the problem solving process depending on its recent activity history. The agent adds new concepts to its slipnet and creates relevant links between new and old concepts. Moreover, history of each learning process is maintained in the agent's case-based memory, which enhances the learning capabilities of the agent in future [4]. As a solution to the action selection mechanism, the authors introduce behavioural learning that helps an agent selects and performs appropriate actions based on the received information. To accomplish this task, the agent may utilize case-based reasoning that adapts solutions of old problems, and apply them to similar perceived information. As an alternative, the agent can communicate with its human supervisor to receive proper instruction. This problem is referred to as a development period. The authors argue that a development phase will be a cost-effective method for the agent to operate in its complex domain. During the development period, the agent obtains the needed knowledge of the domain. This may include observation, conversational interaction and assistance of a human supervisor.

2.2 Learning Adaptive Decision Making Rules

In [5], a real-time self-organisational algorithm is suggested for behavioural learning of an agent in autonomous systems. The authors consider a multi-agent cooperation system model. The action selection mechanism is modelled on a subsumption method [6], where an agent makes an appropriate decision based on the received perception from the environment and evaluates possible actions and their pre-conditions. The researchers only focus on single-rule scenarios and do not investigate the problem of parallel action selection which is a realistic model [5]. They argue that local behaviours and their independence to the final goal allow the agent to self-design in dynamic environments. In their approach the learning process is composed of an adaptive behavioural rules base (ABRB) component, which selects the best match in the list of possible actions by evaluating the pre-conditions. Lastly, the agent will send feedback to ABRB reporting the success of the result with the goal of improving the future cycles. See Fig. 1, redrawn from [5].

The proposed algorithm forms a tree that is composed of a limited number of Boolean expressions representing the pre-conditions. According to the received perceptions, values are assigned to each pre-condition in the tree, in order to select the action that is more likely to provide the best solution to the current perceived states of the environment.

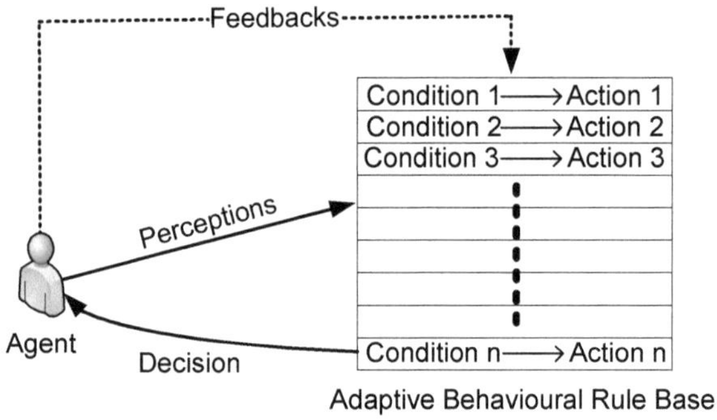

Fig. 1. Adaptive behaviour learning [5]

2.3 Relational Reinforcement Learning

The performance of a real-time learning mechanism in a highly dynamic environment is investigated in [7]. The authors enhance the adaptive logic interpreter (ADLIN) [7], which is a learning algorithm built upon relational reinforcement learning (RRL) [8]. The enhancements are due to the poor performance of ADLIN in time-constrained environments. They propose a real-time learning mechanism that combines ADLIN with a justification-based truth maintenance system (JTMS), a technique for managing the agents' beliefs [9], to enhance the inference process.

Authors argue that logical reasoning mechanisms have to be deployed carefully in intelligent agents due to their high computational complexity. JTMS makes the inference engine more efficient, by storing inferences received through interactions with inference engine, and reducing the number of RRL's states. This way, the previously seen instances that are stored in JTMS do not need to be processed by the inference engine anymore. Through experiments authors have shown that JTMS-based ADLIN outperforms both ADLIN and exhaustive inference systems in learning time. The proposed structure is shown in Fig.2.

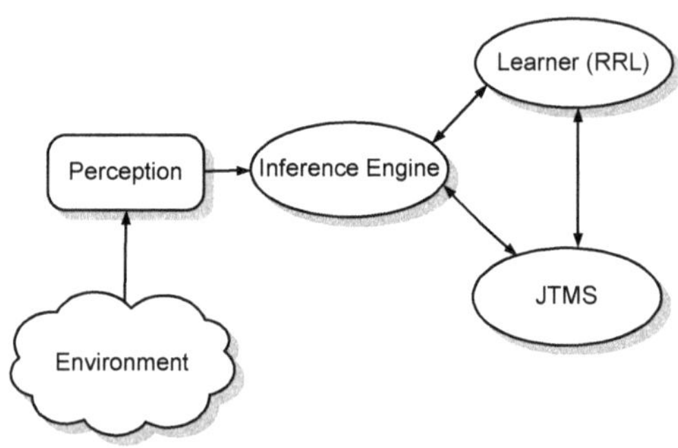

Fig. 2. Schematic of the adaptive logic interpreter (ADLIN) built based on justification-based truth maintenance system (JTMS) [7]

2.4 Participatory Learning

A quite different learning mechanism called a participatory learning mechanism is proposed in [10]. In the approach, the current knowledge of an agent participates in the learning process. This learning model is based on the features of human learning style, where the current beliefs directly affect the acceptance of newly received information. The author formulates the above learning process as a smoothing like algorithm [11], where the current observations from the environment are learned only if they are compatible to some extent with the old beliefs. For this purpose, a compatibility ratio is measured and has to be satisfied in order to consider the current observation valid. In Fig. 3 the upper feedback loop shows the participatory nature of the model, where the old beliefs and theories affect the learning process.

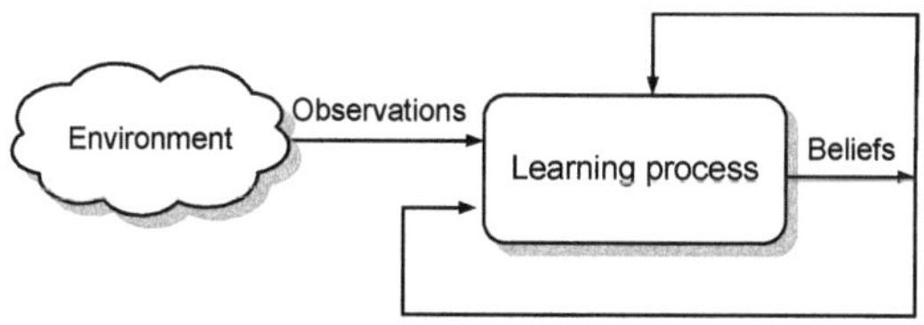

Fig. 3. Participatory learning model [10]

The learning mechanism is formulated as follows [10]:

$$V(j+1) = V(j) + \alpha\rho_i(D(j) - V(j))$$

where $V(j+1)$, $V(j)$ and $D(j)$ are vectors representing new information, old beliefs and current observations, while and ρ_i are the base rate and the compatibility rate, respectively. In [10] the impact of α and ρ_i on learning speed is shown. The author in [10] believes that this learning model is most effective when only a small change or a high compatibility exists between the new observations and the current beliefs; thus only a small change or an update happens to the current beliefs.

2.5 Online Adaptive Fuzzy Learning

In [12] an inference technique for agents' adaptation and learning in ubiquitous environments is proposed. In their model, agents learn and adapt their behaviour from a human (user) model by observing the user interactions with the environment. For this purpose, an adaptive online fuzzy inference system (AOFIS) is presented to model the user's behaviour, via a fuzzy logic controller (FLC), and to provide output actions to the environment. The proposed AOFIS technique is composed of five steps, which are shown in Fig.4.

First, agents observe the user's behaviour while capturing and labelling the inputs (from sensors and actuators) over time. Then, the sampled values are quantized into a set of fuzzy membership functions using a double clustering approach [13]. This algorithm runs iteratively to merge similar data samples based on their observed values until a predefined number of membership functions are created. In their approach, Gaussian membership functions are used as the fuzzy sets. In step 3, rules are extracted from the relationships between the set of inputs and outputs applying the learning from examples [14]. By step 4, the agent is capable of observing and controlling the environment via the learned FLC without the need for human involvement. In case of a new input arrival, the agent evaluates the input value to find which of the previously formed fuzzy sets it belongs to. Next, the proper rule is fired by the agent based on the calculated weight of the rules. In [12], the performance of AOFIS is evaluated in a real test-bed, an intelligent dormitory where 17 sensors were used as inputs and 10 actuators were used as outputs while in an interaction with a human user for five consecutive days. Through experiments it is shown that AOFIS outperforms similar soft-computing based techniques with fewer errors, and less computational complexity in online learning mode.

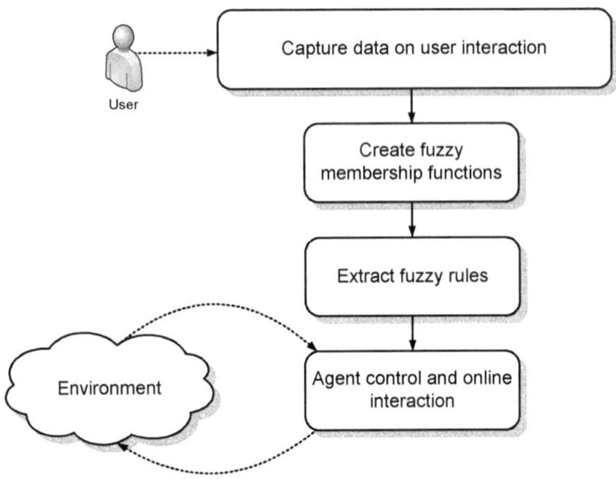

Fig. 4. Flow diagram of an adaptive online fuzzy inference system [12]

2.6 Interactive Artificial Learning

The issue of effectiveness in interactive artificial learning (IAL) is addressed in [15]. It is accomplished by comparing interactive learning method to traditional and conventional learning methods. The effectiveness of a learning method is measured as the ratio of the agent capability over the amount of inputs and skills from a human (designer or end-user). Several other metrics for measuring the quality of a learning method are also investigated in [15]. The authors explain two drawbacks of traditional learning methods as the significant amount of trial and error cycles in order to evaluate the agent's behaviour in a dynamic environment, and the need for a domain expert in addition to a system designer to encode the agent. Furthermore, in conventional learning methods agents operate more independently than in traditional learning methods, yet require human involvement to some extent. Also, conventional learning methods suffer from a slow learning process. Another drawback in conventional

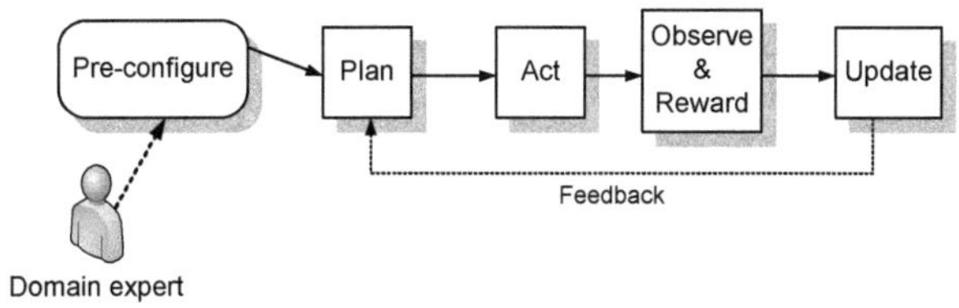

Fig. 5. Conventional learning steps

learning techniques is that the domain expert's role in pre-configuration step, consisting of system's parameter adjustment, reward structure, and learning mechanisms development, is considerably affecting the successfulness of the learning process. Fig. 5 shows a typical conventional learning process.

IAL is a new learning method that recently attracts many researchers' attention. In [15], the authors describe IAL as the learning method in which a human iteratively interacts with the agent during the learning process. The main goal of IAL is to keep the overall human involvement minimal. A general view of IAL learning steps is depicted in Fig. 6. As can be seen the end-user, not necessarily a domain expert, interacts in each step of the learning process thus diminishing the required work on the pre-configuration step [15]. Furthermore, IAL provides mutual understanding and exchange of knowledge between the end-user and the agent which facilitates the end-user's responsibility to provide more efficient inputs to the agent. The authors in [15] discuss the potential benefits of IAL learning on each particular learning step as shown in Fig. 6. Through simulations it is shown that traditional and conventional learning methods require more human involvement in the learning process that leads to a lower learning effectiveness than IAL method.

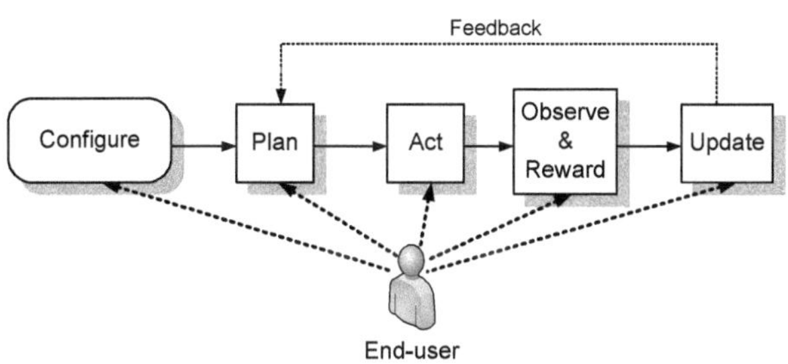

Fig. 6. Interactive artificial learning steps

2.7 Comparison and Discussion

All the discussed mechanisms are efficient learning processes equipped with different capabilities and features. They support various tasks, such as interactive learning, reinforcement learning, fuzzy reasoning, agent's feedback loops, and adaptive decision making. Table 1 provides a comparison of the discussed methods. As can be seen, in all of the mentioned approaches agents are able to adapt online in order to meet real-time needs. Another important observation is that approaches presented in [4] and [15] are the only techniques exploiting continuous human involvement as a vital element during the learning process. In other cases, the agent only relies on the initial knowledge provided by a human.

Table 1. Learning approaches for intelligent agents

Learning mechanisms	Learning model	Multi-agent cooperation	Real-time adaptation	Uncertainty management	Human interaction
Conceptual and behavioural leaning [4]	Conceptual, behavioural learning	No	Yes	No	Yes
Adaptive decision making rules [5]	Subsumption model learning	Yes	Yes	No	No
Relational reinforcement learning [16]	Relational reinforcement learning	No	Yes	No	No
Participatory learning [10]	Participatory learning	No	Yes	No	No
Online adaptive fuzzy learning [12]	Fuzzy inference learning	No	Yes	No	No
Interactive artificial learning [15]	Interactive artificial learning	Yes	Yes	No	Yes

As it can be observed, none of the above methods consider adapting the learning approach due to uncertainty observed by the agent. In fact, uncertainty is a factor that always exists in any complex and dynamic environment. We believe that there exists a relation between the level of agent's uncertainty and the agent's ability to learn. This is due to the fact that levels of uncertainty influence agent's confidence during the learning process.

3 Learning and Uncertainty

The definition of learning – *knowledge or skill acquired by instruction or study* – indicates that learning is a process of assimilating information that contributes to the overall knowledge and experiences of an individual. The pivotal element of the learning process is gaining knowledge. Assimilated knowledge can be evaluated from three different perspectives: its *source*, its *quality*, and its *novelty*. A newly acquired knowledge has to be analyzed in the context of an agent's knowledge base, and then integrated with this base. This process resembles a decision-making activity in which pieces of knowledge are chosen and combined with the existing knowledge. Each of the above perspectives, as well as decision-making mechanisms are potential source of uncertainty. Therefore, the process of learning has to be equipped with procedures suitable for handling uncertainty.

3.1 Knowledge Sources

In general, the web is a large uncensored network to which anyone can contribute by providing truthful as well as false information. Knowledge can be acquired from websites that can have different degrees of reliability. Recently, a lot of attention is dedicated to the issue of trust [17, 18]. In the initial structure of the semantic web [19], the importance of trust is recognized via defining a trust and proof layer as the top layers of the semantic web architecture. Some research activities are focused on different methods for assigning trust values to different sources, as well as methods dedicated to aggregation and inference of trust values. A number of different trust strategies have been proposed to rationale about trust: optimistic, pessimistic, centralized, trust investigation, and trust transitivity [18]. Each of these approaches deals with uncertainty and tries to discover aspects of the environment that are relevant to reduce uncertainty. Overall, the issue of trust in knowledge sources is related to the uncertainty associated with learning processes.

Another important aspect is quality of knowledge. The quality of knowledge relates to the amount of missing or ambiguous information. The quality-based knowledge uncertainty can be divided into three categories: non-specificity (imprecision), fuzziness (vagueness), and strife [20]. Non-specificity is manifested when two or more pieces of information are left unspecified. This may be the result of generalization, simplification, imprecision, or simply time constraints imposed on knowledge collecting processes. Fuzziness is characterized by the lack of definite or sharp distinction among pieces of information and may result from vagueness or any variety of indecisiveness. In some cases, especially for linguistic-based knowledge representation, terms and facts can be ambiguous due to differences in meaning as perceived by authors of the information. Strife or discord is an uncertainty characterized by disagreement in a selection process among pieces of information. This may happen due to dissonance, incongruity, discrepancy, and conflict. There is no doubt that quality of knowledge contributes to the uncertainty associated with the acquired knowledge.

In learning processes, the concept of uncertainty is also associated with novelty of knowledge – new knowledge introduces and changes uncertainty. In general, acquired knowledge can be of different levels of novelty. We can distinguish three scenarios of how acquired knowledge contributes to an agent's knowledge base and how it influences uncertainty.

- Updating existing knowledge – increases confidence in facts, skills and behavioral patterns already known to an agent. The agent's beliefs are modified and its uncertainty about correctness of facts decreases; the information that is "used" for this purpose can be associated with different levels of uncertainty, and it modifies the uncertainty levels of known information to a different degree.
- Modifying existing knowledge – includes changes in facts and skills that an agent currently believes in. The execution of those changes requires the agent's confidence in incoming knowledge; modifications should depend on estimated levels of uncertainty.
- Increasing existing knowledge – means assimilation of acquired knowledge that is new by an agent. This process needs procedures able to handle uncertainty; regulations are required to determine up to what degree of uncertainty an agent accepts new pieces of information.

The mentioned scenarios confirm that uncertainty is a crucial element of a learning process. Agents' learning mechanisms should be properly selected depending on environments. Also, agents should be able to take advantage of new information to increase their knowledge.

Based on presented above aspects, we claim that uncertainty is a part of a learning process and without it the learning would look quite different. Uncertainty is associated with the following issues:

- <u>uncertainty triggers learning:</u> a state of ambiguity forces an individual to search for more information and facts to resolve the vagueness;
- <u>uncertainty enables adaptability:</u> a constant state of not being sure means that an individual has to be prepared for a possible change of his/her opinion, in such a case it is easier to accept a change;
- <u>uncertainty prevents misjudgement:</u> processes of induction and deduction of new facts should have the ability to deal with situations which are not clearly true or false, it is not desirable to simplify everything to those two values;
- <u>uncertainty leads to more accurate models of reality:</u> the real world is not just "black and white", it is full of "gray areas", i.e., vagueness and ambiguity – any models real phenomena should be able to accommodate uncertainty.

3.2 Decision Making

The existence of uncertainty means that any decision-making mechanism has to cope with it. The processes of selecting what actions should be performed or which pieces of information should be integrated with the existing agent's knowledge should use a degree of uncertainty as an input. Decision-making mechanisms should be able to derive a conclusion in the presence of uncertainty, and provide the results that are "labelled" with degrees of uncertainty. Combining uncertainty with decision-making processes is not new. There are a number of different methods and techniques that can be applied here. These methods embrace probabilistic approaches – Bayes nets and Markov Models, possibilistic logic, preference and utility theories, as well as elements of game and auction theories [21-26].

An interesting investigation of decision models and uncertainty has been conducted in [27]. Authors use the bounded rationality concept to describe human decision strategy. They believe in a strong connection between agent rationality and agent model uncertainty. For clarifying this relation, four aspects of decision models are defined where the agent makes the best possible decision based on its knowledge base. These four aspects include: information availability, sampling of alternatives before the decision, the measure of assessment before the decision, and selection of an alternative. Each of those aspects can be associated with different levels of uncertainty. The selected decision will be optimum when the agent has full information with no ambiguity. This will result in full rationality (un-bounded), but it is not a realistic model for making decisions in real-world situations.

The ability to make decisions under uncertainty and to estimate the uncertainty of concluded decisions is a must for an adaptive intelligent agent. The methods and techniques for building and updating agents' knowledge bases with indications about uncertainties of acquired or induced knowledge should be part of learning processes.

4 Uncertainity-Oriented Agent Architecture

Overall, agents should be equipped with multiple learning mechanisms that are utilized depending on the agents' environments and levels of uncertainty associated with acquired information. In order to make it possible we propose an ontology-based uncertainty-oriented architecture for intelligent agents, and a special structure for their knowledge bases, Fig. 7.

Before we describe the architecture in detail, we need to explain the structure and the role of an agent's knowledge base (KB). The base is built based on three different forms of knowledge representations: ontology, causal nets, and belief structures. The ontology provides the basis for expressing facts, their definitions, and different types of relations that can exist among them. One of these relationships is a cause-effect relation. This relation is a fundamental relation of causal nets that are used to express conditional (in)dependence (causal relations) between facts in ontology. Additionally, a belief structure is imposed on ontology facts. It is represented by assessment of beliefs distributed among relevant facts. It can be said that the agent's KB is a multi-dimensional base able to embrace a multi-facet character of information. Furthermore, a number of if-then rules can be built using facts and their definitions contained in the base. The agent's KB has two essential parts: *temporary KB*, and *primary KB*. The *temporary KB* serves as a working memory and is used to store information that is still being retrieved and evaluated. Based on the estimated levels of uncertainty associated with different pieces of information from the *temporary KB*, the information and inferred facts that satisfy pre-defined confidence levels are moved into the *primary KB*. The *primary KB* contains information that has been analyzed via mechanisms of approximate reasoning. There are two parts of the *primary* KB – *facts-part* and *definitions-part.*

- The *facts-part* contains concrete pieces of information; it resembles individuals defined in the semantic web definition of ontology.
- The *definitions-part* contains general knowledge – definitions of things, concepts, and different relations between them; it resembles the definition part of the semantic web ontology. Facts from the *facts-part* and definitions from the *definitions-part* are connected by the "instance-of" relation. Facts and their definitions are associated with belief values that all together constitute a belief structure. Two important elements of the *definitions-part* are relations and rules:
  - *relations* express different types of relations that exist among facts/definitions; each fact is just an instance of a single definition; the relations are built through observing relations among facts and then are generalized to the level of definitions;
  - *rules* are if-then rules of arbitrary complexity built using facts, definitions, and relations between them.

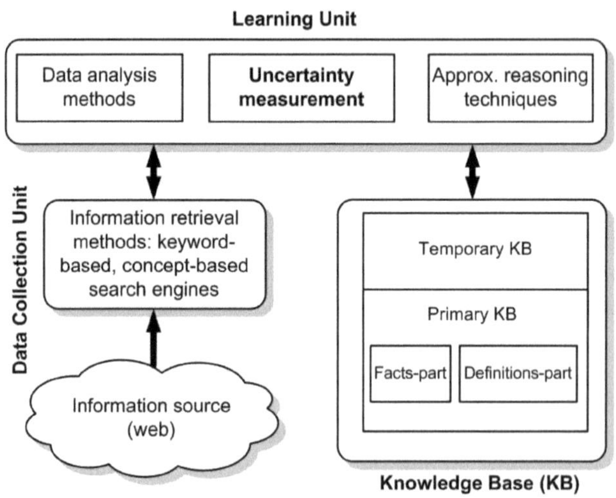

Fig. 7. The proposed uncertainty-oriented agent's architecture

The presented agent's architecture, Fig. 7, can be described in the following way. Information is retrieved from the web using different keyword- and concept- based information retrieval methods. This acquired knowledge is stored in the *temporary KB* and analyzed using variety of methods such as NLP-based pre-processing, different unsupervised and supervised techniques, and ontology-based processing (identification of facts included in ontology, synonyms, and ontology-defined relations). These analyses are integrated with processes leading to estimation of quality- and novelty- based uncertainties – "uncertainty measurement" unit in Fig. 7. The obtained uncertainties are combined with trust values associated with sources of information. This process results in determining belief values that are assigned to the acquired pieces of knowledge. The new information together with uncertainty values is stored in the agent's *primary KB*, and to be more precise, in its *facts-part*.

However, the learning process is far from done. Depending on available data and beliefs assigned, different methods will be used in order to find patterns and rules (association mining, decision tree construction, and supervised learning); to identify groups of items that exhibit similarity (unsupervised and semi-supervised learning); and to award/punish agent's decision and actions (reinforcement learning with possible involvement of a human). These processes will be performed on a regular basis. The levels of uncertainty associated with different facts, definitions and relations will influence the invoked learning mechanism and determine if an additional gathering of information is still needed in order to achieve a satisfying level of uncertainty.

It is worth noting that a human can play a distinct and significant role in almost every part of the agent's architecture by providing raw information to the agent, being involved in decision making processes, assisting the agent in estimating knowledge uncertainty or in building agent's knowledge base.

The importance of dealing with uncertainty and the justification of the proposed architecture can be illustrated with a simple, almost naïve example. Let us define a scenario in which two agents, "A" and "B", operate independently in the same environment. The agent "B" is designed such that it is capable of representing and reasoning under uncertainty, while the agent "A" is not able to do it. The assigned task from an end-user to these two agents is to organize a trip to Disneyland. Firstly, the agents start discovering the location of Disneyland on the web. The agent "A" looks through a number of hits (determined by its configuration parameters) and accepts the results without any doubt. The agent "B" estimates the uncertainty associated with the results, and is able to perform more search, i.e., to find more possibilities related to the trip's destination.

Secondly, once the destination is determined, the agents try to identify the most suitable hotel at a given location. Once again, the agent "A" is more rigid – it only does what is determined by its parameters – number of selection criteria, number of alternative hotels. The agent "B", on the other hand, is more flexible and is able to adapt and modify the selection process in the case the information about alternatives involves imprecision and ambiguity – it increases the search, looks for more criteria that were used in the past. Conclusively, the agent "B" accomplishes the task but the agent "A" struggles to finish it and needs human intervention.

As can be inferred from this example, the agent's ability to understand uncertainty and properly act based on it lead to increased "curiosity" and adaptation in the process of exploring the environment. A large set of available options is evaluated until the agent becomes certain whether its selections and decisions match the preferences of the requested task.

5 Conclusion

Learning is a key feature that converts an ordinary agent into one that intelligently interacts with its environment. This means that an agent is capable of dealing with different situations while adjusting its tactical strategies during operation in a dynamic environment. A brief survey of a number of state-of-the-art learning mechanisms is presented in this paper. The discussed methods address different features of a learning process and types of required knowledge. We have argued that the agent's ability to update its learning schema in the presence of uncertainty is an essential element of learning mechanisms of an intelligent system. This leads to greater flexibility in the agent's functionality in dynamic environments.

We also proposed a new architecture with the focus on uncertainty for intelligent agents. The architecture reflects the importance of: assessing levels of uncertainty, storing the uncertainty values in the agent's knowledge base, and using these values for decision making and learning processes. A new knowledge base structure is also proposed which addresses the issue of knowledge representation with uncertainty. Among different uncertainty management approaches – such as probabilistic models or the Dempster-Schaefer theory – we strongly believe in appropriateness of the fuzzy set theory. This opinion is also shared by the author of [20] who believes that the fuzzy set theory is the most appropriate tool for modeling human decision making processes due to the fact that the fuzzy set theory is inherently suitable for modeling information with imprecision – a situation that is normal for complex environments.

References

[1] Hayes, C.C.: Agents in a nutshell-a very brief introduction. IEEE Transactions on Knowledge and Data Engineering 11, 127–132 (1999)

[2] Edwards, P.: Intelligent agents=learning agents. In: UK Intelligent Agents Workshop, pp. 157–161. SGES Publications, Oxford (1997)

[3] Demiris, Y., Meltzoff, A.: The robot in the crib: a developemental analysis of imitation skills in infants and robots. Infant. Child Dev. 17, 43–53 (2008)

[4] Ramamurthy, U., Negatu, A., Franklin, S.: Learning mechanisms for intelligent systems. In: International Conference on Advances in Infrastructure for Electronic Business, Science, and Education on the Internet (SSGRR 2001), Italy, (2001)

[5] Lemouzy, S., Camps, V., Glize, P.: Towards a self-organising mechanism for learning adaptive decision-making rules. In: IEEE/WIC/ACM International Conference on Web Intelligence and Intelligent Agent Technology, WI-IAT 2008, pp. 616–620 (2008)

[6] Brooks, R.: A robust layered control system for a mobile robot. IEEE Journal of Robotics and Automation 2, 14–23 (1986)

[7] Asgharbeygi, N., Nejati, N., Langley, P., Arai, S.: Guiding Inference Through Relational Reinforcement Learning. In: Kramer, S., Pfahringer, B. (eds.) ILP 2005. LNCS (LNAI), vol. 3625, pp. 20–37. Springer, Heidelberg (2005)

[8] Dzeroski, S., De Raedt, L., Blockeel, H.: Relational reinforcement learning. In: Proceedings of the Fifteenth International Conference on Machine Learning, pp. 136–143 (1998)

[9] Doyle, J.: A truth maintenance system. Artif. Intell. 12, 231–272 (1979)

[10] Yager, R.R.: A model of participatory learning. IEEE Transactions on Systems, Man and Cybernetics 20, 1229–1234 (1990)

[11] Brown, R.G.: Smoothing, Forecasting and Prediction of Discrete Time Series. Prentice Hall, New Jersey (1963)

[12] Doctor, F., Hagras, H., Callaghan, V., Lopez, A.: An adaptive fuzzy learning mechanism for intelligent agents in ubiquitous computing environments. In: Proceedings of the Automation Congress World, pp. 101–106 (2004)
[13] Castellano, G., Fanelli, A.M., Mencar, C.: Generation of interpretable fuzzy granules by a double-clustering technique. Archive of Control Sciences, Special Issue on Granular Computing 12, 397–410 (2002)
[14] Wang, L., Mendel, J.M.: Generating fuzzy rules by learning from examples. In: Proceedings of the 1991 IEEE International Symposium on Intelligent Control, pp. 263–268 (1991)
[15] Crandall, J.W., Goodrich, M.A., Lin, L.: Encoding intelligent agents for uncertain, unknown, and dynamic tasks: From programming to interactive artificial learning. In: AAAI Spring Symposium: Agents that Learn from Human Teachers, pp. 28–35 (2009)
[16] Hamidi, M., Fijany, A., Fontaine, J.: Enhancing inference in relational reinforce-ment learning via truth maintenance systems. In: The Ninth International Conference on Machine Learning and Applications (ICMLA 2010), pp. 407–413 (2010)
[17] Richardson, M., Agrawal, R., Domingos, P.: Trust Management for the Semantic Web. In: Fensel, D., Sycara, K., Mylopoulos, J. (eds.) ISWC 2003. LNCS, vol. 2870, pp. 351–368. Springer, Heidelberg (2003)
[18] O'Hara, K., Alani, H., Kalfoglou, Y., Shadbolt, N.: Trust Strategies for the Semantic Web. In: McIlraith, S.A., Plexousakis, D., van Harmelen, F. (eds.) ISWC 2004. LNCS, vol. 3298, pp. 78–85. Springer, Heidelberg (2004)
[19] Horrocks, I., Parsia, B., Patel-Schneider, P.F., Hendler, J.: Semantic Web Architecture: Stack or Two Towers? In: Fages, F., Soliman, S. (eds.) PPSWR 2005. LNCS, vol. 3703, pp. 37–41. Springer, Heidelberg (2005)
[20] Mccauley-Bell, P.: Intelligent agent characterization and uncertainty management with fuzzy set theory: a tool to support early supplier integration. Journal of Intelligent Manufacturing 10, 135–147 (1999)
[21] Russell, S.J.: Norvig, Artificial Intelligence: A Modern Approach. Prentice Hall (2003)
[22] Pearl, J.: Probabilistic Reasoning in Intelligent Systems: Networks of Plausible Inference. Morgan Kaufmann (1997)
[23] Dubois, D., Lang, J., Prade, H.: A Brief Review of Possibilistic Logic. In: Kruse, R., Siegel, P. (eds.) ECSQAU 1991 and ECSQARU 1991. LNCS, vol. 548, pp. 53–57. Springer, Heidelberg (1991)
[24] Resnik, M.D.: Choices: An Introduction to Decision Theory. University of Minnestoa Press, Minneapolis (1987)
[25] Fudenberg, D., Tirole, J.: Game Theory. MIT Press, Cambridge (1992)
[26] Rosenschein, S., Zlotkin, G.: Rules of Encounter: Designing Conventions for Automated Negotiation among Computers. MIT Press, Cambridge (1994)
[27] Reusser, D.E., Hare, M., Paul-Wostl, C.: Relating choice of agent rationality to agent model uncertainty - an experimental study. In: iIEMSs Complexity and Integrated Resources Management (2004)

Computing with Words Using Weighted Power Mean Aggregation Operators

John T. Rickard[1], Janet Aisbett[2], Ronald R. Yager[3], and Greg Gibbon[2]

[1] Distributed Infinity, Inc., 4637 Shoshone Drive, Larkspur, CO 80118, United States
terry.rickard@gmail.com
[2] Faculty of Science and IT, The University of Newcastle, NSW 2308, Australia
{janet.aisbett,greg.gibbon}@newcastle.edu.au
[3] Machine Intelligence Institute, Iona College, New Rochelle, NY, United States
yager@panix.com

Abstract. Weighted power means are a flexible and powerful family of aggregation functions. The simplest member of this family, the weighted arithmetic mean, previously has been adapted for interval type-2 fuzzy scores and weights. This operator has been termed a "linguistic weighted average," and has been a primary instantiation of a "perceptual computer" in recent literature. We present an algorithm for computing weighted power means of arbitrary power for type-1 or interval type-2 fuzzy inputs and weights, which we call "linguistic weighted power means." We compare the linguistic weighted power mean and the linguistic weighted average on an "investment judgment advisor" example. Our results illustrate the flexibility and range of logical inference provided by this very versatile aggregation operator for computing with words applications.

Keywords: aggregation operators, type-2 fuzzy logic, computing with words, perceptual computing.

1 Introduction

In many applications, a choice has to be made amongst a number of alternatives, such as investment options, candidates for a job, etc. In evaluating alternatives, partial scores are assigned according to the degree to which a candidate satisfies certain properties (variously called attributes, goals, factors, criteria, etc.) that are considered relevant to the decision. Each property may be weighted to reflect its relative importance in the decision context. The partial scores and weights are then aggregated to provide a global score, which can be ranked against the scores of other alternatives. Often this scoring is performed in the context of a hierarchical decision tree, whose leaf nodes represent the most basic properties of a candidate alternative. These basic properties are sub-properties of higher-level attributes in the tree, which in turn may be sub-properties of yet higher-level attributes, all the way up to the root node of the tree. In general, the decision tree may have any number of intermediate levels in its structure.

R.R. Yager et al. (Eds.): *Soft Computing: State of the Art Theory*, STUDFUZZ 291, pp. 145–160.
DOI: 10.1007/978-3-642-34922-5_11

At the heart of modeling evaluation decisions in this context is the selection of an aggregation function L mapping vectors of scores $\mathbf{x} = (x_1, ..., x_n) \in [0,1]^n$ and their associated importance weights $\mathbf{w} = (w_1, ..., w_n)$ into a scalar score $L(\mathbf{x}, \mathbf{w}) \in [0,1]$. The aggregation function may be different for each sub-tree in the overall decision tree. Its output in any instance represents the degree to which a particular candidate satisfies the properties identified for that sub-tree, in a sense determined by the structure of L [1]. This process is a form of data fusion in which scores and weights are generally taken to lie in the unit interval.

The weighted mean of the scores is one of the standard aggregation functions extensively used in decision analysis [1]. More broadly, aggregation is performed with the p-norm [2], which is a generalization of the Euclidean norm. When the power in the p-norm is allowed to vary over the entire real line, the aggregator is called a weighted power mean (WPM). While it is no longer always a norm, the WPM is a very flexible operator that is used in information retrieval and other ranked decision applications [3].

However, the choice of aggregation function L cannot be made without also taking into account imprecision in the decision model. Whether dealing with decision making at the cognitive or perceptual levels, imprecision and instability affect the selection and definition of the properties considered relevant to the decision, affect the assessments of their relative importance, affect the way we score a candidate in terms of these properties, and affect how scores are aggregated [3]-[8].

Powerful techniques for handling imprecision can be obtained by using fuzzy scores, including scores characterized by interval-valued fuzzy sets, which are also known as interval type-2 (IT2) fuzzy sets, and general type-2 fuzzy sets. Aggregation functions should therefore be able to deal with these forms of input. As is well known, even the weighted arithmetic mean does not extend to an analytically closed form when importance weights are interval-valued [9]. Reference [10] presents an efficient algorithm to compute this mean, which is referred to as a *linguistic weighted average* (LWA) since the authors are concerned with aggregation of fuzzy sets representing words or terms. The authors use this operator in their approach to "computing with words" (CWW), a problem area originally suggested by Zadeh [11] that remains of high interest [12]-[16].

The approach to CWW in [10] is motivated by the premise that "words mean different things to different people," and therefore the inputs, computational engine and outputs of a perceptual computer (Per-C) must explicitly take account of this inherent imprecision. Inputs to the Per-C are modeled as interval type-2 (IT2) fuzzy membership functions (MFs) spanning an appropriate range (e.g., a 0-10 scale), which is perhaps the simplest method for representing imprecise knowledge of membership values. However, the LWA Per-C architecture employed in [10] is limited in its ability to represent how humans aggregate preference scores. For examples, it cannot model a mandatory requirement, nor does it allow for control of the degree of conjunction or disjunction. Therefore it is highly desirable to develop more powerful and computationally feasible aggregation operators that can be applied to type-2 inputs.

The primary contribution of this chapter is to provide an efficient algorithm to compute general fuzzy weighted power means, in which we accommodate imprecision of scores, weights and powers. The case of IT2 scores, weights and powers is referred to as a *linguistic weighted power mean* (LWPM), in keeping with the terminology of [10]. This provides an enhanced and powerful aggregation tool for computing with words and other fusion applications involving fuzzy inputs. It is straightforwardly extensible to general type-2 inputs using the z-slices (or α-plane) method [17]. We present a comparison of Per-C computational engines using the LWPM aggregation operator with the LWA Per-C engine using the detailed example of an Investment Judgment Advisor presented in [10] and described below. This example illustrates the flexibility and range of logical inferences available using the LWPM as compared to the LWA. Thus the LWPM represents a significant generalization and extension of the choices of Per-C computational engines available for CWW applications.

The remainder of this chapter is organized as follows: first, the selection of an aggregation function is discussed in terms of characteristics of the properties relevant to the decision, with focus on the weighted power mean. Next, an algorithm is presented to compute fuzzy weighted power means and their De Morgan complements when inputs, weights and scores are intervals rather than scalar values. We outline how this is applied to deal with general fuzzy sets. Then we review the Investment Judgment Advisor (IJA) example of [10]. Next, we compare the IJA outputs using the LWPM versus the LWA, and discuss the advantages of the LWPM. Finally, we describe how even more useful logical aggregation operators can be constructed hierarchically using combinations of LWPM components.

2 Weighted Power Mean as a Partial Conjunction or Partial Disjunction Operator

This section looks at the relationship between a given set of properties and the way that an aggregation function treats those properties, focusing on the weighted power mean. In designing an evaluation system, the aggregation function should be matched to known characteristics of the properties. We deal here only with point value scores and weights; Section 3 will extend these to the case when scores and weights are intervals or more general fuzzy sets.

First, consider what aggregators imply about properties, starting with the special cases of disjunction and conjunction. Disjunction treats properties as *sufficient*, as a full score against any property leads to a global score of unity. Disjunction therefore implies *substitutability* or replaceability of the properties whose scores are being aggregated. Conversely, conjunction implies independence of properties, requiring *simultaneity* of satisfaction [1],[4]. Conjunction treats properties as *mandatory* since a score of zero against any property leads to a zero global score. When an aggregation function L treats a property as neither mandatory nor sufficient, as the weighted arithmetic mean does, the property is said to be *desired* (in the context of L).

When a parameterized aggregation operator ranges between conjunction and disjunction, it is natural to define some measure of similarity between conjunction (or disjunction) and the operator at each parameter setting [18]-[20]. Similarity to conjunction is called the degree of *andness* and similarity to disjunction is called *orness*, where andness is the complement to orness. Andness of an aggregator L for which $x_1 \wedge \cdots \wedge x_n \leq L(\mathbf{x},\mathbf{w}) \leq x_1 \vee \cdots \vee x_n$ everywhere is defined in [4] to be:

$$\alpha_g = \frac{\overline{x_1 \vee \cdots \vee x_n} - \overline{L(\mathbf{x},\mathbf{w})}}{\overline{x_1 \vee \cdots \vee x_n} - \overline{x_1 \wedge \cdots \wedge x_n}} = \frac{n-(n+1)\left(\overline{L(\mathbf{x},\mathbf{w})}\right)}{n-1}. \tag{1}$$

Here, the overbar denotes averaging over the hypercube $[0,1]^n$ in which $\mathbf{x}$ lies.

An aggregator L that is more similar to conjunction than disjunction is referred to as *partial conjunction*, and if vice versa, as *partial disjunction* [4]. Property requirements are partially substitutable if L is a partial disjunction, and partially simultaneous if L is a partial conjunction. We generalize the terminology in [4] to say $|L(\mathbf{x},\mathbf{w}) - x_1 \vee \ldots \vee x_n|$ is the *penalty* and $|L(\mathbf{x},\mathbf{w}) - x_1 \wedge \ldots \wedge x_n|$ is the *reward* at $\mathbf{x} = (x_1,\ldots,x_n)$.

Let us now turn to the case when L is the weighted power mean. Over the full range of the exponent $-\infty \leq r \leq \infty$, satisfaction score vectors $\mathbf{x} = (x_1,\ldots,x_n)$ and normalized weight vectors $\mathbf{w} = (w_1,\ldots,w_n)$ for which $\sum_{i=1}^{n} w_i = 1$ and each $w_i > 0$, the weighted power mean is defined as:

$$L_r(\mathbf{x},\mathbf{w}) = \lim_{q\to r}\left(\sum_{i=1}^{n} w_i x_i^q\right)^{1/q}. \tag{2}$$

Hence $L_0(\mathbf{x},\mathbf{w}) = \prod_{i=1}^{n} x_i^{w_i}$, $L_{-\infty}(\mathbf{x},\mathbf{w}) = x_1 \wedge \cdots \wedge x_n$ and $L_{\infty}(\mathbf{x},\mathbf{w}) = x_1 \vee \cdots \vee x_n$.

De Morgan duality creates an aggregation function satisfying a sufficiency condition on the inputs [3],[4]. We denote this operator $L_r^{(-)}$, defined as

$$L_r^{(-)}(\mathbf{x},\mathbf{w}) = 1 - L_r(\mathbf{1}-\mathbf{x},\mathbf{w}), \text{ where } \mathbf{1}-\mathbf{x} = (1-x_1,\ldots,1-x_n). \tag{3}$$

When the weights and input scores are fixed, $L_r(\mathbf{x},\mathbf{w})$ is a continuous and increasing function of the power r . Therefore from (1), the andness $\alpha_g^{(r)}$ of $L_r(\mathbf{x},\mathbf{w})$ is a monotonically decreasing function of r . It is easy to see that the andness of $L_r^{(-)}$ equals the orness of L_r and vice versa. The ranges of r for which L_r and $L_r^{(-)}$ are partial conjunctions or partial disjunctions are summarized in Table 1, which also shows the form of aggregation that is valid when properties are all mandatory, all sufficient, all desired (neither mandatory nor sufficient) and partially substitutable, or all desired and partially simultaneous [22].

To aggregate a mandatory input x_1 with a desired input x_2, [21] sets up bivariate operators called *conjunctive partial absorption* (see also [4]). Conjunctive partial absorption aggregators apply a partial disjunction to the two inputs, and then take a partial conjunction of that output with the mandatory input. *Disjunctive partial absorption* aggregators defined in [21] and [4] combine a sufficient input x_1 with a

Table 1. Aggregator Properties of Operators as a Function of r

	Partial conjunction	**Partial disjunction**
All properties mandatory	$L_r(\mathbf{x},\mathbf{w})$, $-\infty \le r \le 0$ $L_\infty^{(-)}(\mathbf{x},\mathbf{w})$	
All properties sufficient		$L_\infty(\mathbf{x},\mathbf{w})$ $L_r^{(-)}(\mathbf{x},\mathbf{w})$, $-\infty \le r \le 0$
Properties desired, partially simultaneous	$L_r(\mathbf{x},\mathbf{w})$, $0 < r \le 1$ $L_r^{(-)}(\mathbf{x},\mathbf{w})$, $1 \le r < \infty$	
Properties. desired, partially substitutable		$L_r(\mathbf{x},\mathbf{w})$, $1 \le r < \infty$ $L_r^{(-)}(\mathbf{x},\mathbf{w})$, $0 < r \le 1$

desired input x_2 using a partial conjunction of the two inputs followed by a partial disjunction of that output and the sufficient input.

We can construct such operators from weighted power means as shown in Table 2. Here, $(x_1, L_S(\mathbf{x},\mathbf{w}))$ and $(x_1, L_S^{(-)}(\mathbf{x},\mathbf{w}))$ are both 2-dimensional scores, and $\mathbf{w}'$ is a 2-dimensional weight vector. However $\mathbf{x}$ and $\mathbf{w}$ may be of higher dimension if there are multiple partially substitutable inputs.

Table 2. Conjunctive/Disjunctive Partial Absorption Operators

Conjunctive Partial Absorption Property 1 mandatory, other properties desired and partially substitutable	$L_r((x_1, L_S(\mathbf{x},\mathbf{w})),(w',1-w'))$, $r \le 0,\ 1 \le s < \infty$ $L_r((x_1, L_S^{(-)}(\mathbf{x},\mathbf{w})),(w',1-w'))$, $r \le 0,\ 0 < s \le 1$
Disjunctive Partial Absorption Property 1 sufficient, other properties desired and partially simultaneous	$L_r^{(-)}((x_1, L_S(\mathbf{x},\mathbf{w})),(w',1-w'))$., $r \le 0,\ 0 < s \le 1$ $L_r^{(-)}((x_1, L_s^{(-)}(\mathbf{x},\mathbf{w})),(w',1-w'))$, $r \le 0,\ 1 \le s < \infty$

3 Imprecise Attributes, Weights and Powers

We turn now to the fuzzy case, starting with interval-valued scores $[\underline{x}_i, \overline{x}_i]$ and weights $[\underline{w}_i, \overline{w}_i]$, where for $i \in \{1,\ldots,n\}$, $\overline{x}_i, \underline{x}_i \in [0,1]$ and $\overline{w}_i, \underline{w}_i \ge 0$ (so we allow weight intervals to include zero). Drawing on work in [22], we show how to compute the weighted power mean for a precise power r and then extend this to deal with a power range $[\underline{r}, \overline{r}]$ and thence to an arbitrary fuzzy set on the reals.

A. Bounds of an interval-valued weighted arithmetic mean

Define $\Omega = \{\mathbf{x},\mathbf{w} : x_i \in [\underline{x}_i, \overline{x}_i], w_i \in [\underline{w}_i, \overline{w}_i], i = 1,\ldots,n\}$. Then it is easy to see that $y(\mathbf{x},\mathbf{w}) = \sum_{i=1}^{n} w_i x_i \Big/ \sum_{i=1}^{n} w_i$ achieves its maximum value at $x_i = \overline{x}_i$ and its minimum value at $x_i = \underline{x}$ for $i = 1,\ldots,n$. Reference [23] shows that the bounds of y as $\mathbf{x}$ and $\mathbf{w}$ vary over Ω are given by

$$\underline{y} = y(\underline{\mathbf{x}}, \underline{\mathbf{z}}),\ \text{where}\ \underline{z}_i = \begin{cases} \overline{w}_i, & \underline{x}_i < \underline{y} \\ \underline{w}_i, & \underline{x}_i > \underline{y} \end{cases},\ i \in \{1,\ldots,n\}$$
$$\overline{y} = y(\overline{\mathbf{x}}, \overline{\mathbf{z}}),\ \text{where}\ \overline{z}_i = \begin{cases} \underline{w}_i, & \overline{x}_i < \overline{y} \\ \overline{w}_i, & \overline{x}_i > \overline{y} \end{cases},\ i \in \{1,\ldots,n\} \tag{4}$$

Algorithms for computing these bounds for the fuzzy weighted mean, and extensions to interval type-2 fuzzy sets, are given in [23],[24]. These algorithms are called Enhanced Karnik-Mendel or EKM. Below, we condense and generalize the EKM algorithm to compute fuzzy weighted power means of arbitrary power.

B. Bounds of an interval valued weighted power mean

First we deal with the exceptional cases $r = \pm\infty$, for which inspection of (2) shows that $L_r(\mathbf{x},\mathbf{w})$ is independent of all $w_i > 0$; specifically, $L_{-\infty}(\mathbf{x},\mathbf{w}) = \min_{i:\, w_i > 0}\{x_i\}$, $L_{\infty}(\mathbf{x},\mathbf{w}) = \max_{i:\, w_i > 0}\{x_i\}$. When applied to interval values $[\underline{x}_i, \overline{x}_i]$, $i = 1,\ldots,n$ (for example, using Zadeh's Extension Principle) $L_{-\infty}(\mathbf{x},\mathbf{w})$ and $L_{\infty}(\mathbf{x},\mathbf{w})$ are intervals

$$[\underline{L}_{-\infty}, \overline{L}_{-\infty}] = \left[\min_{x_i \in [\underline{x}_i, \overline{x}_i]}\left\{\min_{i:\, w_i > 0}\{x_i\}\right\},\ \max_{x_i \in [\underline{x}_i, \overline{x}_i]}\left\{\min_{i:\, w_i > 0}\{x_i\}\right\}\right], \tag{5}$$

$$[\underline{L}_{\infty}, \overline{L}_{\infty}] = \left[\min_{x_i \in [\underline{x}_i, \overline{x}_i]}\left\{\max_{i:\, w_i > 0}\{x_i\}\right\},\ \max_{x_i \in [\underline{x}_i, \overline{x}_i]}\left\{\max_{i:\, w_i > 0}\{x_i\}\right\}\right]. \tag{6}$$

Therefore the intervals are determined solely by the x_i interval boundaries, with

$$\underline{L}_{-\infty} = \min_{i:\, w_i > 0}\{\underline{x}_i\},\ \overline{L}_{-\infty} = \min_{i:\, w_i > 0}\{\overline{x}_i\};\ \underline{L}_{\infty} = \max_{i:\, w_i > 0}\{\underline{x}_i\},\ \overline{L}_{\infty} = \max_{i:\, w_i > 0}\{\overline{x}_i\} \tag{7}$$

Next, consider the weighted power mean for a finite power r. The relevant expressions in (2) can be rewritten as

$$L_r(\mathbf{x},\mathbf{w}) = h_r\left(\sum_{i=1}^{n} w_i h_r^{-1}(x_i) \Big/ \sum_{i=1}^{n} w_i\right) = h_r\left(y_r(\mathbf{x},\mathbf{w})\right), \tag{8}$$

where

$$h_r(z) = \begin{cases} z^{1/r}, & r \neq 0 \\ \exp(z), & r = 0 \end{cases} \tag{9}$$

and

$$y_r(\mathbf{x},\mathbf{w}) = \sum_{i=1}^{n} w_i h_r^{-1}(x_i) \Big/ \sum_{i=1}^{n} w_i\ . \tag{10}$$

Note that h_r and h_r^{-1} are strictly increasing on the non-negative reals if $r \geq 0$ and are strictly decreasing if $r < 0$. Note also that (8) can be used for other invertible aggregation functions h, but we restrict attention here to the WPM operator.

With (8), the value of the weighted power mean applied to intervals $[\underline{x}_i, \overline{x}_i]$, $[\underline{w}_i, \overline{w}_i]$ can be written as the interval

$$[\underline{L}_r,\overline{L}_r]=\left[\min h_r\left(\sum_{i=1}^{n}w_ih_r^{-1}(x_i)\Big/\sum_{i=1}^{n}w_i\right),\max h_r\left(\sum_{i=1}^{n}w_ih_r^{-1}(x_i)\Big/\sum_{i=1}^{n}w_i\right)\right] \tag{11}$$

where min and max are taken over Ω.

The lemma below enables the results from [23] to be adapted to this general case. The lemma follows from the elementary expression for the derivative of a composite function, i.e., $(h(g(z)))'=g'(z)h'(g(z))$.

Lemma

Suppose that g and h are real valued differentiable functions on an interval Z and that $f(z)=h(g(z))$ for all z in Z.

(i) If h is strictly increasing then the locations in Z of the minima (resp. maxima) of f coincide with each of the locations of minima $\underline{z}$ (resp. maxima $\overline{z}$) of g, and

$$\min_{z\in Z} f(z)=h(g(\underline{z})),\ \max_{z\in Z} f(z)=h(g(\overline{z}))$$

(ii) if h is strictly decreasing then the locations of the minima (resp. maxima) of f coincide with each of the locations of maxima (resp. minima) of g, and

$$\min_{z\in Z} f(z)=h(g(\overline{z})),\ \max_{z\in Z} f(z)=h(g(\underline{z})).$$

(If Z is unbounded or an open interval, the maximum and minimum may not be attained, in which case the lemma holds with supremum (resp. infimum) substituted for maximum (resp. minimum.)

Applying the lemma to (11) with h_r in the role of h and y_r in the role of g we see that

$$[\underline{L}_r,\overline{L}_r]=\begin{matrix}[h_r(\underline{y}_r),h_r(\overline{y}_r)] & r\ge 0\\ [h_r(\overline{y}_r),h_r(\underline{y}_r)] & r<0\end{matrix}, \tag{12}$$

where

$$[\underline{y}_r,\overline{y}_r]=\left[\min\left\{\frac{\sum_{i=1}^{n}w_ih_r^{-1}(x_i)}{\sum_{i=1}^{n}w_i}\right\},\ \max\left\{\frac{\sum_{i=1}^{n}w_ih_r^{-1}(x_i)}{\sum_{i=1}^{n}w_i}\right\}\right]. \tag{13}$$

Now for any $i\in\{1,\dots,n\}$, $h_r^{-1}(\underline{x}_i)\le h_r^{-1}(\overline{x}_i)$ if h_r is strictly increasing and $h_r^{-1}(\underline{x}_i)\ge h_r^{-1}(\overline{x}_i)$ if h_r is strictly decreasing. So to compute (13), we apply Wu and Mendel's results (4) using the input intervals $[h_r^{-1}(\underline{x}_i),h_r^{-1}(\overline{x}_i)]$ when $r\ge 0$ and $[h_r^{-1}(\overline{x}_i),h_r^{-1}(\underline{x}_i)]$ when $r<0$. Then we apply h_r to the resulting interval in accordance with (17). The upper and lower bounds are transposed at this step if h_r is strictly decreasing. Thus we see that in all cases, the bounds in (12) are given by:

$$\underline{L}_r = h_r\left(\sum_{i=1}^{n} \underline{z}_i h_r^{-1}(\underline{x}_i)\Big/\sum_{i=1}^{n}\underline{z}_i\right),\quad \underline{z}_i = \begin{cases}\overline{w}_i, & \underline{x}_i < h_r\left(y_r\left(\underline{\mathbf{x}},\underline{\mathbf{z}}\right)\right) = \underline{L}_r \\ \underline{w}_i, & \underline{x}_i > h_r\left(y_r\left(\underline{\mathbf{x}},\underline{\mathbf{z}}\right)\right) = \underline{L}_r\end{cases}$$
$$\overline{L}_r = h_r\left(\sum_{i=1}^{n} \overline{z}_i h_r^{-1}(\overline{x}_i)\Big/\sum_{i=1}^{n}\overline{z}_i\right),\quad \overline{z}_i = \begin{cases}\underline{w}_i, & \overline{x}_i < h_r\left(y_r\left(\overline{\mathbf{x}},\overline{\mathbf{z}}\right)\right) = \overline{L}_r, \\ \overline{w}_i, & \overline{x}_i > h_r\left(y_r\left(\overline{\mathbf{x}},\overline{\mathbf{z}}\right)\right) = \overline{L}_r\end{cases} \tag{14}$$

C. WPMEKM Algorithm for computing fuzzy weighted power means

The EKM algorithms [23],[24] can be adapted to compute the interval bounds

$$\left[\underline{y},\overline{y}\right] = \left[\min h\left(\frac{\sum_{i=1}^{n} w_i h^{-1}(x_i)}{\sum_{i=1}^{n} w_i}\right), \max h\left(\frac{\sum_{i=1}^{n} w_i h^{-1}(x_i)}{\sum_{i=1}^{n} w_i}\right)\right]$$

for any strictly monotonic function h [17]. We call the resulting algorithm WPMEKM since it is targeted at computing the special case of weighted power means, while recognizing the more general application. For conciseness we also conflate what are essentially two identical algorithms that the earlier works used to compute the upper and lower bounds.

<u>The WPMEKM Algorithm</u>

1) If the minimum $L_r(\mathbf{x},\mathbf{w}) = \underline{L}_r$ is to be computed, set $k = \left[n/2.4\right]$ where $\left[x\right]$ is the nearest integer to x, and set $x_i = \underline{x}_i$, $u_i = \underline{w}_i$ and $v_i = \overline{w}_i$ for each $i = 1,\ldots,n$. If the maximum $L_r(\mathbf{x},\mathbf{w}) = \overline{L}_r$ is to be computed, set $k = \left[n/1.7\right]$, $x_i = \overline{x}_i$, $u_i = \overline{w}_i$ and $v_i = \underline{w}_i$ for $i = 1,\ldots,n$.
2) Sort the x_i for $i = 1,\ldots,n$ in ascending order so that $x_{i_1} \le x_{i_2} \le \cdots \le x_{i_n}$ where i_j, $j = 1,\ldots,n$ represents the original index values. Rearrange the weight interval endpoints u_i and v_i using the same permutation of index values i_j, $j = 1,\ldots,n$ that result from the x_i sort. Now refer to x_i, u_i, v_i, $i = 1,\ldots,n$ as the sorted sets of inputs and weights.
3) Compute
$$a = \sum_{i=1}^{k} h^{-1}(x_i)v_i + \sum_{i=k+1}^{n} h^{-1}(x_i)u_i,\ \ b = \sum_{i=1}^{k} v_i + \sum_{i=k+1}^{n} u_i,\ \ y = \left(\frac{a}{b}\right).$$
4) Find $k' \in \{0,1,\ldots,n\}$ such that $x_{k'} \le h(y) < x_{k'+1}$. Check if $k' = k$. If yes, set $L_r(\mathbf{x},\mathbf{w}) = h(y)$ and stop. If no, continue.
5) Set $s = sign(k'-k)$, and compute
$$a' = a + s\sum_{i=\min(k,k')+1}^{i=\max(k,k')} h^{-1}(x_i)(v_i - u_i),$$
$$b' = b + s\sum_{i=\min(k,k')+1}^{i=\max(k,k')} (v_i - u_i),\ \ y = \left(\frac{a'}{b'}\right).$$

6) Set $a = a'$, $b = b'$ and $k = k'$. Go to step 4).

Note that because h is strictly monotonic, the sort on x_i results in either an ascending or descending sort on $h^{-1}(x_i)$ depending on whether h is increasing or decreasing.

With the WPMEKM algorithm, the weighted power mean interval boundaries can be computed for any value of r from given input and weight intervals. These interval-valued results enable us directly to compute arbitrarily accurate approximations to the weighted power mean membership function for scores and weights which are general fuzzy sets, using the α-cut representation theorem [25] as described in [23] and [24]. In this case the interval computation is applied for each α-cut of the input variables to determine the corresponding α-cut of the output variable. To further extend to the interval type-2 case, this computation is applied to both the upper and lower bounding functions for the scores and weights, to determine the corresponding upper and lower bounding functions of the global score. To extend to the general type-2 case, we apply the latter approach to each z-slice.

D. Allowing for Imprecise Powers in the Weighted Power Mean

Imprecise weights impart imprecision to the way that input satisfaction scores are aggregated into a global score. Imprecision is also involved in the choice of power to use in the weighted power mean. Again, we first model this imprecision as an interval, that is, we suppose that the power lies in a range $[\underline{r},\overline{r}]$. As mentioned earlier, the weighted power mean is known to be a continuous increasing function of the power r as r varies over the real line. When the power is an interval rather than a point, aggregation of scalar input scores for a given set of scalar weights is therefore the interval

$$L_{[\underline{r},\overline{r}]}(\mathbf{x};\mathbf{w}) = \left[L_{\underline{r}}(\mathbf{x};\mathbf{w}), L_{\overline{r}}(\mathbf{x};\mathbf{w})\right]. \tag{15}$$

If satisfaction scores are intervals $[\underline{x}_i,\overline{x}_i], i \in \{1,\ldots,n\}$ (in vector notation, $[\underline{\mathbf{x}},\overline{\mathbf{x}}]$), the aggregate global preference becomes the wider interval

$$L_{[\underline{r},\overline{r}]}([\underline{\mathbf{x}},\overline{\mathbf{x}}],\mathbf{w}) = \left[L_{\underline{r}}(\underline{\mathbf{x}},\mathbf{w}), L_{\overline{r}}(\overline{\mathbf{x}},\mathbf{w})\right]. \tag{16}$$

If also weights are fuzzy intervals $[\underline{w}_i,\overline{w}_i]$, $i \in \{1,\ldots,n\}$ the WPMEKM algorithm is used to determine the weight vector that gives the smallest value of $L_{\underline{r}}(\underline{\mathbf{x}},\mathbf{w})$ (call it $\mathbf{w}'$) and the weight vector that gives the largest value of $L_{\overline{r}}(\overline{\mathbf{x}},\mathbf{w})$ (call it $\mathbf{w}''$). Then

$$L_{[\underline{r},\overline{r}]}([\underline{\mathbf{x}},\overline{\mathbf{x}}],[\underline{\mathbf{w}},\overline{\mathbf{w}}]) = \left[L_{\underline{r}}(\underline{\mathbf{x}},\mathbf{w}'), L_{\overline{r}}(\overline{\mathbf{x}},\mathbf{w}'')\right] \tag{17}$$

From the expression (3) for the De Morgan complement, it is then easy to see that

$$L^{(-)}_{[\underline{r},\overline{r}]}([\underline{\mathbf{x}},\overline{\mathbf{x}}],[\underline{\mathbf{w}},\overline{\mathbf{w}}]) = \left[1 - L_{\overline{r}}(\mathbf{1}-\underline{\mathbf{x}},\mathbf{w}''), 1 - L_{\underline{r}}(\mathbf{1}-\overline{\mathbf{x}},\mathbf{w}')\right] \tag{18}$$

where the weight vector $\mathbf{w}'$ is computed to find a minimum with respect to the vector score $\mathbf{1}-\overline{\mathbf{x}}$ and the weight vector $\mathbf{w}''$ is computed to find a maximum with respect to the vector score $\mathbf{1}-\underline{\mathbf{x}}$. The extension from intervals to fuzzy sets is made with repeated application of the α-cut representation theorem.

4 Investment Example

The IJA example in [10] originated with Tong and Bonissone [26], who considered the decision confronting an individual with a moderately large amount of capital who wishes to select among five investment sector alternatives, using four investment criteria. The alternatives and criteria are as shown in Table 3. The first three criteria concern quantities and the fourth is a quality. The investor rates each criterion for importance and then rates each investment with respect to the criteria. The ratings are then aggregated into a decision statistic that can be used to guide capital allocation.

Through a process of interval analysis [27] using inputs from real subjects, Mendel and Wu derive three vocabularies associated with the concepts: *quantity*, *quality* and *importance*. The descriptors are represented using IT2 MFs, whose UMF and LMF are trapezoids disposed on the domain interval $[0,10]$. The extreme words in each vocabulary are typically represented by left- or right-shoulder MFs, while the intermediate words are represented by mid-range MFs. (For details, see [10].) An eight-term vocabulary describes *quantity*, namely None to Very Little (NVL), Very Low (VL), Low (L), More or Less Low (MLL), Fair to More or Less High (FMLH), More or Less High (MLH), High (H) and Extremely High (EH). A seven-term vocabulary described *quality*: Very Bad (VB), Bad (B), Somewhat Bad (SB), Fair (F), Somewhat Good (SG), Good (G) and Very Good (VG). A four-word vocabulary describes *importance*: Unimportant (U), More or Less Unimportant (MLU), More or Less Important (MLI) and Very Important (VI).

Table 3 (drawn from [10]) presents three hypothetical word evaluation matrices grading each investment alternative from the perspectives of speculative, conservative and in-between investors, respectively. Mendel and Wu use the LWA to aggregate the criteria evaluations and importance evaluations for each alternative into corresponding output IT2 MFs, which are then ranked to determine the order of appropriateness of the investment categories for each of the three types of investors.

We next generalize these results using the LWPM with different power exponents on the same input data, in order to illustrate the versatility of this aggregation operator [28]. In keeping with [10], a simple *antonym* $\mu_{10-\tilde{A}}(x)$ of the respective IT2 MF is used for criteria 1 and 2 in the aggregation, i.e., $\mu_{10-\tilde{A}}(x)=\mu_{\tilde{A}}(10-x)$. Alternative forms for the membership functions antonyms which exhibit more intuitive behavior are discussed in [29].

Table 3. Linguistic Ratings of Alternatives and Weights WRT Criteria

Alternative/weight	*Criterion* **Risk of capital loss**	**Inflation vulnerability**	**Investment return**	**Liquidity**
	Speculative Investor			
Criterion importance	U	MLU	VI	MLU
Commodities	FMLH	FMLH	EH	G
Stocks	MLH	FMLH	H	SG
Gold	VL	L	VL	SB
Real Estate	MLL	MLH	MLH	SG
Long Term Bonds	NVL	MLL	VL	VB
	Conservative investor			
Criterion importance	**VI**	**MLI**	**VI**	**MLI**
Commodities	EH	MLH	MLL	F
Stocks	H	H	MLH	SG
Gold	MLL	FMLH	FMLH	F
Real Estate	MLH	MLH	H	F
Long Term Bonds	NVL	MLL	MLH	SG
	In-between investor			
Criterion importance	**MLI**	**MLI**	**VI**	**MLI**
Commodities	H	H	MLH	SB
Stocks	FMLH	FMLH	H	VG
Gold	L	MLL	MLL	SB
Real Estate	FMLH	FMLH	H	SG
Long Term Bonds	L	L	L	F

The following sequences of figures show the output IT2 MF footprint of uncertainty (FOU) of the LWPM for different values of the power exponent r. In each figure we also plot the FOU of the weighted average ($r=1$, with andness $\alpha_g^{(1)}=0.5$). The "+" crosshatch corresponds to the LWA MF, while the "x" crosshatch corresponds to the LWPM MF; the UMF is shown as a solid line, while the LMF is shown as a dashed line. The solid vertical line indicates the defuzzified centroid of the LWPM IT2 MF, and the two dashed vertical lines indicate the left and right endpoints of the LWPM centroid interval. Note that our purpose in the choices of alternatives shown below is to illustrate the cases where the LWPM is qualitatively the *most* similar and the *least* similar to the LWA results for a given value of the parameter r.

Consider first the extreme cases of $r=\pm\infty$, where the andness of the LWPM is 1 ($r=-\infty$) or 0 ($r=+\infty$). Figs. 1-3 are for $r=+\infty$, with andness $\alpha_g^{(\infty)}=0$ (pure disjunction), where the LWPM MF lies generally to the right of the LWA MF. The left plot correspond to the alternatives for which the LWPM MF FOU is most similar qualitatively to that of the LWA, and the right correspond to alternatives for which it is most dissimilar to the LWA. We observe a dramatic range of differences between the LWPM and LWA over different alternatives, which is a function of the *dispersion* of the evaluation words' IT2 MFs over the 0-10 scale with respect to the various criteria. The LWA tends to smooth out these differences, resulting in an interior trapezoidal IT2 MF in all cases, while the LWPM for $r=+\infty$ often results in a right-shoulder trapezoidal IT2 MF.

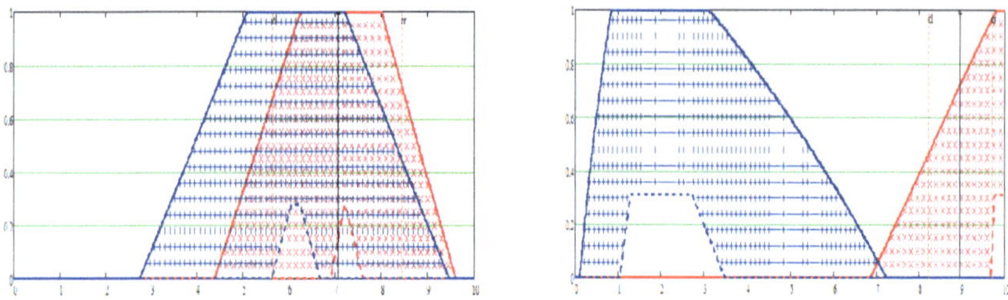

Fig. 1. $r = +\infty$ Speculative investor: FOUs for Real Estate (left) and Long Term Bonds (right)

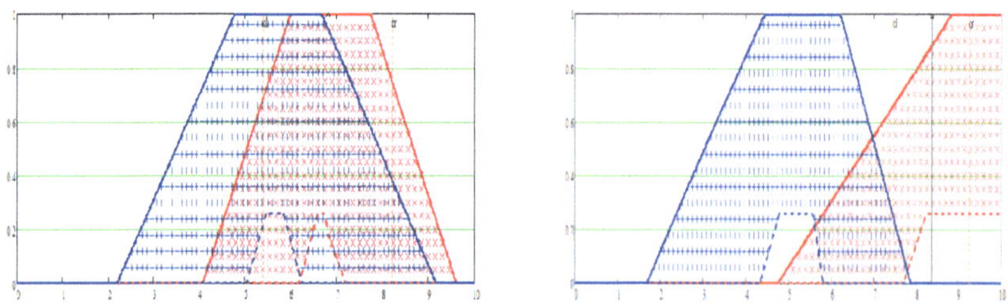

Fig. 2. $r = +\infty$ Conservative investor: FOUs for Gold (left) and Real Estate (right)

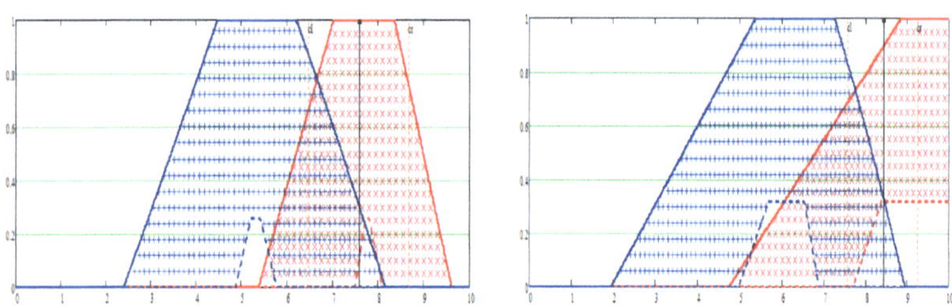

Fig. 3. $r = +\infty$ In-between investor: FOUs for Long Term Bonds (left) and Real Estate (right)

Figs. 4-6 show analogous results for the case $r = -\infty$ (pure conjunction), which corresponds to andness $\alpha_s^{(-)} = 1$. Here, the LWPM MF lies generally to the left of the LWA MF, since the LWPM for $r = -\infty$ selects the *minimum* left and right endpoints over the input MFs for each α-cut. Again we observe a substantial range of differences between the LWPM and LWA IT2 MFs, with the LWPM often transitioning to a left-shoulder trapezoidal IT2 MF.

These examples serve to illustrate the substantially greater aggregation versatility of the LWPM over the LWA. By the choice of r, we impart different perspectives to the fuzzy IT2 decision process, ranging from the most optimistic view having an orness of 1, to the most pessimistic view having an andness of 1, with a continuous range of andness/orness between these extremes. This compares with the single value of andness = orness = 0.5 available with the LWA, which averages out the more extreme inputs and thus might be termed a "consensus" view.

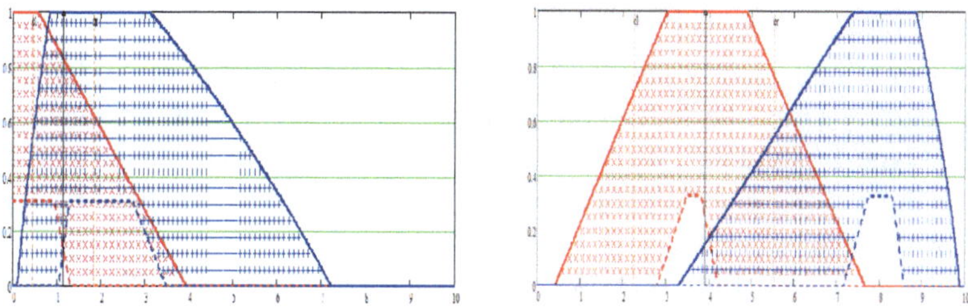

Fig. 4. $r = -\infty$ Speculative investor: FOUs for Long Term Bonds (left) and Commodities (right)

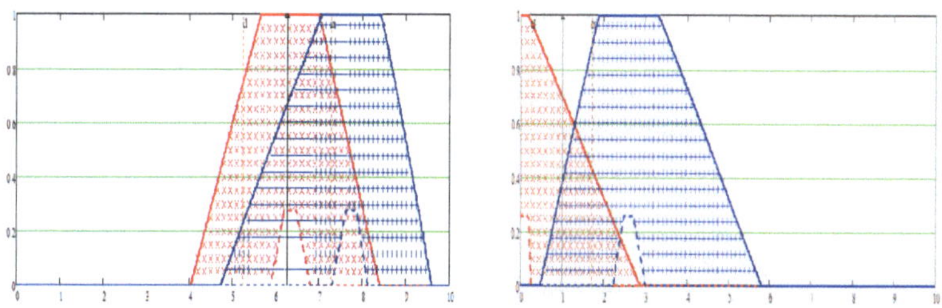

Fig. 5. $r = -\infty$ Conservative investor: FOUs for Long Term Bonds (left) and Commodities (right)

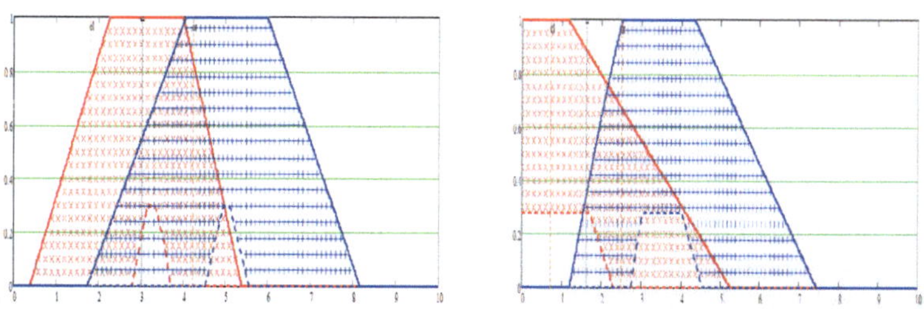

Fig. 6. $r = -\infty$ In-between investor: FOUs for Gold (left) and commodities (right)

Conjunctive and Disjunctive Partial Absorption aggregators can be used to aggregate subcategories of criteria scores into higher-level criteria (Table 2). To do so further generalizes the use of hierarchical LWAs in a decision tree [10]. Suppose the speculative investor places a mandatory (in a type-1 context) requirement on investment return (see Table 3), with the remaining criteria desirable. Let x_1 be the score on the mandatory criterion and x_2 be the aggregate of the scores of the desirable criteria under a WPM of power -1 (i.e. the harmonic mean), which is a moderately conjunctive combination of desired inputs. Now apply a conjunctive partial absorption aggregator as in Table 2 to combine the mandatory with the desirable scores:

$$\text{CPA}= L_r\left(\left(x_1, L_S\left(\mathbf{x},\mathbf{w}\right)\right),\left(w',1-w'\right)\right) \tag{18}$$

where $\mathbf{x} = (x_1, x_2), \mathbf{w} = (w, 1-w)$. Suppose that we specify a maximum reward of +15%, with a maximum penalty of -25%. From the methods of [4],[21], we set $r = -10$ and $s = 1.449$ (the latter corresponding to an andness of 7/16) in (18) and compute type-1 weights $w = 0.347$ and $w' = 0.935$, which achieve these reward/penalty values, and then calculate the LWPMs in (18) on the IT2 input variables, with singleton weights.

The plot of the left of Fig. 7 shows the CPA IT2 FOU (X crosshatch) and the LWA FOU (+ crosshatch) for investment alternative Commodities. The plot on the right used a LWPM on all four inputs with $r = -1$. We observe that the CPA FOU is substantially narrower, and is disposed over considerably higher values, than the LWPM FOU. This is understandable from inspection of Table 3, where we observe that the potential for high investment returns in commodities is graded as "extremely high" by a speculative investor. Thus we would expect an enhanced score for commodities when this criterion is elevated to mandatory.

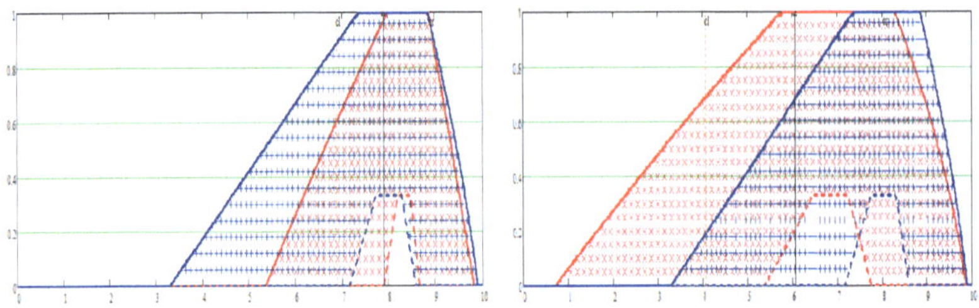

Fig. 7. Speculative investor, Commodities: CPA when Investment Return a mandatory criteria (left) and FOU with $r = -1$ (right)

On the other hand, Fig. 8 shows the same comparison for investing in Long Term Bonds, whose potential for high returns is graded "very low" by the speculative investor. The CPA FOU differs relatively little from the comparable LWPM FOU, since the alternative of Long Term Bonds scored relatively low with respect to *all* criteria by the speculative investor.

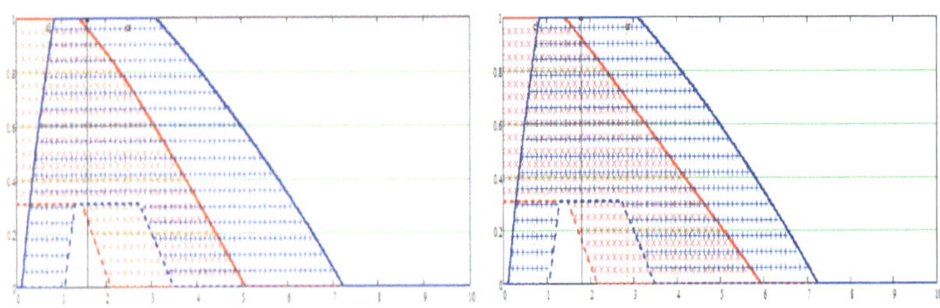

Fig. 8. Speculative investor, Long Term Bonds: CPA with Investment Return a mandatory criteria (left) and FOU with $r = -1$ (right)

5 Conclusion

We introduce a new family of Per-C computational engines based on the LWPM, and compare it with the LWA using an extended example described in [10]. The WPMEKM algorithm is presented for the computation of LWPM aggregations with interval input and weight variables. This can be extended to inputs and weights described by IT2 MFs using the α-cut Representation Theorem. The LWPM is shown to be a much more versatile aggregation operator, and thus it provides a significant extension to the suite of tools available for CWW applications. We also show how LWPM modules can be used to produce conjunctive partial absorption operators, and illustrate their differences with respect to a straightforward LWPM. These operators enable more general hierarchical logic functions to be constructed than are available with simple hierarchies of LWAs.

References

[1] Dubois, D., Prade, H.: A review of fuzzy set aggregation connectives. Inf. Sci. 36, 85–121 (1985)

[2] Salton, G., Fox, E.A., Wu, H.: Extended Boolean information retrieval. Communications of the ACM 26(12), 1022–1036 (1983)

[3] Hong, W.S., Chen, S.J., Wang, L.H., et al.: A new approach for fuzzy information retrieval based on weighted power-mean averaging operators. Computers & Mathematics with Applications 53(12), 1800–1819 (2007)

[4] Dujmović, J.: Continuous preference logic for system evaluation. IEEE Trans. on Fuzzy Systems 15(6), 1082–1099 (2007)

[5] Gärdenfors, P.: Conceptual Spaces: The Geometry of Thought. MIT Press (2000)

[6] Goldstone, R.: Influences of categorization on perceptual discrimination. J. Experimental Psychology: General 123, 178–200 (2000)

[7] Schyns, P., Goldstone, R., Thibaut, J.-P.: The development of features in object concepts. Behavioural and Brain Science 21, 1–54 (1998)

[8] Gigerenzer, G., Todd, P., and the ABC Research Group: Simple Heuristics That Make Us Smart. Oxford University Press, NY (1999)

[9] Liu, F., Mendel, J.M.: Aggregation using the fuzzy weighted average as computed using the Karnik-Mendel algorithms. IEEE Trans. Fuzzy Syst. 17(6) (2009)

[10] Mendel, J.M., Wu, D.: Perceptual Computing. John Wiley & Sons, Inc., Hoboken (2010)

[11] Zadeh, L.A.: Fuzzy logic = computing with words. IEEE Trans. Fuzzy Syst. 4, 103–111 (1996)

[12] Special Section on Computing with Words. IEEE Trans. Fuzzy Syst. 18 (June 2010)

[13] Mendel, J.M.: Computing with words: Zadeh, Turing, Popper and Occam. IEEE Computational Intelligence Magazine 2, 10–17 (2007)

[14] Yager, R.R.: Approximate reasoning as a basis for computing with words. In: Zadeh, L.A., Kacprzyk, J. (eds.) Computing with Words in Information/Intelligent Systems I: Foundations, pp. 50–77. Physica-Verlag, Heidelberg (1999)

[15] Yager, R.R.: On the retranslation process in Zadeh's paradigm of computing with words. IEEE Trans. on Systems, Man, and Cybernetics—Part B: Cybernetics 34, 1184–1195 (2004)

[16] Zadeh, L.A.: Toward human-level machine intelligence—Is it achievable? The need for a new paradigm shift. IEEE Computational Intelligence Magazine 3, 11–22 (2008)
[17] Wagner, C., Hagras, H.: Toward general type-2 fuzzy logic systems based on zSlices. IEEE Trans. on Fuzzy Systems 18(4), 637–660 (2010)
[18] Yager, R.R.: On ordered weighted averaging aggregation operators in multi-criteria decision making. IEEE Trans. Syst. Man Cybern. 18, 183–190 (1988)
[19] Larssen, H.: Efficient andness-directed importance weighted averaging operators. Int. J. Uncertainty, Fuzziness, Knowledge-Based Syst. 12, 67–82 (2003)
[20] Dujmović, J., Larsen, H.L.: Generalized conjunction/disjunction. Int. J. Approx. Reas. 46, 423–446 (2007)
[21] Dujmović, J.: Partial absorption functions. J. Univ. Belgrade, EE Dept. Ser. Math & Physics (659), 156–163 (1979)
[22] Rickard, J.T., Aisbett, J., Yager, R., Gibbon, G.: Fuzzy weighted power means in evaluation decisions. In: Proc. World Symposium on Soft Computing, Paper #100, San Francisco, CA (2011)
[23] Wu, D., Mendel, J.M.: Aggregation using the linguistic weighted average and interval type-2 fuzzy sets. IEEE Trans. on Fuzzy Systems 15(6), 1145–1161 (2007)
[24] Wu, D., Mendel, J.M.: Corrections to 'Aggregation using the linguistic weighted average and interval type-2 fuzzy sets'. IEEE Trans. on Fuzzy Systems 16(6), 1664–1666 (2008)
[25] Klir, G., Yuan, B.: Fuzzy Sets and Fuzzy Logic. Prentice-Hall, Upper Saddle River (1995)
[26] Tong, R.M., Bonissone, P.P.: A linguistic approach to decision making with fuzzy sets. IEEE Trans. Syst., Man, Cybern. SMC-10, 716–723 (1980)
[27] Liu, F., Mendel, J.M.: Aggregation using the fuzzy weighted average as computed using the Karnick-Mendel algorithms. IEEE Trans. Fuzzy Syst. 17 (December 2009)
[28] Rickard, J.T., Aisbett, J., Yager, R., Gibbon, G.: Linguistic weighted power means: comparison with the linguistic weighted average. In: Proc. FUZZ-IEEE 2011, 2011 World Congress on Computational Intelligence, Taipei, Taiwan, pp. 2185–2192 (June 2011)
[29] Trillas, E., Moraga, C., Guadarrama, S., Cubillo, S., Castiñeira, E.: Computing with Antonyms. In: Nikravesh, M., Kacprzyk, J., Zadeh, L.A. (eds.) Forging New Frontiers: Fuzzy Pioneers I. STUDFUZZ, vol. 217, pp. 133–153. Springer, Heidelberg (2007)

Querying Possibilistic Databases: Three Interpretations

Patrick Bosc and Olivier Pivert

IRISA–ENSSAT, Lannion, France
{bosc,pivert}@enssat.fr

1 Introduction

Many authors have made proposals to model and handle databases involving uncertain data. In particular, the last two decades have witnessed a blossoming of researches on this topic (cf. e.g., [3,4,19] for some recent ones). Even though most of the literature about uncertain databases uses probability theory as the underlying uncertainty model, some approaches rather rest on possibility theory [26]. The initial idea consisting in applying possibility theory to this issue goes back to the early 80's [24]. More recent advances on this topic can be found in [10]. In contrast with probability theory, one expects the following advantages when using possibility theory:

- the qualitative nature of the model makes easier the elicitation of the degrees attached to candidate values;
- in probability theory, the fact that the sum of the degrees from a distribution must equal 1 makes it difficult to deal with incompletely known distributions.

Our aim is not to claim (nor to demonstrate) that the possibility-theory-based framework is "better" than the probabilistic one at modeling uncertain databases, but that it constitutes an interesting alternative inasmuch as it captures a different kind of uncertainty (of a qualitative nature). An example is that of a person who witnesses a car accident and is not sure about the model of the car involved. In such a case, it seems reasonable to model the uncertain value by means of a possibility distribution, e.g., {1/Mazda, 1/Toyota, 0.7/Honda} — where 0.7 is a numerical encoding in a usually finite possibility scale — rather than with a probability distribution which would be artificially normalized.

The rest of the paper is organized as follows. Section 2 is devoted to a reminder about basic notions concerning the interpretation of an uncertain database in terms of a set of possible worlds. In Section 3, two models of uncertain databases founded on possibility theory are presented. Then, in Section 4, three fairly different families of queries are proposed, that have quite different meanings. Section 5 concludes the paper and opens some lines for future works.

2 Basic Notions

2.1 The Possible Worlds Semantics

The possible worlds model is founded on the fact that uncertainty in data makes it impossible to define what precisely the real world is. One can only describe the set of

R.R. Yager et al. (Eds.): *Soft Computing: State of the Art Theory*, STUDFUZZ 291, pp. 161–176.
DOI: 10.1007/978-3-642-34922-5_12

possible worlds which are consistent with the available information. As far as a table T conveys some imprecision/uncertainty, several interpretations (I) can be drawn from T and the set of all the interpretations of T is denoted by $rep(T)$. The notation $rep(D)$ extends naturally to an uncertain database D involving several tables. A regular database is nothing but a special case of an uncertain one which has only one interpretation. From a semantic point of view, such an uncertain database D can be interpreted in terms of a set of usual databases, also called worlds W_1, ..., W_p, and $rep(D) = \{W_1, ..., W_p\}$. In the following, we consider the case where $rep(D)$ is finite. Any world W_i is obtained by choosing a candidate value in each set appearing in a relation T_j pertaining to D. One of these (regular) databases, let us say W_k, is supposed to correspond to the actual state of the universe modeled. The assumption of independence between the sets of candidates is usually made and then any world W_i corresponds to a conjunction of independent choices (thus the degree associated to a world is based on a conjunction operator, e.g., "min" or "product").

Example 1. Let us consider the uncertain database D involving a single relation *im* whose schema is *IM*(#*i*, *airc*, *date*, *place*). Relation *im* is assumed to describe satellite images of aircrafts. Each image, numbered (#*i*), was taken on a certain location (*place*) a given day (*date*) and it is supposed that it includes a single aircraft (*airc*). With the extension of *im* depicted in Table 1 six worlds can be drawn, W_1, W_2, W_3, W_4, W_5 and W_6 since there are three candidates for *date* in the first tuple and two candidates for *airc* in the second one. Two of the worlds associated with the uncertain relation *im* are represented in Table 1. ◇

Table 1. An extension of *im* (top) and two worlds associated with it (bottom)

#i	airc	date	place
i_1	a_1	$\{d_1, d_3, d_7\}$	c_1
i_3	$\{a_3, a_4\}$	d_1	c_2

#i	airc	date	place
i_1	a_1	d_1	c_1
i_3	a_3	d_1	c_2

#i	airc	date	place
i_1	a_1	d_7	c_1
i_3	a_4	d_1	c_2

2.2 Strong Representation Systems and Compact Calculus

When dealing with an uncertain database D, a very important issue is that of the efficiency of the querying process. A naive way of doing would be to make explicit all the interpretations of D (at least when they are finite) in order to query each of them. Such an approach is intractable in practice and it is of prime importance to find a more realistic alternative. To this end, the notion of a representation system has been introduced — initially by Imielinski and Lipski [22] — and discussed in [1]. The basic idea is to look for a way for representing both initial tables and those resulting from queries so that the representation of the result of a query q against any database D (made of tables T_1, ..., T_p) denoted by $q(D)$, is equivalent (in terms of interpretations, or worlds) to the set of results obtained by applying q to every interpretation of D, i.e.:

$$rep(q(D)) = q(rep(D)) \quad \text{(P1)}$$

where $q(rep(D)) = \{q(W) \mid W \in rep(D)\}$. If property P1 holds for a representation system ρ and a subset σ of the relational algebra, ρ is called a *strong representation system* for σ. From a querying point of view, P1 enables a direct (or compact) calculus of a query q, which then applies to D itself without making the worlds explicit (see Figure 1). So doing, provided that relational operations are defined over tables of the system considered, reasonable performances can be expected.

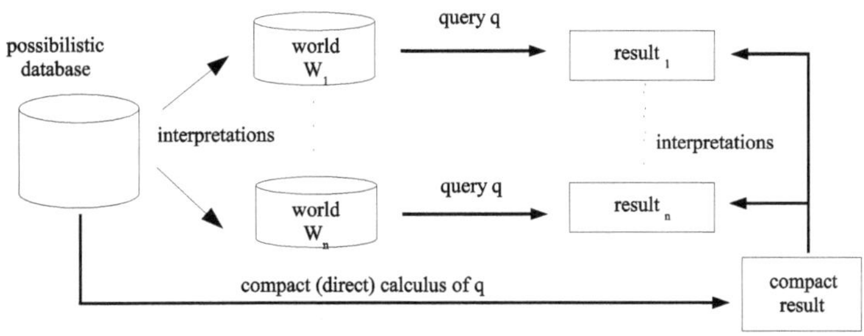

Fig. 1. Compact query evaluation

3 Two Uncertain Database Models Based on Possibility Theory

3.1 Full Possibilistic Model

In the "full possibilistic model" [10], any attribute value can be a possibility distribution which acts as a restriction over the values that are more or less preferred for a considered attribute (a precise value is an extreme case where only one candidate is possible). Besides, there is a need for expressing that some tuples may not be represented in some worlds. Indeed, a selection may lead to discard candidate values from a distribution, but one must be able to compute the degree of any world of the answer, including those in which some tuples are not represented. A simple solution is to introduce a new attribute, denoted by N, which states whether or not it is legal to build worlds where the corresponding tuple has no representative, and, if so, the influence of this choice in terms of possibility degree. N expresses the certainty of the presence of a representative of the tuple in any world. By doing so, it is possible to generate the worlds in which a tuple is not represented, by taking into account the degree of possibility of its absence, which, according to possibility theory, is given by $(1-N)$. A tuple is denoted as a pair N/t where N equals 1 for tuples of initial possibilistic relations as well as when no alternative has been discarded. A second aspect is related to the fact that it is sometimes necessary to express dependencies between candidate values of different attributes of a same tuple. For instance, let A and B be two attributes whose respective candidates in a given tuple t are $\{a_1, a_2\}$ and $\{b_1, b_2, b_3\}$. If, according to a given selection criterion, the only legal associations are (a_1, b_1) and (a_2, b_3), one cannot call on a Cartesian product of subsets of $t.A$ and $t.B$. In other words, A and B values cannot be kept separate (which would mean that they are independent) and the correct associations can be explicitly represented if the model incorporates attribute values defined as possibility distributions over several domains. In this context, candidates can be (weighted) tuples in a model based on the concept of nested relations. Besides, let us emphasize two particular aspects, both connected with the fact that nested relations are used to support possibility distributions: i) tuples of nested relations are weighted since any element of a possibility dis-tribution is assigned a level of preference and ii) the extension of a nested relation has a disjunctive meaning according to the semantics of a possibility distribution. The notation

$$R(A_1, \dots, A_m, X_1(A_p, \dots, A_q), \dots, X_n(A_k, \dots, A_r))$$

stands for a schema in which A_1 to A_m are elementary attributes (also called level-one attributes) whose values are either precise or possibility distributions and $X_i(A_h, \dots, A_j)$ represents a "structured" attribute X_i whose values are possibility distributions made of tuples built over attributes A_h to A_j which are called "nested" attributes. Obviously, such relations have an interpretation in terms of worlds as it is the case for ordinary possibilistic relations. When one moves to a given world, a structured candidate value is split into atomic values and the schema becomes unnested. The idea is to use the extended model to represent the result of intermediate operations in a correct fashion.

Table 2. An extension of relation r

A	B	X: C	X: D	X: E	X: π	N
a_1	$\{\pi_1/b_1, \pi_2/b_4\}$	c_2	d_1	e_3	π_3	1
a_2	b_3	c_1	d_1	e_2	π_4	0.4
		c_3	d_2	e_3	π_5	
		c_2	d_4	e_2	π_6	
		c_2	d_1	e_3	π_7	

Example 2. Let us consider the intermediate relation of schema $R(A, B, X(C, D, E))$ represented by Table 2 where the π_i's denote possibility degrees. Five possibilities exist as to the second tuple since it may be absent ($N < 1$). Consequently, ten worlds can be derived from this imprecise relation. The world containing only the tuple $\langle a_1, b_4, c_2, d_1, e_3 \rangle$, in which the second tuple is not represented, is associated with the degree:

$$min(min(1, \pi_2, \pi_3), 1 - 0.4).$$

The world with the two tuples $\langle a_1, b_1, c_2, d_1, e_3 \rangle$ and $\langle a_2, b_3, c_3, d_2, e_3 \rangle$ can also be drawn and its degree is:

$$min(min(1, \pi_1, \pi_3), min(1, 1, \pi_5)).\diamond$$

3.2 Certainty-Based Model

3.2.1 Main Features of the Model

In the certainty-based model [13,15], a possibility distribution is "synthetized" by keeping only its most plausible elements. So, to each uncertain value a of an attribute A is attached a certainty degree α. The underlying possibility distribution associated with an uncertain attribute value (a, α) is $\{1/a, (1-\alpha)/\omega\}$ where ω denotes $domain(A)\backslash\{a\}$ (due to the duality necessity/possibility: $N(a) \geq \alpha \Leftrightarrow \Pi(\omega) \leq 1-\alpha$ [21]). For instance, let us assume that the domain of attribute City is $\{$*Newton*, *Quincy*, *Boston*$\}$. The uncertain attribute value (*Newton*, α) is assumed to correspond to the possibility distribution $\{$1/*Newton*, $(1-\alpha)$/*Quincy*, $(1-\alpha)$/*Boston*$\}$. More generally, the model can deal with disjunctive values, and the underlying possibility distributions are of the form $\{max(\mu_S(x_1), 1-\alpha)/x_1, \ldots, max(\mu_S(x_p), 1-\alpha)/x_p\}$ where S is an α-certain subset of the attribute domain and $\mu_S(x_i)$ equals 1 if $x_i \in S$, 0 otherwise [20]. Let us notice that, in general, there is not a strict equivalence between an initial possibility distribution (e.g., $\{$1/*Newton*, 1/*Malden*, 0.6/*Quincy*, 0.2/*Boston*$\}$ and the distribution ($\{$1/*Newton*, 1/*Malden*, 0.6/*Quincy*, 0.6/*Boston*$\}$) derived from its synthetized form (*Newton* $\vee$ *Malden*, 0.4).

Moreover, since some operations may create "maybe tuples" (e.g., the selection as in the full possibilistic model), each tuple t from an imprecise relation r has to be associated with a degree N expressing the certainty that t exists in r. It will be denoted by N/t.

Example 3. Let us consider the relation r of schema (#id, Name, City) containing tuple $t_1 = \langle 1, John, (Boston, 0.8)\rangle$, and the query "find the persons who live in Boston". Let the domain of attribute City be $\{$*Newton*, *Quincy*, *Boston*$\}$. The answer contains $0.8/t_1$ since it is 0.8 certain that t_1 satisfies the requirement, while the result of the query "find the persons who live in Boston, Newton or Quincy" contains $1/t_1$ since it is totally certain that t_1 satisfies the condition.$\diamond$

To sum up, a tuple $\alpha/\langle 37, John, (Boston, \beta)\rangle$ from relation r means that it is α certain that person 37 exists in the relation, that it is totally sure that the name of that person is *John*, and that it is β certain that 37 lives in *Boston* (independently from the fact that it is or not in relation r).

Given a query, only answers that are *somewhat certain* are considered of interest (in contrast with those that are just possible), which makes the approach much simpler. Consider the relations r and s from Table 3 and a query asking for the persons who live in a city with a flea market. *John* will be retrieved with a certainty level equal to $min(\alpha, \beta)$ (in agreement with the calculus of necessity measures [20]). Although it is not impossible that *Mary* lives in a city with a flea market, she does not belong to the answer because this is just possible.

As mentioned above, it is also possible to handle cases of disjunctive information in this setting. For instance, ⟨3, Peter, (Gardner $\vee$ Fitchburg, 0.8)⟩ represents the fact that it is 0.8-certain that the person number 3 named Peter lives in Gardner or in Fitchburg.

Table 3. Relations r (left) and s (right)

#id	*Name*	*City*	*N*
1	John	(Newton, α)	1
2	Mary	(Norwood, δ)	1

City	*Flea Market*	*N*
Newton	(yes, β)	1
Norwood	(no, γ)	1

3.2.2 Strong Representation System

Let us now examine what becomes of property P1 in such a context. Let us denote by D an imprecise database involving certainty levels, $poss(D)$ the corresponding imprecise database involving the simplified possibility distributions of Subsection 3.2.1 (i.e., those associated with values that are somewhat certain), q an algebraic query, and q_c the compact version of q. The counterpart of property P1 is:

$$q_c(D) = \psi(q(rep(poss(D)))) \quad \text{(P2)}$$

where $\psi(r')$ denotes the certainty-based relation which gathers the tuples somewhat certainly in the intersection of all the (more or less) possible worlds from the set r' (each world from r' represents a possible result of q applied to D).

Table 4. Extension of *im* for Example 4

#i	*airc*	*date*	*place*
7	$\{1/\text{MiG31}, 0.8/\text{MiG29}\}$	96/03/02	$\{1/v_1, 0.2/v_2\}$
9	$\{1/\text{Su27}, 0.3/\text{Su30}, 0.5/\text{MiG31}\}$	92/12/01	v_1
17	MiG31	96/09/27	$\{1/v_2, 0.4/v_1\}$
5	$\{1/\text{MiG29}, 1/\text{Su7}\}$	95/06/09	v_2
34	MiG31	95/10/01	v_1

Table 5. Result of the query of Example 4

#i	*airc*	*date*	*place*	Π	*N*
7	$\{1/\text{MiG31}, 0.8/\text{MiG29}\}$	96/03/02	$\{1/v_1, 0.2/v_2\}$	1	0.2
9	$\{1/\text{Su27}, 0.3/\text{Su30}, 0.5/\text{MiG31}\}$	92/12/01	v_1	0.5	0
17	MiG31	96/09/27	$\{1/v_2, 0.4/v_1\}$	0.4	0
34	MiG31	95/10/01	v_1	1	1

4 Three Families of Query Semantics

Though it would make sense to envisage fuzzy queries (i.e., involving preferences expressed through fuzzy predicates), for space reasons, we only focus on Boolean queries.

4.1 Event-Oriented Querying

The corresponding model and query language were first introduced in [24] where it was possible to issue fuzzy queries against a possibilistic database. First, it is important to notice that this approach is not related to the possible worlds semantics. The idea is rather to see a query as a way to build facts (or events) as tuples using algebraic operations. Each tuple is assigned a pair of grades Π, N expressing the possibility and necessity of the corresponding event. The central operator is the selection for which ouput tuples are input tuples (kept unchanged) accompanied by the two grades Π and N mentioned before. In the presence of a Boolean selection condition ϕ applying to attribute A, the value of Π for tuple t (inside which A is represented as the possibility distribution $\pi_{t.A}$) is defind as:

$$sup_{d \in domainA}\, min(\pi_{t.A}(d), \phi(d)).$$

It equals 1 if there is (at least) one value in the core of $\pi_{t.A}$ that satisfies ϕ and 0 if no value of the support of $\pi_{t.A}$ matches ϕ. Of course, other values of the unit interval can be taken (see Example 4). Similarly, the necessity degree is given by:

$$1 - sup_{d \in domain(A)}\, min(\pi_{t.A}(d), \neg\phi(d)) = inf_{d \in domain(A)}\, max(1 - \pi_{t.A}(d), \phi(d)).$$

It equals 1 if any somewhat possible value of $\pi_{t.A}$ satisfies ϕ and 0 if a completely possible value of $\pi_{t.A}$ does not comply with ϕ. Of course, one has the property: $\Pi < 1 \Rightarrow N = 0$, as illustrated in the next example.

Example 4. Let us consider the relation *im* whose schema is given in Example 1 with the extension of Table 4. The query looking for images of "*MiG31*" taken in city v_1 returns the relation of Table 5.◇

It is worth noticing that, in such an approach, the composition of operations is problematic since input tuples are not "updated". For instance, the query looking for persons whose age is between 28 and 32 would reject $\langle John, \{1/25, 1/35\}\rangle$ whereas this tuple is selected if two successive selections are used.

4.2 Possible Worlds

4.2.1 Queries in the Full Possibilistic Model

Let us first point out some difficulties raised by the presence of disjunctive values. Let us consider the following relations $r(A, B)$ and $s(B, C)$:

$$r = \{\langle\{\alpha/a_1, \beta/a_2, \gamma/a_3\}, b\rangle\};\ s = \{\langle b, c_1\rangle, \langle b, c_2\rangle\}$$

where incompleteness is only due to the fact that the actual value of A in the tuple of r is either a_1, or a_2, or a_3. The natural join of r and s leads to a relation $t(A, B, C)$ involving two tuples, but it is mandatory to guarantee that only three possible worlds can be drawn from t (and not 3^2), since attribute A should take the same value in each of the two tuples, for property P1 to hold. Now, let us perform the natural join of the following relations:

$$r = \{\langle a, \{\alpha/b_1, \beta/b_2, \gamma/b_3\}\rangle\} \text{ and } s = \{\langle b_1, c_1\rangle, \langle b_3, \{\eta/c_2, \delta/c_3\}\rangle\}.$$

Here, the resulting relation is either empty, or made of a single tuple among three possible: $\langle a, b_1, c_1\rangle$, $\langle a, b_3, c_2\rangle$ and $\langle a, b_3, c_3\rangle$. It is then necessary to express that these four situations are exclusive. This implies using a sophisticated data model such as c-tables introduced by Imielinski and Lipski [22], which in turn raises important complexity issues.

Binary relational operations may be categorized the following way. Type 1 (resp. type 2) operations are such that any tuple from an operand relation can take part in the generation of at most one (resp. several) tuple(s) in the resulting relation. An example of a type 1 operation is the union. Type 2 operations include the intersection, the difference, the Cartesian product and the join (in their most general forms). However, in some particular cases linked to the presence of keys, an operation that is in general of type 2 can behave as a type 1 one (for instance the join operation when the join attributes are precise and constitute the keys of the operand relations or the foreign-key join detailed later). To summarize, let us say that in a strict relational framework, it is not possible to define a strong representation system allowing to deal with an operation of type 2 in the presence of imprecise values [9].

We now give an overview of four operators which define a language for which the full possibilistic model is an SRS. The reader will find more details and examples in [9] and [10]. In the following, because of space limits, we consider the case where input relations only include level-one attributes.

Selection

The usual selection keeps the tuples of a relation which satisfy a given predicate. Here, the idea is to retain only candidate values complying with the selection criterion. We review the various cases of selection conditions and examine their impact on the structure of the result.

When the condition is of the form "*att* θ *constant*" ($\theta \in \{=, \neq, >, <, \geq, \leq\}$), the structure of the result is the same as that of the input relation. If the schema of the input relation r is $R(A, B)$, the condition concerns attribute A and $scv(t.A)$ denotes the non-weighted set of candidate values appearing in $t.A$, the selection is defined as:

$$select(r, \theta(A, v)) = \{N'/\langle restrict(t.A, \theta(A,v)), t.B\rangle \mid N/t \in r \wedge \\ N' = min(N, 1 - sup_{x \in scv(t.A) \mid \neg\theta(x,v)} \pi_{t.A}(x))\}$$

with

$$restrict(t.A, \theta(A,v)) = \{\ldots + \pi/a + \ldots\} \text{ s.t. } a \in scv(t.A) \wedge \theta(a,v) \wedge \pi = \pi_{t.A}(a).$$

This formula says that, in any tuple t, only the elements of the distribution $t.A$ which satisfy the condition are retained in the resulting tuple. Moreover, the degree of certainty associated with this tuple ($t.N$) is updated according to the highest possible value which is discarded. It is proven in [9] that property P1 holds with this definition.

Let us now consider with selection conditions of the form "$A_1 \theta A_2$" or "$cond_1(A_1)$ or $cond_2(A_2)$". In both cases, if A_1 and A_2 are imprecise attributes, it is necessary to gather their candidate values in a nested relation so that only the correct pairs of values are kept in the result. The corresponding definition is given in [9]. The way the operator works is illustrated in the following example by a condition involving a disjunction.

Example 5. Let us consider the schema (#c, name, city, mileage) of an intermediate relation *ac* describing cars with their number, name, city of last owner and mileage. The condition (brand = "C*" or city = "Paris") applied to:

$$\{0.7/\langle 1, \{1/Camry, 0.4/Taurus\}, \{1/Madrid, 0.7/Paris\}, 75000\rangle\}$$

leads to the result:

$$\{0.6/\langle 1, \{1/\{Camry, Madrid\}, \\ 0.7/\{Camry, Paris\}, \\ 0.4/\{Taurus, Paris\}\}, 75000\rangle\}.$$

The necessity degree 0.6 attached to the tuple corresponds to $min(0.7, 1-\rho)$ where $\rho = 0.4$ is the possibility degree of the most possible pair of candidates that does not satisfy the selection criterion, i.e., $\langle Taurus, Madrid\rangle$ here. This way of doing guarantees the validity of property P1. ◇

Other Operators

As stated before, the classical join cannot apply in general for possibilistic relations due to the disjunctive nature of possibility distributions. However, we point out a specific type of join, called fk-join [9], where this problem does not appear since the tuples resulting from the join are independent in terms of their interpretation.

The operation $fk\text{-}join(r, s, (U, V))$ composes a possibilistic relation *r* whose schema is $R(U, Y)$ with a regular relation *s* (whose schema is $S(V, Z)$ where V is compatible with U) describing the graph of the functional dependency $V \rightarrow Z$. The fk-join computes the image of any imprecise U-value present in *r* by means of the function. In order to keep the elementary associations between antecedents and images of the FD $V \rightarrow Z$, it is mandatory to place U and Z candidate values inside a same nested relation. Let us consider the case where the schema of *r* is $R(A, B, G)$ with $U = \{A, B\}$ and the schema of *s* is $S(C, D, E)$ with $V = \{C, D\}$. The schema of the result is $Res(X(A, B, E), G)$.

Contrary to the usual case, the projection of a possibilistic relation does not entail any duplicate removal. One proceeds so that it is impossible to get a world after projection which would be more possible than the corresponding one before projection. This means that, for a given tuple, the possibility of the most possible candidate of the attributes which are removed becomes the upper bound of any interpretation of the tuple issued from the projection [7].

It is also possible to show that this model constitutes a strong representation system for the union operator provided that input relations are independent. Under this assumption, the union gathers the tuples of the two input relations and produces a result where the tuples are independent.

About Generalized Yes-No Queries

Queries addressed to an imprecise database may raise the problem of the interpretability of their results by an end-user. Indeed, even when "simple" models based on relations without any conditions are used — such as that presented above —, it appears difficult for an end-user to grasp the content of a relation that may include nested subrelations, distributions of possible values and necessity degrees. This is why several authors have considered a class of queries which are more specialized (or targeted) to fit user needs. This is the case, for instance, of S. Abiteboul who studied such queries in the context of Codd-tables and tables with conditions [2]. In the context of possibilistic databases, the queries considered in [11,12] are basically yes-no questions about some properties possessed (or not) by some of the worlds of an imprecise database. Their general query format is: "to what extent is it possible {and/or} certain that the answer to q fulfills condition C?" where q is a (constrained) relational algebraic query which may include only the operators for which the model is an SRS, i.e. projection, selection, fk-join and union (cf. above). More precisely, the following types of queries are considered:

- vacuity-based yes-no queries: to what extent is it possible and certain that the answer to q is non-empty?
- tuple-membership-based yes-no queries: to what extent is it possible and certain that tuple t belongs to the answer to q?
- cardinality-based yes-no queries: to what extent is it possible and certain that the answer to q contains at least (resp. at most, exactly) k items?
- inclusion-based yes-no queries: to what extent is it possible and certain that the answer to q contains the set of tuples $\{t_1, ..., t_k\}$?

For each of these queries, the authors show that the processing obeys the following three step scheme:

1. pre-processing in order to eliminate the unnecessary attributes (and, for tuple-membership-based queries, to remove from the relations the tuples that cannot generate the target tuple);
2. evaluation of q, which yields a resulting possibilistic relation *res*;
3. post-processing aimed at computing the final possibility and certainty degrees Π and N.

The four previous types of queries can be clustered into two categories: those which require only a sequential scan of the result of q (vacuity and tuple-membership-based queries) and those for which it is necessary to use a "trial and error" type of algorithm (cardinality and inclusion-based queries).

4.2.2 Queries in the Certainty-Based Model

We now outline the compact version of the relational algebraic operators in the certainty-based database model [13,15]. The only limitation with respect to the usual algebraic framework consists in the fact that the operands of union, Cartesian product and join must be independent relations. Indeed, the presence of non-independent relations (for instance stemming from two selections on the same relation or a self join) might induce dependencies between uncertain values in a same tuple of the result, which cannot be handled in the model.

Selection

Let us consider a relation r of schema (A, X) where A is an attribute and X is a set of attributes, and a selection condition ϕ on A. Let us denote by $scv(t.A)$ the disjunctive set of values — which may be a singleton — somewhat certain for attribute A in tuple μ/t, and by $cl(t.A)$ the associated certainty level. Let us first deal with the case where ϕ writes $A\,\theta\,v$ where θ denotes a comparator and v a constant.

$$\begin{aligned} select(r, A\,\theta\,v) = \{\mu'/t \mid \exists \mu/t \in r \text{ s.t. } \forall a_i \in scv(t.A),\, a_i\,\theta\,q \,\wedge \\ \mu' = min(\mu, 1) = \mu \text{ if } \forall a_i \in domain(A),\, a_i\,\theta\,v; \\ \mu' = min(\mu, cl(t.A)) \text{ otherwise}\}. \end{aligned}$$

The proof that this definition of the selection satisfies property P2 can be found in [13]. The case of a condition ϕ of the form $A_1\,\theta\,A_2$ where A_1 and A_2 denote two attributes is dealt with in [13] but is omitted here for space reasons.

Example 6. Let us consider the database D made of the sole relation *emp* of schema (#id, name, city, job). Let us suppose that *emp* only contains tuple t = 0.9/⟨17, *John*, (*Boston*, 0.8), (*Engineer*, 0.7)⟩ and let us consider the query:

q = select(emp, city = 'Paris' and job = 'Engineer').

Its compact result is 0.7/⟨17, *John*, (*Boston*, 0.8), (*Engineer*, 0.7)⟩. Let us show that property P2 is satisfied. Identifier 17 is present in every completely possible world of the result. The most possible world of *emp* where 17 is not present in the result of the selection is made of the tuple ⟨17, *John*, *Boston*, ε)⟩ (where $\varepsilon \in \omega = domain(job) \setminus \{Engineer\}$) and has the possibility degree $min(1,\, 1-0.7) = 0.3$. Hence, the certainty degree attached to 17 in the result is 1 – 0.3 = 0.7. The most possible world where 17 has a *city* value different from *Boston* in the result has the possibility degree $1-0.8 = 0.2$. Hence, the certainty degree attached to the *city* value *Boston* in the tuple identified by 17 in the result is $1-0.2 = 0.8$. The most possible world where 17 has a *job* value different from *Engineer* in the result has the possibility degree $1-0.7 = 0.3$. Hence the certainty degree attached to the *job* value *Engineer* in the tuple identified by 17 in the result is $1-0.3 = 0.7$. The compact calculus is thus correct. ◇

Join

The compact definition of the join is:

$$\begin{aligned} join(r_1, r_2, A = B) = \{min(\alpha, \beta, \chi, \delta)/t_1 \oplus t_2 \mid \exists \alpha/t_1 \in r_1, \exists \beta/t_2 \in r_2 \text{ s.t.} \\ card(scv(t_1.A)) = 1 \wedge card(scv(t_2.B)) = 1 \wedge \\ scv(t_1.A) = scv(t_2.A) \wedge cl(t_1.A) = \chi \wedge cl(t_2.B) = \delta\} \end{aligned}$$

where $\oplus$ denotes the concatenation and *card* returns the cardinality of a set. Notice that only the tuples whose value for the join attribute is non-disjunctive (i.e., is a singleton) can participate in the result: for the other ones, one cannot be certain at all that they match a tuple from the other relation. Indeed, for a tuple t_1 of r whose join attribute value $t_1.A$ is disjunctive, it is always possible to find a completely possible interpretation such that the (equi-)join condition is false, whatever the tuple t_2 from s. Note that property P2 would not hold in the case of a θ-join where θ is not equality. In [13], it is shown that the usual equivalence between a semi-join and a join followed by a projection: $r_1 \ltimes r_2 \equiv (r_1 \bowtie r_2)[X]$ where X denotes the attributes of r_1, is not valid anymore in the context of the certainty-based model. However, the semi-join can be defined in a sound way in this framework, see [13]. The key to the fact that join (and semi-join) can be easily handled in this model lies in the property that a tuple involving disjunctive values can produce at most one tuple in the result (due to the semantics of certainty).

Projection

Let r be a relation of schema (X, Y). The projection operation is straightforwardly defined as follows:

$$project(r, X) = \{\alpha/t.X \mid \alpha/t \in r \wedge \nexists \alpha'/t' \text{ s.t. } posbs(\alpha'/t'.X, \alpha/t.X)\}.$$

The only difference with respect to the definition of the projection in a classical database context concerns duplicate elimination, which is here based on the concept of "possibilistic subsumption" (using predicate *posbs*). Intuitively, an X-value of a tuple t is kept in the result if there is no other tuple t' with the same candidate values and a higher certainty level. More formally, letting $X = \{A_1, \ldots, A_n\}$, predicate *posbs* is defined as follows:

$$posbs(\alpha'/t'.X, \alpha/t.X) \equiv \forall i \in 1..n,\ scv(t.A_i) = scv(t'.A_i) \wedge cl(t.A_i) \leq cl(t'.A_i) \wedge \alpha \leq \alpha' \wedge ((\exists i \in 1..n,\ cl(t.A_i) < cl(t'.A_i)) \vee \alpha < \alpha').$$

The validity of the result before duplicate removal is guaranteed by the satisfaction of P2. As to the duplicate removal step, its soundness relies on the axioms of possibility theory. The definitions of the other relational algebraic operators in the certainty-based model can be found in [15].

4.3 Representation-Based Querying

The main motivation underlying the representation-based querying approach is to be able to exploit at a query level all the information available concerning the qualification of imperfectness in the data. In other words, one wants to be able to express conditions on the *descriptions* of ill-known data. Hereafter, we present a framework that was introduced in [5]. Representation-based queries can notably be used to:

- express conditions on specified sets of candidates (the specified set being a subset of a distribution representing an ill-known attribute value). The generic query is: "find the tuples such that all the elements of a specified subset of the candidate values (for a given attribute) satisfy a given condition",
- compute aggregates on the weighted sets corresponding to the representations of ill-known data (e.g., the cardinality of a specified subset of candidate values for a given attribute) and to use these aggregates inside conditions,

- compare a piece of data with a given vague pattern. In the representation-based querying framework, the comparison is based on the notion of synonymy of representations, contrary to the "value-based" framework where the comparison is founded on the notion of possibility/necessity of matching.

It is important to notice that representation-based queries are not just value-based queries expressed another way, but that they are queries of a different nature. A value-based criterion applying to an ill-known value has to be evaluated on each possible world associated with the attribute value (even though the explicit computation of those worlds is not always necessary, cf. Section 4.2.1), while a representation-based condition does not at all refer to worlds.

Example 7. We consider again a database containing aerial images of aircrafts (each image is supposed to represent a single aircraft), described by the set of attributes: (*#id*, *location*, *date*, *type*). The attributes *#id*, *location*, and *date* are supposed to take precise values whereas the attribute *type* describing the type of aircraft present in the picture will generally take imperfect values due to ambiguities in image interpretations. Examples of conditions involving one representation are:

- find the images which represent more likely a MiG29 than a MiG23,
- find the images such that all the candidates which are possible over 0.3 are of the type MiG,
- find the images for which at most 2 types of airplane are considered possible over 0.3,
- find the images for which the only best candidate is 'MiG29',
- find the images representing airplanes whose type is not precisely known (i.e., there are more than one candidate).◇

A language for representation-based conditions is described in [5]. In this framework, conditions involving two representations deserve a particular attention. Several methods have been proposed to compare possibility distributions or fuzzy sets and one can distinguish among two families of approaches. In the first family, a measure is used to evaluate the possibility degree of (approximate) equality between two imprecise values [17,18,24]. In the second family, what is measured is the extent to which two representations are globally close to each other [6,16,23,25]. In the representation-based querying framework, it is quite clear that only the second family of approach makes sense. Let us consider an attribute A and two items x and y whose A-values are ill-known. Let us denote by $\pi_{A(x)}$ and $\pi_{A(y)}$ the possibility distributions to be compared. Let D be the domain of attribute A. First, let us recall the expression of strict equality:

$$\forall d \in D,\ \pi_{A(x)}(d) = \pi_{A(y)}(d).$$

Several authors have proposed to relax the preceding measure into a measure of approximate equality. Raju and Majumdar [25] define the fuzzy equality measure, denoted EQ, in the following way:

$$\mu_{EQ}(\pi_{A(x)}, \pi_{A(y)}) = min_{u \in D}\ \psi(\pi_{A(x)}(u),\ \pi_{A(y)}(u))$$

where ψ is a resemblance relation (i.e., reflexive and symmetric) over [0, 1]. An alternative approach consists in defining the similarity of two fuzzy sets (two possibility distributions in our case) A and B as a function of $A \cap B$, $B - A$ and $A - B$. This approach is studied in particular by Bouchon-Meunier *et al.* [16] where different kinds of measures of comparison are considered.

Example 8. Let us consider the following description of an image I_1:

$$I_1 = \langle 27, \mathit{Krasnoyarsk}, \text{1992-12-01}, \{1/Su27, 1/Su30, 0.7/Mig29, 0.2/Yak130\}\rangle.$$

Let us consider the query: "find the pictures taken over Krasnoyarsk in 1992 representing an airplane similar to the one in image I_1" and let us assume that the database contains notably the following description:

$$I_2 = \langle 51, \mathit{Krasnoyarsk}, \text{1992-04-15}, \{1/Su30, 0.9/Mig29, 0.8/Su27, 0.4/Mig23\}\rangle.$$

If strict equality were used, it is clear that image I_2 would not belong to the result. Using Raju-Majumdar's measure of approximate equality with $(a, b) = 1 - |a - b|$, the matching degree between I_1 and I_2 is equal to:

$$EQ(I_1.type, I_2.type) = min(0.8\,(Su27), 1\,(Su30), 0.8\,(Mig29),$$
$$0.8\,(Yak130), 0.6\,(Mig23)) = 0.6. \diamond$$

On the other hand, these measures can be used to compare an ill-known attribute value D with a linguistic label P. The basic idea is the same: one evaluates the extent to which the value and the linguistic label represent the same concept. For example, let us consider a possibility distribution D representing John's age and a linguistic label P = "middle-aged" (represented by a fuzzy set). While the value-based querying approach aims at assessing the extent to which John is possibly (resp. necessary) middle-aged, the representation-based approach can be used to measure the extent to which the description of John's age and the linguistic label "middle-aged" are close to each other. This approach is especially useful in the context of applications where user queries can be conveniently expressed by means of linguistic terms defined on continuous domains. Lastly, the concept of representation-based comparison can be used to define the notions of representation-based intersection, union and difference in a straightforward manner.

5 Conclusion

In this paper, we have reviewed different types of queries that can be addressed to a database containing imprecise values represented in the possibilistic framework. We have distinguished three main lines: i) the initial approach proposed by Prade and Testemale which is intended for building "events" and their associated possibility and necessity degrees from data, ii) works based on the possible worlds semantics with two data models: the full possibilistic model where queries are constrained and the certainty-based model which offers the richness of the entire relational algebra, iii) queries where the conditions bear on the representation of imprecise data. The focus has been put on the semantic aspects, not on implementation and performances due to space limits. However, most of the operators proposed are very similar to those defined in regular database systems and reasonable performances can be expected. Among others, future works could concern queries involving preferences in the spirit of [8,14].

References

1. Abiteboul, S., Hull, R., Vianu, V.: Foundations of Databases. Addison-Wesley, Reading (1995)
2. Abiteboul, S., Kanellakis, P., Grahne, G.: On the representation and querying of sets of possible worlds. Theoretical Computer Science 78, 159–187 (1991)
3. Antova, L., Jansen, T., Koch, C., Olteanu, D.: Fast and simple processing of uncertain data. In: Proc. of 24th Int. Conf. on Data Engineering (ICDE 2008), pp. 983–992 (2008)
4. Benjelloun, O., Das Sarma, A., Halevy, A., Widom, J.: ULDBs: Databases with uncertainty and lineage. In: Proc. of VLDB 2006, pp. 953–964 (2006)
5. Bosc, P., Duval, L., Pivert, O.: Value-based and representation-based querying of possibilistic databases. In: Bordogna, G., Pasi, G. (eds.) Recent Research Issues on Fuzzy Databases, pp. 3–27. Physica-Verlag, Heidelberg (2000)
6. Bosc, P., Pivert, O.: On the comparison of imprecise values in fuzzy databases. In: Proc. of the 6th IEEE International Conference on Fuzzy Systems (FUZZ-IEEE 1997), Barcelona, Spain, pp. 707–712 (1997)
7. Bosc, P., Pivert, O.: About the projection operator in a possibilistic database framework. In: Proc. of the 21st Int. Conf. of the North American Fuzzy Information Processing Society (NAFIPS 2002), New Orleans, Louisiana, USA, pp. 371–376 (2002)
8. Bosc, P., Pivert, O.: From boolean to fuzzy algebraic queries in a possibilistic database framework. In: Proc. of the 13th IEEE International Conference on Fuzzy Systems (FUZZ-IEEE 2004), Budapest, Hungary (2004)
9. Bosc, P., Pivert, O.: On a strong representation system for imprecise relational databases. In: Proc. of the 10th International Conference on Information Processing and Management of Uncertainty in Knowledge-Based Systems (IPMU 2004), Perugia, Italy, pp. 1759–1766 (2004)
10. Bosc, P., Pivert, O.: About projection-selection-join queries addressed to possibilistic relational databases. IEEE Trans. on Fuzzy Systems 13(1), 124–139 (2005)
11. Bosc, P., Pivert, O.: Vacuity-oriented generalized yes/no queries addressed to possibilistic databases. In: Bordogna, G., Psaila, G. (eds.) Flexible Databases Supporting Imprecision and Uncertainty, pp. 55–74. Springer (2006)
12. Bosc, P., Pivert, O.: About yes/no queries against possibilistic databases. International Journal of Intelligent Systems 22, 691–722 (2007)
13. Bosc, P., Pivert, O., Prade, H.: A Model Based on Possibilistic Certainty Levels for Incomplete Databases. In: Godo, L., Pugliese, A. (eds.) SUM 2009. LNCS, vol. 5785, pp. 80–94. Springer, Heidelberg (2009)
14. Bosc, P., Pivert, O., Prade, H.: A possibilistic logic view of preference queries to an uncertain database. In: Proc. of the 19th IEEE International Conference on Fuzzy Systems (FUZZ-IEEE 2010), Barcelona, Spain, pp. 379–384 (2010)
15. Bosc, P., Pivert, O., Prade, H.: An uncertain database model and a query algebra based on possibilistic certainty. In: Proc. of the 2nd International Conference on Soft Computing and Pattern Recognition (SoCPaR 2010), Cergy-Pontoise, France (2010)
16. Bouchon-Meunier, B., Rifqi, M., Bothorel, S.: Towards general measures of comparison of objects. Fuzzy Sets and Systems 84, 143–153 (1996)

17. Chen, G., Kerre, E., Vandenbulcke, J.: A general treatment of data redundancy in a fuzzy relational data model. Journal of the American Society for Information Science 43, 304–311 (1992)
18. Cubero, J., Vila, M.: A new definition of fuzzy functional dependency in fuzzy relational databases. International Journal of Intelligent Systems 9, 441–448 (1994)
19. Dalvi, N., Suciu, D.: Management of probabilistic data: Foundations and challenges. In: Proc. of PODS 2007, pp. 1–12 (2007)
20. Dubois, D., Prade, H.: Necessity measures and the resolution principle. IEEE Trans. Syst., Man and Cyber. 17(3), 474–478 (1987)
21. Dubois, D., Prade, H.: Possibility Theory. Plenum, New York (1988)
22. Imielinski, T., Lipski, W.: Incomplete information in relational databases. J. of the ACM 31(4), 761–791 (1984)
23. Liu, W.: The fuzzy functional dependency on the basis of the semantic distance. Fuzzy Sets and Systems 59, 173–179 (1993)
24. Prade, H., Testemale, C.: Generalizing database relational algebra for the treatment of incompleteuncertain information and vague queries. Information Sciences 34, 115–143 (1984)
25. Raju, K., Majumdar, A.: Fuzzy functional dependencies and lossless join decomposition of fuzzy relational database systems. ACM Transactions on Database Systems 13, 129–166 (1988)
26. Zadeh, L.: Fuzzy sets as a basis for a theory of possibility. Fuzzy Sets and Systems 1(1), 3–28 (1978)

The Theory and Applications of Generalized Complex Fuzzy Propositional Logic

Dan E. Tamir[1], Mark Last[2], and Abraham Kandel[3]

[1] CS Department, Texas State University, 601 University Drive, San Marcos, Texas 78666
dt19@txstate.edu

[2] Faculty of Engineering Sciences, Department of Information Systems Engineering, Ben-Gurion University of the Negev, Beer Sheva Israel
mlast@bgu.ac.il

[3] CSE Department, University of South Florida, 4202 E. Fowler Ave, Tampa, Florida 33620
kandel@cse.usf.edu

Abstract. The current definition of complex fuzzy logic has two limitations. First, the derivation uses complex fuzzy relations; hence, it assumes the existence of complex fuzzy sets. Second, current theory is based on a restricted polar representation of complex fuzzy proposition, where only one component of a complex fuzzy proposition carries fuzzy information. In this chapter we present a novel form of complex fuzzy logic. The new theory, referred to as generalized complex fuzzy logic, overcomes the limitations of the current theory and provides several advantages. First, the derivation of the new theory is based on axiomatic approach and does not assume the existence of complex fuzzy sets or complex fuzzy classes. Second, the new form supports Cartesian and polar representation of complex logical propositions with two components of fuzzy information. Hence, the new form significantly improves the expressive power and inference capability of complex fuzzy logic. Finally, the new form is compatible with (yet independent of) the definition of complex fuzzy classes; thereby providing further improvement in the expressive power and inference capability. The chapter surveys the current state of complex fuzzy sets, complex fuzzy classes, and complex fuzzy logic; and provides a new and generalized complex fuzzy propositional logic theory. The new theory has potential for usage in advanced complex fuzzy logic systems and latent for extension into multidimensional fuzzy propositional and predicate logic. Moreover, it can be used for inference with type 2 (or higher) fuzzy sets. Furthermore, the introduction of complex logic can be used for analysis of periodic temporal fuzzy processes where the period is fuzzy.

1 Introduction

In 1965, L. A. Zadeh has established the theory of fuzzy sets [1]. In fuzzy set theory the degree of membership of an item in a set can get any value in the interval [0, 1], rather than the two values $\{\notin, \in\}$; where higher values denote higher degree of membership [1,2]. Fuzzy logic, introduced later, is a multilevel extension of classical logic where propositions can get truth values in the interval [0, 1], and are not limited

R.R. Yager et al. (Eds.): *Soft Computing: State of the Art Theory*, STUDFUZZ 291, pp. 177–192.
DOI: 10.1007/978-3-642-34922-5_13

to one of the two values {True, False} (or the equivalent Boolean logic values of {0, 1}) [3]. The four decades that followed his pioneering work has shown a multitude of research work and applications related to, control theory [4], inference [5], data mining [6], and reasoning [7,8]. In 1975 Zadeh introduced the concept of linguistic variable and the induced concept of type-2 (type-n) fuzzy sets [9]. In recent years, type-1 and type-2 along with interval type1/type-2 fuzzy logic and fuzzy systems have been applied in many areas including classification [10], fuzzy neural networks [11], fuzzy clustering [12], data base management and data mining [13, 14], and software testing [11].

Ramot et al. propose an extension of fuzzy set theory and fuzzy logic where the range of degrees of membership and the range of truth values is the complex unit circle [15,16]. The definition of complex fuzzy logic provided by Ramot, however, has two constraints. First, the derivation uses complex fuzzy relations thereby presumes the existence of complex fuzzy sets. Second, under the formalism proposed by Ramot, complex fuzzy membership functions and complex fuzzy propositions are represented in polar coordinates where only the absolute value of the complex membership function conveys fuzzy information.

Tamir and Kandel provide further generalization of the concept of complex fuzzy membership function and use a Cartesian complex fuzzy membership function where both the real part and the imaginary part can be fuzzy functions. Alternatively, polar representation where both the absolute value and the phase value of the complex membership function convey fuzzy information can be utilized [17]. Furthermore, they provide a new interpretation of complex fuzzy grades of membership as a representation of a complex fuzzy class along with complex fuzzy set / class operations.

In this chapter we present a new formal definition of complex fuzzy logic referred to as generalized complex fuzzy logic. The new theory resolves most of the limitations of the current complex fuzzy logic theory. Moreover, it presents several advantages. Including:

- The derivation of the new formalism is based on axiomatic approach that does not assume the existence of complex fuzzy sets or complex fuzzy classes. Hence, complex fuzzy logic is independent of complex fuzzy classes and vice versa. On the other hand, the new formalism, presented here, is compatible with previous work.
- The new theory supports Cartesian as well as polar representation of complex logical fuzzy propositions with two components of ambiguous information. Hence, this form significantly improves the expressive power and inference capability of complex fuzzy logic.
- The compatibility with the definition of complex fuzzy classes provides for further improvement in the expressive power and inference capability.

The chapter reviews the current state of complex fuzzy sets theory, provides a brief overview of complex fuzzy classes, and complex fuzzy logic; and introduces a new and generalized complex fuzzy propositional logic theory. The new formalism can be used in advanced complex fuzzy logic systems and provides ways for extension into multidimensional fuzzy propositional and first order logic. Furthermore, it can be used for inference with type 2 (or higher) fuzzy sets. Finally, the new formalism can be used as the basis for new definition of fuzzy temporal logic.

The rest of the chapter is organized in the following way: section II introduces the concepts of complex fuzzy set and complex fuzzy classes [15-18]. Section 3 provides the definition of generalized fuzzy complex logic along with interesting examples, and section 4 includes conclusions and directions for further research.

2 Complex Fuzzy Sets and Classes

This section reviews the basic concepts and operations of complex fuzzy sets and complex fuzzy classes [15-17, 19].

2.1 Complex Fuzzy Sets

A complex fuzzy set S on a universe of discourse U is defined by a complex-valued grade of membership function $\mu_S(x)$ [15,16]:

$$\mu_S(x) = r_S(x)e^{j\omega_S(x)} \tag{1}$$

Where $j = \sqrt{-1}$. The function $\mu_S(x)$ maps U into the unit disc of the complex plane. This definition utilizes polar representation of complex numbers along with conventional fuzzy set definition; where $r_S(x)$, the amplitude part of the grade of membership, is a fuzzy function defined in the interval [0, 1]. And $\omega_S(x)$ is a real number, derived from a real function ($\omega_S()$), standing for the phase part of the grade of membership.

Basic Operations on Complex Fuzzy Sets

Ramot et al. propose a directional complex (DC) fuzzy complement which is a combination of phase rotation and traditional fuzzy complement of the amplitude. Suppose that S is a complex fuzzy set, and its degree of membership is defined by the function $\mu_S(x) = r_S(x)e^{j\omega(x)}$. The rotation of S by θ radians is defined as:

$$Rot_\theta(\mu_S(x)) = r_S(x)e^{j(\omega_S(x)+\theta)} \tag{2}$$

The DC fuzzy complement follows the traditional "axioms" of complement and induces the following membership function

$$c(\mu_S(x)) = c(r_S(x))e^{j(\omega_S(x)+\theta)} \tag{3}$$

Note that the set of axioms listed in [15,16,19] is actually a set of theorems that can be proved using axiomatic fuzzy set theory [7,20-23].

The union function, ∪, of two complex fuzzy sets A and B returns a complex-valued grade of membership of the elements in the set $A \cup B$. The membership function of $A \cup B$ is defined to be:

$$\mu_{A\cup B}(x) = [r_A(x) \oplus r_B(x)] \cdot e^{j\omega_{A\cup B}(x)} \tag{4}$$

where $\oplus$ represents a *t-conorm* function of [24]; and $\omega_{A\cup B}(x)$ is a real function.

The definition of intersection of complex fuzzy sets is analogous to the definition of the union operation. The intersection function ∩, of two complex fuzzy sets *A* and *B*, returns a complex-valued grade of membership of the element in the set $A \cap B$. The membership function of $A \cap B$ is defined to be:

$$\mu_{A\cap B}(x) = [r_A(x) \odot r_B(x)] \cdot e^{j\omega_{A\cap B}(x)} \quad (5)$$

where $\odot$ represents a *t-norm* [24]; and $\omega_{A\cup B}(x)$ is a real function.

Current and Related Research on Complex Fuzzy Sets

Ramot et al. propose several applications for the concept of fuzzy set, and demonstrate the use of the theory for inference about periodic time sequences [15,16]. In this sense, the concept can be used for fuzzy temporal reasoning. Nevertheless, they have limited their focus to periodic signals with a fixed period. Hence, only the amplitude term of their complex fuzzy membership function is a fuzzy function. The generalization provided in this chapter, however, enables dealing with periodic processes with fuzzy period. These types of processes, e.g., the behaviour of the stock market, occur in numerous applications of interest. Dick expands the research on complex fuzzy sets [19,25-30]. It is interesting to note that the idea of complex grade of membership and the utilization of the concept in complex fuzzy sets as well as complex fuzzy classes shares common features with complex neural networks where the excitation, the outputs, and weights can get complex values [31-35]. It should be noted that there is a substantial difference between the definitions of complex fuzzy numbers given by J. Buckley [36-40] and the concept of complex fuzzy sets or complex fuzzy logic proposed by Ramot et al. [15,16], and generalized in this chapter. Buckley is concerned with generalizing number theory; while the papers by Ramot as well as the current paper are concerned with the generalization of fuzzy set theory and fuzzy logic [15,16]. Complex fuzzy numbers have been utilized in several numerical applications [41-43].Yet, the concept of a complex fuzzy number is different than the concept of complex fuzzy sets or complex fuzzy classes.

2.2 New Interpretation of Complex Fuzzy Classes

Tamir et al. introduced a new interpretation of complex fuzzy membership grade and derive the concept of complex fuzzy classes [17,18]. This section introduces the concept of a complex fuzzy grade of membership, the interpretation of this concept as the denotation of a fuzzy class, and the basic operations on fuzzy classes. In addition, the section outlines coordinate transformations. To distinguish between Ramot's interpretation of complex fuzzy sets and classes and Tamir's interpretation of the same concepts we refer to the later as "pure."

Pure Complex Fuzzy Membership Grade

To distinguish between classes, sets, and elements of a set we use the following notation: a class is denoted by an upper case Greek letter, a set is denoted by an upper case Latin letter, and a member of a set is denoted by a lower case Latin letter.

The Cartesian representation of the complex grade of membership is given in the following way:

$$\mu(V,z) = \mu_r(V) + j\mu_i(z) \quad (6)$$

Where $\mu_r(V)$ and $\mu_i(z)$, the real and imaginary components of the complex fuzzy grade of membership, are real value fuzzy grades of membership. That is, $\mu_r(V)$ and $\mu_i(z)$ can get any value in the interval $[0,1]$. The polar representation of the complex grade of membership is given by:

$$\mu(V,z) = r(V)e^{j\sigma\phi(z)} \quad (7)$$

Where $r(V)$ and $\phi(z)$, the amplitude and phase components of the complex fuzzy grade of membership, are real value fuzzy grades of membership. That is, they can get any value in the interval $[0,1]$. The scaling factor, σ is in the interval $(0,2\pi]$. It is used to control the behavior of the phase within the unit circle according to the specific application. Typical values of σ are $\{1,\frac{\pi}{2},\pi,2\pi\}$. Without loss of generality, for the rest of the discussion in this section we assume that $\sigma = 2\pi$.

The difference between generalized complex fuzzy grades of membership proposed here and the complex fuzzy grade of membership proposed by Ramot et al. [15,16], is that both components of the membership grade are fuzzy functions that convey information about a fuzzy set. This entails different interpretation of the concept as well as a different set of operations and a different set of results obtained when these operations are applied to complex grades of membership. This is detailed in the following sections.

Pure Complex Fuzzy Class

A fuzzy class is a finite or infinite collection of objects and fuzzy sets that can be defined in an unambiguous way and complies with class theory [23], the axioms of fuzzy sets given by Tamir et al. [7], and the axioms of fuzzy classes given by Běhounek, Montagna, Hájek, Cintula, and others [21,22,44-47]. While a general fuzzy class can contain individual objects as well as fuzzy sets, a *fuzzy class of order one* can contain only fuzzy sets. In other words, individual objects cannot be members of a fuzzy class of order one. A fuzzy class of order M is a collection of fuzzy classes of order $M-1$. We define a *Complex Fuzzy Class* Γ to be a fuzzy class of order one i.e., a fuzzy set of fuzzy sets. That is, $\Gamma = \{V_i\}_{i=1}^{\infty}$; or $\Gamma = \{V_i\}_{i=1}^{N}$ where V_i is a fuzzy set and N is a finite integer. Note that despite the fact that we use the notation $\Gamma = \{V_i\}_{i=1}^{\infty}$ we do not imply that the set of sets $\{V_i\}$ is enumerable. The set of sets $\{V_i\}$ can be finite, countably infinite, or uncountably infinite. The use of the notation $\{V_i\}_{i=1}^{\infty}$ is just for convenience.

The class Γ is defined over a universe of discourse T. It is characterized by a complex membership function $\mu_\Gamma(V,z)$ that assigns a complex-valued grade of membership in Γ to any element $z \in U$ (where U is the universe of discourse). The values that $\mu_\Gamma(V,z)$ can receive lie within the unit square or the unit circle in the complex plane, and are in one of the following forms:

$$\mu_\Gamma(V,z) = \mu_r(V) + j\mu_i(z) \quad (8a)$$

$$\mu_\Gamma(z,V) = \mu_r(z) + j\mu_i(V) \tag{8b}$$

Where $\mu_r(\alpha)$ and $\mu_i(\alpha)$, are real functions with a range of [0,1].

Alternatively:

$$\mu_\Gamma(V,z) = r(V)e^{j\theta\phi(z)} \tag{9a}$$

$$\mu_\Gamma(z,V) = r(z)e^{j\theta\phi(V)} \tag{9b}$$

Where $r(\alpha)$ and $\phi(\alpha)$, are real functions with a range of [0, 1] and $\theta \in (0,2\pi]$.

In order to provide a concrete example we define the following fuzzy class. Let the universe of discourse be the set of all the stocks that were available for trade on the opening of the New York stock exchange (NYSE) market on January 3, 2012 along with a set of attributes related to historical price performance of each of these stocks. Let M_i denote the set of NYSE stocks in the portfolio of an NYSE tradable Mutual Fund i, on the same day. Furthermore consider a function (f_1) that associates a number between 0 and 1 with each subset of stocks, such as the portfolio of each NYSE tradable mutual fund. For example, this function might reflect the performance of the portfolio in the last quarter. In addition, consider a second function (f_2) that associates a number between 0 and 1 with each tradable stock. For example, this function might be a normalized value of volatility of this stock. The functions (f_1, f_2) can be used to define a fuzzy class of order one. A compound of the two functions in the form of a complex number can represent the degree of membership in the fuzzy class of "volatile stocks in the portfolio of high performance mutual funds."

Formally, let U be a universe of discourse and let 2^U be the power set of U. Let f_1 be a function from 2^U to [0, 1] and let f_2 be a function that maps elements of U to the interval [0, 1]. For $V \in 2^U$ and $z \in U$ define $\mu_\Gamma(V,z)$ to be:

$$\mu_\Gamma(V,z) = \mu_r(V) + j\mu_i(z) = f_1(V) + jf_2(z) \tag{10}$$

Then, $\mu_\Gamma(V,z)$ defines a fuzzy class of order one, where for every $V \in 2^U$, and for every $z \in U$,

$\mu_\Gamma(V,z)$; is the degree of membership of z in V and the degree of membership of V in Γ. Hence, a complex fuzzy class Γ can be represented as the set of ordered triples:

$$\Gamma = \{V, z, \mu_\Gamma(V,z) | V \in 2^U, z \in U\} \tag{11}$$

Depending on the form of $\mu_\Gamma(\alpha)$ (Cartesian or polar), $\mu_r(\alpha)$, $\grave{\imath}_i(\alpha)$, $r(\alpha)$, and $\phi(\alpha)$ denote the degree of membership of z in V and / or the degree of membership of V in Γ. Without loss of generality, however, we assume that $\mu_r(\alpha)$ and $r(\alpha)$ denote the degree of membership of V in Γ for the Cartesian and the polar representations respectively. In addition, we assume that $\mu_i(\alpha)$ and $\phi(\alpha)$ denote the degree of membership of z in V for the Cartesian and the polar representations respectively. Throughout this chapter, the term "complex fuzzy class" refers to a fuzzy class with complex-valued membership function, while the term "fuzzy class" refers to a traditional fuzzy class such as the one defined by Běhounek [22].

Coordinate Transformations

Changing the coordinate system from Cartesian to polar and vice versa according to the classical transformation, is straightforward. Nevertheless, due to the nonlinear nature of the transformation it might produce a completely different and often meaningless class. Moreover, depending on the value of θ, a scaling may be required. On the other hand, as shown in the next paragraph, it is very simple to define a set of transformations that maintains the same semantics.

Cartesian to polar coordinate transformation:

$$\mu_\Gamma(V,z) = T_r(\mu_r(V) + \mu_i(z)) = \mu_r(V)e^{j\theta\phi(x)} \tag{12}$$

Polar to Cartesian Coordinate Transformation:

$$\mu_\Gamma(V,x) = T_r(r(V)e^{j\theta\phi(x)}) = r(V) + j\theta\phi(x) \tag{13}$$

Where $T_r(\alpha)$ denotes the coordinate transformation function. Note that this definition assumes that: $\mu_r(V)$ and $r(V)$ denote the degree of membership of V in Γ and that $\mu_i(z)$ and $\phi(z)$ denote the degree of membership of z in V.

Figure 1 illustrates the dual Cartesian representations of complex fuzzy classes. Each stripe that is parallel to the Y axis in the left image represents a fuzzy set. The x coordinate of each pixel in the left image represent the degree of membership of the fuzzy set that contains this pixel in the complex fuzzy class, and the gray level of the pixel (normalized to (0,1)) represents the value $\mu_i(z)$; that is, the degree of membership of the pixel in the fuzzy set that contains this pixel. In the right image, the role of the X axis and the Y axis is interchanged.

Fig. 1. An illustration of the Cartesian representation of complex fuzzy classes

Figure 2 shows the two polar coordinate system based representation interpretations. Assuming that $\sigma = 2\pi$. Each ring around the origin in the left image represents a fuzzy set; for each pixel in that image, the distance of the pixel from the origin represents the degree of membership of the fuzzy set that contains this pixel, in the complex fuzzy class; and the normalized gray level of the pixel represents the value $\phi(z)$; that is, the degree of membership of the pixel in the fuzzy set that contains this pixel.

In the right image, the role of the radius r and the angle θ is interchanged. Several additional linear transformations that maintain the same class semantics can be considered. Without loss of generality, the class operations given in the next sections use only the Cartesian form.

Degree of Membership of Order N

The traditional fuzzy grade of membership is a real number that defines a fuzzy set. It can be considered as degree of membership of order 1. The complex degree of membership defined in this chapter is a complex number that defines a fuzzy class. That is, a fuzzy set of fuzzy sets. This degree of membership can be considered as degree of membership of order 2 and the class defined can be considered as a fuzzy class of order 1. Additionally, one can consider the definition of a fuzzy set (a class of order 0) as a mapping into a one dimensional space and the definition of a fuzzy class (a class of order 1) as a mapping into a two dimensional space. Hence, it is possible to consider a degree of membership of order N as well as a mapping into an N-dimensional space. The following is a recursive definition of a fuzzy class of order N. Note that part 2 of the definition is not really necessary; it is given in order to connect the terms complex fuzzy grade of membership and the term grade of membership of order 2.

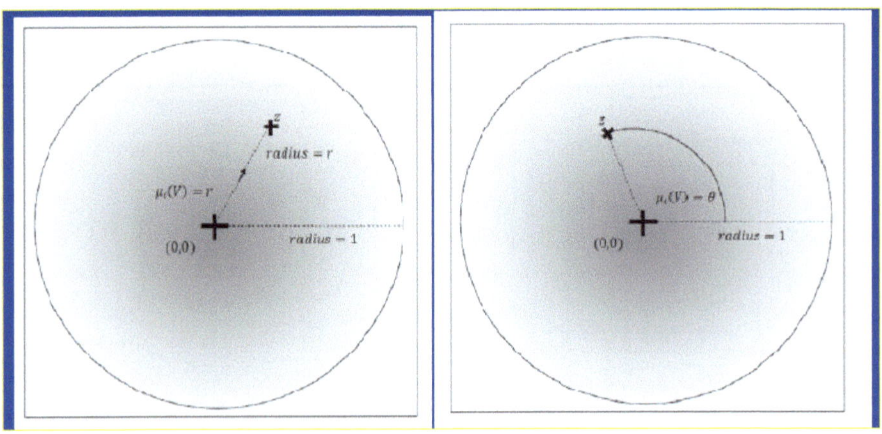

Fig. 2. An illustration of the polar representation of complex fuzzy classes

Definition:

1) A fuzzy class of order 0 is a fuzzy set; it is characterized by a degree of membership of order 1 and a mapping into a one dimensional space.
2) A fuzzy class of order 1 is a fuzzy class; that is, set of fuzzy sets. It is characterized by a complex degree of membership. Alternatively, it can be

characterized by a degree of membership of order two and a mapping into a two dimensional space.

3) A fuzzy class of order N is a fuzzy set of fuzzy classes of order N-1; it is characterized by a degree of membership of order $N+1$ and a mapping into an $(N+1)$-dimensional space.

Operations on Pure Complex Fuzzy Classes

This section defines the three basic operations on complex fuzzy classes; complement, union, and intersection. The definitions are provided in terms of operations on grades of membership and the implication of these operations on the underlying complex fuzzy classes. Note that the definitions utilize the Cartesian form. This is general enough due to the fact that one can change the role of the real and imaginary parts and / or perform coordinate transform without changing the semantics of complex fuzzy grade of membership.

The Complement of a Pure Complex Fuzzy Class

Let $\Gamma = \{V, z, \mu_\Gamma(V, z) | V \in 2^U, z \in U\}$ be the class defined by: $\mu_\Gamma(V, z) = \mu_r(V) + j\mu_i(z)$, where $\mu_r(\alpha)$, and $\mu_i(\alpha)$ stand for the real and imaginary parts of $\mu_\Gamma(V, z)$. The complement operation $(c(\alpha))$ on Γ is defined via complement operations on $\mu_\Gamma(\alpha)$ and denoted as $c(\alpha)$.

$$c\,(\mu_\Gamma(V, z)) = c(\mu_r(V)) + jc(\mu_i(z)) \tag{14}$$

That is, the complement function operates on the sets that comprise the class Γ as well as on the individual members of each of these sets. It changes the degree of membership of each set V in Γ and each element z in the set V. There are several ways to define the operation $(c(\alpha))$. A simple and straightforward way would be to use the classical fuzzy definition. That is $c(\mu_x(y)) = 1 - \mu_x(y)$; where x is either r (real part) or i (imaginary part) and y is either V or z.

Pure Complex Fuzzy Class Union Operation

Consider the two complex fuzzy classes:

1. $\Gamma = \{V, z, \mu_\Gamma(V, z) | V \in 2^U, z \in U\}$
2. $\Psi = \{T, z, \mu_\Psi(T, z) | T \in 2^U, z \in U\}$

Where V and T are fuzzy sets. Assume that Γ and Ψ are defined over a universe of discourse U and let 2^U denote the power set of U. Further assume that the degree of membership of an object $z \in V$, and an object $y \in T$ is given by: $\mu_\Gamma(V, z) = \mu_{\Gamma r}(V) + j\mu_{\Gamma i}(z)$; $\mu_\Psi(T, y) = \mu_{\Psi r}(T) + j\mu_{\Psi i}(y)$ respectively, where $\mu_{\Gamma r}(\alpha), \mu_{\Psi r}(\alpha), \mu_{\Gamma i}(\alpha)$, and $\mu_{\Psi i}(\alpha)$ stand for the real and imaginary parts of $\mu_\Gamma(V, z)$ and $\mu_\Psi(T, y)$. Finally, let $W \in 2^U$, and let $\oplus$ denote a t-conorm operation. The union function $\Gamma \cup \Psi$ is defined to be:

$$\mu_{\Gamma \cup \Psi}(W, z) = \left(\mu_{\Gamma r}(V) \oplus \mu_{\Psi r}(T)) + j(\mu_{\Gamma i}(z) \oplus \mu_{\Psi i}(z))\right) \tag{15}$$

Pure Complex Fuzzy Class Intersection Operation

The discussion on complex fuzzy intersection is analogous to the discussion on union. Consider be two complex fuzzy classes:

1. $\Gamma = \{V, z, \mu_\Gamma(V, z) | V \in 2^U, z \in U\}$
2. $\Psi = \{T, z, \mu_\Psi(T, z) | T \in 2^U, z \in U\}$

Where V and T are fuzzy sets. Assume that Γ and Ψ are defined over a universe of discourse U and let 2^{U} denote the power set of U. Further assume that the degree of membership of an object $z \in V$, and an object $y \in T$ is given by: $\mu_\Gamma(V, z) = \mu_{\Gamma r}(V) + j\mu_{\Gamma i}(z)$ and $\mu_\Psi(T, y) = \mu_{\Psi r}(T) + j\mu_{\Psi i}(y)$ respectively, where $\mu_{\Gamma r}(\alpha), \mu_{\Psi r}(\alpha), \mu_{\Gamma i}(\alpha)$, and $\mu_{\Psi i}(\alpha)$ stand for the real and imaginary parts of $\mu_\Gamma(V, z)$ and $\mu_\Psi(T, y)$. Finally, let $W \in 2^{\mathrm{U}}$, and let $\odot$ denote a t-norm operation. The intersection function $\Gamma \cap \Psi$ is defined as:

$$\mu_{\Gamma \cap \Psi}(W, z) = \left(\mu_{\Gamma r}(V) \odot \mu_{\Psi r}(T) + j\left(\mu_{\Gamma i}(z) \odot \mu_{\Psi i}(z)\right)\right) \tag{16}$$

3 Generalized Complex Propositional Fuzzy Logic

There are several ways to define fuzzy logic, fuzzy inference, and fuzzy logic system (FLS). One of these ways is to use fuzzy set theory to define fuzzy relations, and then define logical operations, such as implication and negation, as well as inference rules, as special types of relations on fuzzy sets. Alternatively, fuzzy logic can be formalized as a direct generalization of classical logic. Under this "traditional" approach, notions that relate to the syntax and semantics of classical logic, such as propositions, interpretation, and inference are used to define fuzzy logic. Although the relations based definition can be carefully formalized it is generally less rigorous than the traditional approach.

Ramot et al. use the first approach [16]. They use the definition of complex fuzzy relations to define complex fuzzy logic via the definition of logical operations. In the current paper, however, we use the traditional approach. This section provides the definition of generalized complex propositional fuzzy logic using direct generalization of fuzzy logic.

A fuzzy proposition P can get any truth value in the real interval [0, 1]. The intuitive interpretation of truth values in the range [0, 1], referred to in this chapter as fuzzy truth values, is that $'0'$ denotes "False," and $'1'$ denotes "True". Furthermore, the relations $\leq$, over the set of real numbers in the interval $[0, 1]$ implies a monotonically increasing ordering on the truth values associated with every $x, y \in [0, 1]$. Formally a fuzzy interpretation of a proposition P is an assignment of fuzzy truth value to P.

A *linguistic variable* is a variable whose domain of values is formal or natural language words [9]. Generally, a linguistic variable is related to a fuzzy set such as: $\{very\ young\ male, young\ male, old\ male, very\ old\ male\}$ and can get any value from the set. A linguistic constant has a fixed and unmodified linguistic value i.e., a single word or phrase from formal or natural language.

One form of a fuzzy proposition is: "x ... A ...;" where A is a linguistic variable such as "young male", or "tall person", and '...' denote natural language constants such as "Moses", "is a," "portfolio," "mutual fund" etc.

For example, under one interpretation, the fuzzy truth value associated with the fuzzy proposition: $P =$ "Moses is a $young\ male$" can be 0.3, and under another interpretation, the fuzzy truth value associated with the proposition P can be 0.9. In this case, the linguistic variable is "young $male$", and it is distinguished from the Fuzzy constants "Moses," and "is" by its Italics font. Set theory and fuzzy logic can be connected through fuzzy membership functions. For example, let A denote the fuzzy set of *young males,* and let f_A be a specific fuzzy membership function of A, then f_A can be used as the basis for interpretations of P.

Following the definition of propositions (syntax) and interpretation (semantics), one may want to define the syntax and semantics of fuzzy logical operations (connectives). Table 1 includes a specific definition of connectives along with their interpretation. In this table P, Q, and R denote fuzzy propositions and $f(R)$ denotes the fuzzy interpretation of R. We use the fuzzy Łukasiewicz logical system as the basis for the definitions [22]. Hence, the max t-norm is used for conjunction and the min t-conorm is used for disjunction. Nevertheless, other logical systems such as Gödel fuzzy systems can be used [47].

Table 1. Basic Propositional Fuzzy Logic Connectives

Operation	Interpretation
Negation	$f('P) = 1 - f(P)$
Implication	$f(P \longrightarrow Q) = \min(1{,}1 - f(P) + f(Q))$
Conjunction	$f(P \otimes Q) = \min(f(P), f(Q))$
Disjunction	$f(P \oplus Q) = \max(f(P), f(Q))$

We assume the following set of axioms [22, 47]:

A1: $(P \rightarrow Q) \rightarrow ((Q \rightarrow R) \rightarrow (P \rightarrow R))$
A2: $(P \otimes Q) \rightarrow P$
A3: $(P \otimes Q) \rightarrow (Q \otimes P)$
A4: $(P \otimes (P \rightarrow Q)) \rightarrow (Q \otimes (Q \rightarrow P))$
A5a: $(P \rightarrow (Q \rightarrow R)) \rightarrow ((P \otimes Q) \rightarrow R)$
A5b: $((P \otimes Q) \rightarrow R) \rightarrow (P \rightarrow (Q \rightarrow R))$
A6: $((P \rightarrow Q) \rightarrow R) \rightarrow (((Q \rightarrow P) \rightarrow R) \rightarrow R)$
A7: $0 \rightarrow P$

In order to completely specify propositional fuzzy complex logic, it is enough to specify a universal set of operators along with a rule of inference. A commonly used inference rule is Modus ponens defined as:

$$A \wedge (A \longrightarrow B) \longrightarrow B \tag{17}$$

That is, using degree of confidence as a synonym for fuzzy logical truth value, and given the propositions and the interpretations provided above we can infer B from A and $(A \longrightarrow B)$ with the implied fuzzy degree of confidence; calculated using table

1. In the next section, we define generalized complex fuzzy logic using generalized complex fuzzy connectives and generalized complex fuzzy Modus ponens [22, 47].

Generalized Complex Fuzzy Logic

A general form of a complex fuzzy proposition is: "x ... A ... B ..." where A and B are values assigned to linguistic variables and '...' denotes natural language constants. A complex fuzzy proposition P can get any pair of truth values from the Cartesian interval $[0,1] \times [0,1]$ or the unit circle. Formally a fuzzy interpretation of a complex fuzzy proposition P is an assignment of fuzzy truth value of the form $p_r + jp_i$, or of the form $r(p)e^{j\theta(p)}$, to P. In this case, assuming a proposition of the form "x ... A ... B ...," then p_r $(r(p))$ is assigned to the term A and p_i $(\theta(p))$ is assigned to term B.

For example, under one interpretation, the complex fuzzy truth value associated with the complex fuzzy proposition: "x is a *volatile stock* in a *strong portfolio*," can be $0.1 + j0.5$. Alternatively, in another context, the same proposition can be interpreted as having the complex truth value $0.3e^{j0.2}$. As in the case of traditional propositional fuzzy logic we use the tight relation between complex fuzzy classes / complex fuzzy membership to determine the interpretation of connectives. For example, let C denote the complex fuzzy set of volatile stocks in a strong portfolio, and let $f_C = c_r + jc_i$, be a specific fuzzy membership function of C, then f_C can be used as the basis for interoperations of P. Next we define several connectives along with their interpretation.

Table 2 includes a specific definition of connectives along with their interpretation. In this table $\boldsymbol{P}$, $\boldsymbol{Q}$, and $\boldsymbol{S}$ denote complex fuzzy propositions and $\boldsymbol{f(S)}$ denotes the complex fuzzy interpretation of $\boldsymbol{S}$. We use the fuzzy Łukasiewicz logical system as the basis for the definitions [22,47]. Hence, the max t-norm is used for conjunction and the min t-conorm is used for disjunction. Nevertheless, other logical systems such as Gödel fuzzy systems can be used [22,47] .

Table 2. Basic Complex Propositional Fuzzy Logic Connectives

Operation	Interpretation
Negation	$f('P) = 1 + j1 - f(P)$
Implication	$f(P \rightarrow Q) = \min(1, 1 - p_r + q_r) + j \times \min(1, 1 - p_i + q_i)$
Conjunction	$f(P \otimes Q) = \min(p_r, q_r) + j \times \min(p_i, q_i)$
Disjunction	$f(P \oplus Q) = \max(p_r, q_r) + j \times \max(p_i, q_i)$

The same axioms used for fuzzy logic are used for complex fuzzy logic, and Modus ponens is the rule of inference.

Complex Fuzzy Propositions and Connectives Examples

Consider the following propositions(P and Q respectively):

1) "x is a *volatile stock* in a *strong portfolio*."
2) "x is a stock in a *decline trend* in a *strong portfolio*."

Hence, P is of the form: "x is a A in a B, "and Q is of the form"x is a stock C in B ." In this case, *"volatile stock," "a strong portfolio,"* and *"a decline trend"* are values assigned to the linguistic variables $\{A, B, C\}$. Assume that the complex fuzzy interpretation (i.e., degree of confidence or complex fuzzy truth value) of P is $p_r + jp_i$, while the complex fuzzy interpretation of Q is $q_r + jp_i$ $(q_i = p_i)$. Thus, the truth value of *"x is a volatile stock,"* is p_R, the truth value of *"x is in a strong portfolio,"* is p_i , and the truth value of *"x is in a decline trend,"* is q_r. Suppose that the term "*non* – volatile" stands for *"not volatile,"* the term *"weak"* stands for *"not strong,"* and the term *"rising"* stands for *"not declining."* Note that this is not the only way to define these linguistic terms and it is used to exemplify the expressive power and the inference power of the logic. Then, the complex fuzzy interpretation of the following composite propositions is:

1) $f('P) = (1 - p_r) + j(1 - p_I)$

That is, $'P$ denotes the proposition "x is a non – volatile stock in a weak portfolio." The confidence level in $'P$ is $(1 - p_r) + j(1 - p_i)$; where the fuzzy truth value of the term "x is a non – volatile stock," is $(1 - p_r)$ and the fuzzy truth value of the term "weak portfolio," is $(1 - p_i)$

2) $'P \rightarrow 'Q = \min(1, q_r - p_r) + j \times \min(1, q_i - p_I)$

Thus, $('P \rightarrow' Q)$ denotes the proposition "IF x is a *non – volatile stock* in a *weak portfio*, THEN x is a stock in a *rising trend* in a *strong portfolio."* The truth values of individual terms, as well as the truth value of $'P \rightarrow 'Q$ are calculated according to table 2.

3) $f('P \otimes Q) = \min(1 - p_r, q_r) + j \times \min(1 - p_i, q_i)$.

That is, $('P \otimes Q)$ denotes the proposition "x is a volatile stock in a *strong portfolio*" AND " x is a stock in a *rising trend* in a *strong portfolio."* The truth values of individual terms, as well as the truth value of $'P \otimes Q$ are calculated according to table 2.

4) $f(P \oplus 'Q) = \max(p_r, 1 - q_r) + j \times \max(p_i, 1 - q_i)$.

That is, $(P \oplus' Q)$ denotes the proposition "x is a volatile stock in a strong *portfolio*" OR " x is a stock in a *rising trend* in a *wek portfolio."* The truth values of individual terms, as well as the truth value of $P \oplus 'Q$ are calculated according to table 2.

Complex Fuzzy Inference Example

Assume that the degree of confidence in the proposition $R = 'P$ defined above is $r_r + jr_i$, let $S = 'Q$ and assume that the degree of confidence in the fuzzy implication $T = R \rightarrow S$ is $t_r + jt_i$. Then, using Modus ponens:

$$\frac{\begin{array}{l} R \\ R \rightarrow S \end{array}}{S}$$

one can infer S with a degree of confidence $\min(r_r, t_r) + j \times \min(r_i, t_i)$.

In other words using:
"x is a non − *volatile stock* in a *weak portfolio.*"
IF "x is a non volatile stock in a *weak portfolio,*" THEN
"x is a stock in a *rising trend* in a *strong portfolio.*"
"x is a stock in a *rising trend* in a *strong portfolio*"

we can infer "x is a stock in a *rising trend* in a *strong portfolio*" with a degree of confidence $\min(r_r, t_r) + j \times \min(r_i, t_i)$.

4 Conclusions and Directions for Further Research

A new and innovative formal definition of complex fuzzy logic, referred to as generalized complex fuzzy logic, is presented in this chapter. The new form significantly improves the expressive power and inference capability of complex fuzzy logic. In the future, we plan to extend the theory to multidimensional fuzzy propositional and predicate logic; explore the utility of the theory for fuzzy temporal logic; and further explore its potential for usage in advanced complex fuzzy logic systems as well as inference with type 2 (or higher) fuzzy sets.

References

1. Zadeh, L.A.: Fuzzy sets. Information and Control 8, 338–353 (1965)
2. Kandel, A.: Fuzzy Mathematical Techniques with Applications. Addison Wesley, New York (1987)
3. Zadeh, L.A.: Fuzzy Logic and Its Application to Approximate Reasoning. In: Proceedings ofthe IFIP Congress, pp. 591–594 (1974)
4. Drianko, D., Hellendorf, H., Reinfrank, M.: An introduction to fuzzy control. Springer (1993)
5. Halpern, J.Y.: Reasoning about uncertainty. MIT Press (2003)
6. Last, M., Klein, Y., Kandel, A.: Knowledge Discovery in Time Series Databases. IEEE Transactions on Systems, Man, and Cybernetics 31(Part B,1), 160–169 (2001)
7. Tamir, D.E., Kandel, A.: An axiomatic approach to fuzzy set theory. Information Sciences 52, 75–83 (1990)
8. Tamir, D.E., Kandel, A.: Fuzzy semantic analysis and formal specification of conceptual knowledge. Information Sciences, Intelligent systems 82(3-4), 181–196 (1995)
9. Zadeh, L.A.: The concept of a linguistic variable and its application to approximate reasoning - Part I. Information Sciences 7, 199–249 (1975)
10. Ronen, M., Shabtai, R., Guterman, H.: Hybrid model building methodology using unsupervised fuzzy clustering and supervised neural networks. Biotechnology and Bioengineering 77(4), 420–429 (2002)
11. Agarwal, D., Tamir, D.E., Last, M., Kandel, A.: A comparative study of software testing using artificial neural networks and Info-Fuzzy networks. Accepted for publication in the IEEE, Transactions on Man Machine and Cybernetics, Part A (2012)
12. Tamir, D.E., Kandel, A.: The Pyramid Fuzzy C-means Algorithm. International Journal of Computational Intelligence in Control (December 2011)

13. De, S.P., Krishna, R.P.: A new approach to mining fuzzy databases using nearest neighbor classification by exploiting attribute hierarchies. International Journal of Intelligent Systems, 2004 19(12), 1277–1290 (2004)
14. Qiu, T., Chen, X., Liu, Q., Huang, H.: Granular computing approach to finding association rules in relational database. International Journal of Intelligent Systems 25(2), 165–179 (2010)
15. Ramot, D., Milo, R., Friedman, M., Kandel, A.: Complex fuzzy sets. IEEE Transactions on Fuzzy Systems 2002 10(2), 171–186 (2002)
16. Ramot, D., Friedman, M., Langholz, G., Kandel, A.: Complex fuzzy logic. IEEE Transactions on Fuzzy Systems 11(4), 450–461 (2003)
17. Tamir, D.E., Lin, J., Kandel, A.: A New Interpretation of Complex Membership Grade. International Journal of Intelligent Systems (2011)
18. Tamir, D.E., Kandel, A.: Axiomatic Theory of Complex Fuzzy Logic and Complex Fuzzy Classes (invited paper). Int. J. of Computers, Communications & Control VI(3) (2011)
19. Dick, S.: Towards complex fuzzy logic. IEEE Transaction on Fuzzy Systems 13, 405–414 (2005)
20. Cintula, P.: Weakly implicative fuzzy logics. Archive for Mathematical Logic 45(6), 673–704 (2006)
21. Baaz, M., Hajek, P., Montagna, F., Veith, H.: Complexity of t-tautologies. Annals of Pure and Applied Logic 113(1-3), 3–11 (2002)
22. Běhounek, L., Cintula, P.: Fuzzy class theory. Fuzzy Sets and Systems 154(1), 34–55 (2005)
23. Fraenkel, A.A., Bar-Hillel, Y., Levy, A.: Foundations of set theory, 2nd edn. Elsevier, Amsterdam (1973)
24. Montagna, F.: On the predicate logics of continuous t-norm BL-algebras. Archives of Mathematical Logic 44, 97–114 (2005)
25. Moses, D., Degani, O., Teodorescu, H., Friedman, M., Kandel, A.: Linguistic coordinate transformations for complex fuzzy sets. In: 1999 IEEE International Conference on Fuzzy Systems, Seoul, Korea (1999)
26. Zhang, G., Dillon, T.S., Cai, K., Ma, J., Lu, J.: Operation properties and delta equalities of complex fuzzy sets. International Journal on Approximate Reasoning 50(8), 1227–1249 (2009)
27. Nguyen, H.T., Kandel, A., Kreinovich, V.: Complex fuzzy sets: towards new foundations. In: Proceedings, 2000 IEEE International Conference on Fuzzy Systems, San Antonio, Texas (2000)
28. Deshmukh, A.Y., Bavaskar, A.B., Bajaj, P.R., Keskar, A.G.: Implementation of complex fuzzy logic modules with VLSI approach. International Journal on Computer Science and Network Security 8, 172–178 (2008)
29. Chen, Z., Aghakhani, S., Man, J., Dick, S.: ANCFIS: A Neuro-Fuzzy Architecture Employing Complex Fuzzy Sets. IEEE, Transactions on Man Machine and Cybernetics, Part A 19(2) (2011)
30. Man, J., Chen, Z., Dick, S.: Towards inductive learning of complex fuzzy inference systems. In: Proceedings of the International Conference of the North American Fuzzy Information Processing, San Diego, CA (2007)
31. Hirose, A.: Complex-valued neural networks. Springer (2006)
32. Michel, H.E., Awwal, A.A.S., Rancour, D.: Artificial neural networks using complex numbers and phase encoded weights-electronic and optical implementations. In: Proceedings of the International Joint Conference on Neural Networks, Vancouver, BC (2006)

33. Noest, A.J.: Discrete-state phasor neural nets. Physics Review A 38, 2196–2199 (1988)
34. Leung, S.H., Hyakin, S.: The complex back propagation algorithm. IEEE Transactions on Signal Processing 39(9), 2101–2104 (1991)
35. Iritani, N.T., Sakakibara, K.: Improvements of the traffic signal control by complex-valued Hopfield networks. In: Proceedings of the International Joint Conference on Neural Networks, Vancouver, BC,
36. Buckley, J.J.: Fuzzy complex numbers. Fuzzy Sets and Systems 33, 333–345 (1989)
37. Buckley, J.J., Qu, Y.: Fuzzy complex analysis I: differentiation. Fuzzy Sets and Systems 41, 269–284 (1991)
38. Buckley, J.J.: Fuzzy complex analysis II: integration. Fuzzy Sets and Systems 49, 171–179 (1992)
39. Buckley, J.J., Qu, Y.: Solving linear and quadratic fuzzy equations. Fuzzy Sets and Systems 38, 43–59 (1990)
40. Buckley, J.J., Qu, Y.: Solving fuzzy equations: a new solution concept. Fuzzy Sets and Systems 41, 291–301 (1991)
41. Zhang, G.: Fuzzy limit theory of fuzzy complex numbers. Fuzzy Sets and Systems 46(2), 227–2352 (1992)
42. Wu, C., Qiu, J.: Some remarks for fuzzy complex analysis. Fuzzy Sets and Systems 106, 231–238 (1999)
43. Ma, S., Peng, D., Li, D.: Fuzzy complex value measure and fuzzy complex value measurable function. In: Proceedings of the Third Annual Conference on Fuzzy Information and Engineering, Haikou, China, pp. 187–192 (2008)
44. Mundici, D., Cignoli, R., D'Ottaviano, I.M.L.: Algebraic foundations of many-valued reasoning. Kluwer Academic Press (1999)
45. Hájek, P.: Fuzzy logic and arithmetical hierarchy. Fuzzy Sets and Systems 3(8), 359–363 (1995)
46. Casasnovas, J., Rosselló, F.: Scalar and fuzzy cardinalities of crisp and fuzzy multisets. International Journal of Intelligent Systems 24(6), 587–623 (2009)
47. Cintula, P.: Advances in LΠ and LΠ1/2 logics. Archives of Mathematical Logic 42, 449–468 (2003)

Coherent Conditional Probability, Fuzzy Inclusion and Default Rules

Giulianella Coletti[1], Romano Scozzafava[2], and Barbara Vantaggi[2]

[1] University of Perugia, Italy
coletti@dmi.unipg.it
[2] University "La Sapienza", Rome, Italy
{romscozz,barbara.vantaggi}@sbai.uniroma1.it

Abstract. We adopt the interpretation of fuzzy set as coherent conditional probability, and we study coherent enlargement of a probability distribution (on a random variable) and of a membership function on "fuzzy conditional events". We consider a family of fuzzy sets closed with respect to the union and the intersection, whose membership functions are ruled by Frank t-norms and t-conorms. We study the concept of degree of fuzzy inclusion by focusing in particular on inclusion of degree 1, which can be regarded as a default rule. We get also a default logic with the relevant inference rules.

1 Introduction

We refer to the interpretation of a fuzzy set E^*_φ as a pair $(E_\varphi, \mu_\varphi(x))$ with $\mu_\varphi(x) = P(E_\varphi|X = x)$, where X is a variable with range $\mathcal{C}_X$, φ any property related to X, E_φ the event "You claim that X has the property φ" and P a coherent conditional probability (see Section 3, for details, see, e.g., [5–7]).

In this context it has been proved in [5] that, under the hypothesis of logical independence between E_φ and E_ψ, the membership functions $\mu_{\varphi\vee\psi}$ and $\mu_{\varphi\wedge\psi}$ of the fuzzy sets $E^*_\varphi \cup E^*_\psi$ and $E^*_\varphi \cap E^*_\psi$ can be coherently computed as extensions of the coherent conditional probabilities μ_ψ and μ_φ related to a t-conorm and a t-norm of the Frank class. In this context logical independence is not a very strong condition: for instance E_φ and $E_{\neg\varphi}$ (which differs from $\neg E_\varphi$) are logical independent, in fact You can claim both X is φ and X is $\neg\varphi$ or claim only one of them or none.

Starting from a family of fuzzy sets $\mathcal{C} = \{E^*_{\varphi_i}\}$ related to a variable X, with E_{φ_i} logical independent, let us consider the class $< \mathcal{C} >$ obtained as closure of $\mathcal{C}$ with respect to the union and intersection of fuzzy sets ruled by a Frank t-norm and t-conorms. Our main aim is to study, given $\mathcal{C}$ and a probability distribution on X, coherent conditional probability on the "conditional fuzzy events" $A|B$, where A and B are the events related to the elements of $< \mathcal{C} >$. The possibility of defining this conditional probability assessment in the class of interest is assured by coherence.

In this paper we study the concept of "degree of inclusion" (introduced in [23]) as a fuzzy relation on the family $< \mathcal{C} >$ of fuzzy subsets E^*_π of $\mathcal{C}_X$, which should measure "how much" a given element of $< \mathcal{C} >$ is contained into another element of $< \mathcal{C} >$.

In particular, considering a set $\mathcal{D}$ of (ordered) pairs (E^*_φ, E^*_ψ) of fuzzy sets in $< \mathcal{C} >$ with degree of inclusion (of E^*_φ in E^*_ψ) equal to 1, we show that $\mathcal{D}$ can be seen as a "generalized" set of default rules.

R.R. Yager et al. (Eds.): *Soft Computing: State of the Art Theory*, STUDFUZZ 291, pp. 193–208.
DOI: 10.1007/978-3-642-34922-5_14

Moreover, we get also the corresponding inference rules of default logic (as given, for example, by D. Lehmann and M. Magidor [17]).

2 Preliminaries

We first recall the basic concepts of coherent conditional probability theory necessary to give a rigorous formulation of our interpretation of fuzzy sets, fuzzy event and fuzzy inclusion.

2.1 Coherent Conditional Probability

The approach to probability adopted here is based on *coherence* (a notion that goes back to de Finetti [10]). The starting point is conditional probability as a function of two variable ruled by a set of axioms.

Definition 1. *Let $\mathcal{E} = \mathcal{B} \times \mathcal{H}$, with $\mathcal{B}$ a Boolean algebra and $\mathcal{H} \subseteq \mathcal{B}$ an additive set (i.e., closed with respect to finite logical sums), not containing $\emptyset$. A function $P : \mathcal{E} \to [0,1]$ is a conditional probability if the following conditions hold:*

(C1) $P(H|H) = 1$, *for every* $H \in \mathcal{H}$,
(C2) for any $H \in \mathcal{H}$ *the function* $P(\cdot|H)$ *is a (finitely additive) probability on* $\mathcal{B}$,
(C3) for every $A \in \mathcal{B}$, $E \wedge H \in \mathcal{H}$,

$$P(E \wedge A|H) = P(E|H)P(A|E \wedge H).$$

We recall an easy consequence of the above axioms, i.e. the *disintegration formula* for the probability of an event $E|H$ with respect to a partition of an event H

$$P(E|H) = \sum_{k=1}^{N} P(H_k|H)P(E|H_k) \tag{1}$$

Actually, when $P_0(\cdot) = P(\cdot|\Omega)$ is strictly positive on $\mathcal{H}$, any conditional probability can be derived as a ratio by means of this unique "unconditional" probability P_0; while otherwise (when P_0 is not strictly positive on $\mathcal{H}$) to get a similar representation we need to resort to a sequence of unconditional probabilities (see [5, 2, 3]), each of them defined where the previous one is equal to zero.

The above definition of (conditional) probability is strictly based on the Boolean structure of the domains. Actually, in real problems, the logical conditions on the domain can be unrealistic: in fact, the expert (or decision maker) usually has information and interest only on a "bunch" of (conditional) events.

The concept of coherence, introduced by de Finetti [10] in probability theory, has the fundamental role to manage partial assessments of an uncertainty measure and its enlargements. In other words, coherence is a tool to check whether a function defined on an arbitrary set of (conditional) events is consistent with a probability and to make inference in the general sense, that is to extend this function to new conditional events.

Definition 2. *Given an arbitrary set $\mathcal{F} = \{E_i|H_i\}$ of conditional events, a real function P on $\mathcal{F}$ is a coherent assessment if there exists a conditional probability $P'(\cdot|\cdot)$ extending P on $\mathcal{E} = \mathcal{B}\times\mathcal{H}$, with $\mathcal{B}$ the Boolean algebra spanned by the events $\{E_i, H_i\}$ and $\mathcal{H}$ the additive set spanned by the events $\{H_i\}$.*

In the literature many characterizations of a coherent conditional probability assessment are present, we recall here one of them (see [5]), that is given in terms of solutions of finite sequences of linear systems for every finite subset of the assessment (slightly different versions are given in [2, 3, 15]). This condition is based on the notion of atoms generated by a finite subset of conditional events. Given a family of conditional events $E_1|H_1, ..., E_n|H_n$ the relevant atoms are all the possible events obtained by the following conjunctions $E_1^* \wedge H_1^* \wedge ... \wedge E_n^* \wedge H_n^*$ where E_i^* (similarly for H_i^*) stands for either E_i or $\neg E_i$.

Theorem 1. *Let $\mathcal{C}$ be an arbitrary family of conditional events and $\mathcal{F}$ any finite subfamily. Denote by $\mathcal{A}_o^{\mathcal{F}}$ the set of the relevant atoms A_r generated by $\mathcal{F}$. For a real function P on $\mathcal{C}$ the following three statements are equivalent:*

a) P is a coherent conditional probability on $\mathcal{C}$;
b) for every finite subfamily

$$\mathcal{F} = \{E_{j_1}|H_{j_1}, \ldots, E_{j_n}|H_{j_n}\} \subseteq \mathcal{C}$$

there exists a probability distribution P' defined in $A_o^{\mathcal{F}}$ such that for every $E_{j_i}|H_{j_i} \in \mathcal{F}$, $P(E_{j_i}|H_{j_i})$ is solution of the equation

$$x \cdot P'(H_{j_i}) = P'(E_{j_i} \wedge H_{j_i}),$$

and the solution is unique for at least one element of $\mathcal{F}$.
c) for every finite subfamily $\mathcal{F} \subseteq \mathcal{C}$, there exists a solution of the following systems S with unknowns $x_r = P_0(A_r)$, $A_r \in \mathcal{A}_o^{\mathcal{F}}$,

$$\begin{cases} \displaystyle\sum_{A_r \subseteq E_{j_i} \wedge H_{j_i}} x_r = P(E_{j_i}|H_{j_i}) \sum_{A_r \subseteq H_{j_i}} x_r & \left[\text{for all } E_{j_i}|H_{j_i} \in \mathcal{F}\right] , \\ \displaystyle\sum_{A_r \subseteq H_o} x_r = 1 & \end{cases}$$

where $H_o = \bigvee_{E_i|H_i \in \mathcal{F}} H_i$.

The above result states that coherence for an infinite set of conditional events can be reduced to check the coherence on any finite subset. In the finite case, the above result can be reformulated [2] (see also [5, 3]) in a way to avoid to check the coherence in any subset. In fact, the following result (in particular condition *(c)*) gives an operative tool to check coherence by solving a sequence of linear systems where the unknowns are probabilities of atoms. This result provides a procedure to prove coherence in a finite set of conditional events:

Theorem 2. *Let $\mathcal{C} = \{E_{j_1}|H_{j_1}, \ldots, E_{j_n}|H_{j_n}\}$ be an arbitrary finite family of conditional events and by $\mathcal{A}_o$ denote the set of the relevant atoms A_r. For a real function P on $\mathcal{C}$ the following three statements are equivalent:*

a) *P is a coherent conditional probability on $\mathcal{C}$;*
b) *there exists (at least) a family of probabilities $\mathcal{P} = \{P_0, ..., P_k\}$, $(k \leq n)$ each probability being defined on a suitable subset $\mathcal{A}_\alpha \subseteq \mathcal{A}_o$ (with $\mathcal{A}_0 = \mathcal{A}_o$ and $\mathcal{A}_\alpha = \{E \in \mathcal{A}_{\alpha-1} : P_{\alpha-1}(E) = 0\}$), for $\alpha = 1, ..., k$, such that for any $E_{j_i}|H_{j_i} \in \mathcal{F}$ there exists a unique P_α such that $P(E_{j_i}|H_{j_i})$ is solution of the equations*

$$x \cdot P_\beta(H_{j_i}) = P_\beta(E_{j_i} \wedge H_{j_i})$$

for every $\beta \leq \alpha$ and $P(E_{j_i}|H_{j_i})$ is the unique solution for $\beta = \alpha$;
c) *there exists a sequence of compatible systems S_β ($\beta = 0, \ldots, k$), with unknowns $x_r^\beta = P_\alpha(A_r)$, $A_r \in \mathcal{A}_\beta$ for $\beta = 0, 1, 2, \ldots, k \leq n$ $(\mathcal{A}_0 = \mathcal{A}_o^{\mathcal{F}}, \mathcal{A}_\beta = \{E \in \mathcal{A}_{\beta-1} : \sum_{A_r \subseteq E} \mathbf{x}^{\beta-1} = 0\})$,*

$$\begin{cases} \sum_{A_r \subseteq E_{j_i} \wedge H_{j_i}} x_r^\beta = P(E_{j_i}|H_{j_i}) \sum_{A_r \subseteq H_{j_i}} x_r^\beta \ \left[\text{for } E_{j_i}|H_{j_i} \in \mathcal{C} \text{ s.t. } \sum_{A_r \subseteq H_{j_i}} \mathbf{x}_r^{\beta-1} = 0\right] \\ \sum_{A_r \subseteq H_o^\beta} x^{\beta_r} = 1 \end{cases}$$

(put, for all H_{j_i}'s, $\sum_{A_r \subseteq H_{j_i}} \mathbf{x}_r^{-1} = 0$), where $H_o^o = H_o = H_1 \vee \ldots \vee H_n$, and $\mathbf{x}_r^{\beta-1}$ denotes a solution of $(S_{\beta-1})$ and H_o^β is, for $\beta \geq 1$, the logical sum of the H_{j_i}'s such that $\sum_{A_r \subseteq H_{j_i}} \mathbf{x}_r^{\beta-1} = 0$.

The two previous theorems allow to easily prove some sufficient conditions assuring the coherence on suitable sets of conditional events:

Corollary 1. *Let $\mathcal{C} = \{E|H_i\}_{i \in I}$, where $card(I)$ is arbitrary and the events H_i's are a partition of Ω. Then any function $P : \mathcal{C} \to [0,1]$ such that $P(E_i|H_i) = 0$ if $E_i \wedge H_i = \emptyset$ and $P(E_i|H_i) = 1$ if $H_i \subseteq E_i$ (and taking otherwise any value in the interval $[0,1]$) is a coherent conditional probability.*

Moreover, if the only coherent conditional probability assessment is such that $P(E|H_i) \in \{0,1\}$ for any H_i with $i \in I$, then $H_i \wedge E = \emptyset$ for every $H_i \in \mathcal{H}_o$, and $H_i \subseteq E$ for every $H_i \in \mathcal{H}_1$, where $\mathcal{H}_r = \{H_i : P(E|H_i) = r\}$, $r = 0, 1$.

We recall now that the events E_i with $i = 1, ..., n$, are *logically independent* if the cardinality of the set of relevant atoms is 2^n. Given a partition $\{H_j\}$ the events E_i with $i = 1, ..., n$, are said *logically independent with respect to* $\{H_j\}$ if, for any H_j, $E_i \wedge H_j \neq \emptyset$ for any $i = 1, ..., n$, then $\wedge_{i=1}^n E_i \wedge H_j \neq \emptyset$. Note that logical independence with respect to a partition is stronger than logical independence.

Corollary 2. *Let $\mathcal{C} = \{E_j|H_i\}_{j=1,\ldots,n; 1=i,\ldots,m}$ be a set of conditional events such that the events H_i are a partition of Ω and the events E_j are logically independent with respect to $\{H_i\}$. Let $p(\cdot)$ be a probability on the partition $H_1, ..., H_m$. Then for every function $P : \mathcal{C} \to [0,1]$ with $P(E_j|H_i) = 0$ if $E_j \wedge H_i = \emptyset$ and $P(E_j|H_i) = 1$ if $H_i \subseteq E_j$, the global assessment $\mathcal{P} = \{P(E_j|H_i), p(H_i)\}_{j=1,\ldots,n; 1=i,\ldots,m}$ is a coherent conditional probability assessment.*

Proof - Coherence follows from Theorem 3, in particular condition c). Since the events E_j's are logically independent with respect to $\{H_i\}$, it follows that any atom of the form $\bigwedge_j E_j \wedge H_i$ is different from $\emptyset$ whenever there is no E_j with $E_j \wedge H_i = \emptyset$. By considering the relevant system, the subsystem related to each H_i (without the last equation) admits solution.

Moreover, since the events H_i's are a partition, the relevant system has independent equations related to H_i. So it admits a solution which is given by a suitable vector obtained by concatenation of the solutions related to H_i. In fact the last equation of the system is satisfied, since $p(\cdot)$ on $H_1, ..., H_m$ is a probability distribution.

Remark 1. *Corollary 2 can be stated in a way to hold also by considering the assessment*

$$\mathcal{P}' = \{P(E_j|H_i), p(H_i), p(H_i|B)\}$$

(with B event of the algebra spanned by the events H_i). In fact, if $p(H_i) > 0$, for every i, $p(H_r|B)$ is univocally determined by the $p(H_i)$'s and so the whole assessment carries the same information of the assessment $\{P(E_j|H_i), p(H_i)\}$ and it is coherent only if $p(H_i|B)$ can be computed from $p(H_i)$'s. On the contrary, if $p(H_i) = 0$ for some i, the proof goes along the same line of that of Corollary 2. In fact, the equations related to the events $H_i|B$ with $p(B) = 0$ have no role in the first system but just in systems S_β ($\beta > 0$), where the variable related to some of the H_i's contained in B have positive value. Then there is no constraint for them and so the coherence of assessment $\{p(H_r|H_r \vee H_s)\}$ is sufficient to assure global coherence.

2.2 Extending Coherent Conditional Probability Assessments

Concerning coherence, we recall the following fundamental result for conditional probability (essentially due to de Finetti [10]):

Theorem 3. *Let $\mathcal{C} = \{E_i|H_i\}$ be any family of conditional events, and take an arbitrary family $\mathcal{C}' \supseteq \mathcal{C}$. Let P be an assessment on $\mathcal{C}$; then there exists a (possibly not unique) coherent extension of P to $\mathcal{C}'$ if and only if P is coherent on $\mathcal{C}$.*

In particular, if $\mathcal{C}' = \mathcal{C} \cup \{E|H\}$, then the possible coherent values $p = P(E|H)$ are all the values of a suitable closed interval $[\underline{p}, \overline{p}] \subseteq [0, 1]$, with $\underline{p} \leq \overline{p}$. If moreover $\mathcal{C}$ is finite, then it is possible to compute the bounds of the interval $\underline{p} \leq \overline{p}$ by solving a linear programming problem (see e.g. [5]).

We recall here the procedure, related to this last problem, only for the case of events $E \wedge H$ and H logically dependent on $\{E_i, H_i\}$ (i.e. they are union of some atoms generated by $\{E_i, H_i\}$), since they are the only of interest in this paper. The problem is to find the minimum and maximum value of

$$P(E|H) = \frac{P_\alpha(E \wedge H)}{P_\alpha(H)}$$

with α such that $P_\alpha(H) > 0$, for every class $\mathcal{P}$ agreeing with P (in the sense of condition *b)* of Theorem 2). Actually, the problem can be solved by adding to systems $\mathcal{S}_\alpha$

(with $\alpha \geq 0$) of condition *c)* of Theorem 2 the constraint $\sum_{C_r \subseteq H} x_r^\alpha = 0$ till the system is compatible. If for $\overline{\alpha}$ the system $\mathcal{S}_{\overline{\alpha}}$ with the above constraint has no solution, then all possible solutions of system $\mathcal{S}_{\overline{\alpha}}$ give positive probability to H. Then, the minimum and maximum coherent value for $P(E|H)$ coincides with

$$\min / \max \sum_{C_r \subseteq E \wedge H} y_r^{\overline{\alpha}}$$

under $\mathcal{S}'_{\overline{\alpha}}$ that is

$$\begin{cases} \sum\limits_{C_r \subseteq E_i \wedge H_i} y_r^{\overline{\alpha}} = P(E_i|H_i) \sum\limits_{C_r \subseteq H_i} y_r^{\overline{\alpha}} \text{ if } P_{\overline{\alpha}-1}(H_i) = 0 \\ \sum\limits_{C_r \subseteq H} y_r^{\overline{\alpha}} = 1 \\ y_r^{\overline{\alpha}} \geq 0 \qquad\qquad C_r \in \mathcal{C}_{\mathcal{F}} \cap \mathcal{A}_\alpha \end{cases}$$

Note that the solutions $\mathbf{x}_r^{\overline{\alpha}}$ and $\mathbf{y}_r^{\overline{\alpha}}$ are linked by a normalization constant, $\mathbf{x}_r^{\overline{\alpha}} = \dfrac{\mathbf{y}_r^{\overline{\alpha}}}{\sum\limits_{C_r \in \mathcal{C}_{\mathcal{F}} \cap \mathcal{A}_\alpha} \mathbf{y}_r^{\overline{\alpha}}}$.

3 Fuzzy Sets as Coherent Conditional Probabilities

We adopt the interpretation of fuzzy sets in terms of coherent conditional probabilities, introduced in [5–7]. We briefly recall here the main concepts.

3.1 Main Definition

Let X be a (not necessarily numerical) variable, with range $\mathcal{C}_X$, and, for any $x \in \mathcal{C}_X$, let us indicate by x the event $\{X = x\}$, for every $x \in \mathcal{C}_X$.

Let φ be any *property* related to the variable X .

Let us refer to the state of information of a real (or fictitious) person that will be denoted by "You". It is natural to think that You have some information about possible values of X, which allows You to refer to a suitable *membership function* of the fuzzy subset of elements of $\mathcal{C}_X$ with the property φ.

It follows that You may assign to each of these conditional events a degree of belief $P(E_\varphi|x)$, without any syntactical restriction. Corollary 1 assures in fact that any assessment $\{P(E|x)\}_{x \in X}$ is a coherent conditional probability assessment.

Moreover, this conditional probability $P(E_\varphi|x)$ is directly introduced as a function on the set of conditional events (and without assuming any given algebraic structure).

From the above considerations, it follows that the coherent conditional probability $P(E_\varphi|\cdot)$ comes out to be a natural interpretation of the membership function $\mu_\varphi(\cdot)$, according to [6] (see also [5, 7]).

Definition 3. *For any variable X with range $\mathcal{C}_X$ and a related property φ, the fuzzy subset E_φ^* of $\mathcal{C}_X$ is the pair*

$$E_\varphi^* = \{E_\varphi \, , \, \mu_{E_\varphi}\},$$

with $\mu_{E_\varphi}(x) = P(E_\varphi|x)$ for every $x \in \mathcal{C}_X$.

Note that, by the Corollary 1, a fuzzy subset E_φ^* is a crisp set when the property φ is such that, for every $x \in \mathcal{C}_X$, either $E_\varphi \wedge x = \emptyset$ or $x \subseteq E_\varphi$.

In the relevant literature there are several papers (see e.g. [16, 18, 12, 19]) containing approaches which are similar to that expounded here.

3.2 Operations

By referring to [6] we recall the operations between fuzzy subsets: under the hypothesis of logical independence between E_φ and E_ψ with respect to X (i.e. with respect the partition $\{(X = x)\}_{x \in \mathcal{C}_X}$), the binary operations of union and intersection and that of complementation can be obtained directly by using the rules of coherent conditional probability.

For this aim let us denote by $\varphi \vee \psi$, $\varphi \wedge \psi$, respectively, the properties "φ or ψ", "φ and ψ".

Define

$$E_{\varphi \vee \psi} = E_\varphi \vee E_\psi \, , \tag{2}$$

$$E_{\varphi \wedge \psi} = E_\varphi \wedge E_\psi \, . \tag{3}$$

Let us consider two fuzzy subsets E_φ^*, E_ψ^*, related to the same variable X, with the events E_φ, E_ψ logically independent with respect to X. As proved in [6], for any given x in the range of X, the assessment $P(E_\varphi \wedge E_\psi|x) = v$ is coherent if and only if takes values in the interval

$$\max\{P(E_\varphi|x) + P(E_\psi|x) - 1, 0\} \leq v \leq \min\{P(E_\varphi|x), P(E_\psi|x)\}.$$

From probability rules, given a value to $P(E_\varphi \wedge E_\psi|x)$, we get also the value of $P(E_\varphi \vee E_\psi|x)$. Then, as proved in [6] the t-norms and the t-conorms of the class of Frank ([14]) give rise to coherent assessments for the probability assessments $P(E_{\varphi \wedge \psi})$ and $P(E_{\varphi \vee \psi})$.

Then, we put

$$E_\varphi^* \cup E_\psi^* = \{E_{\varphi \vee \psi} \, , \, \mu_{\varphi \vee \psi}\} \;\; , \;\; E_\varphi^* \cap E_\psi^* = \{E_{\varphi \wedge \psi} \, , \, \mu_{\varphi \wedge \psi}\} \; ,$$

with

$$\mu_{\varphi \vee \psi}(x) = P(E_\varphi \vee E_\psi|x) \;\; , \;\; \mu_{\varphi \wedge \psi}(x) = P(E_\varphi \wedge E_\psi|x) \, .$$

Moreover, denoting by $E_{\neg\varphi}^*$ the complementary fuzzy set of E_φ^*, the relation $E_{\neg\varphi} \neq (\neg E_\varphi)$ holds, since the propositions "You *claim* $\neg\varphi$" and "You *do not claim* φ" are logically independent with respect to X. Then, while

$$E_\varphi \vee (\neg E_\varphi) = \mathcal{C}_X \, ,$$

we have instead

$$E_\varphi \vee E_{\neg\varphi} \subseteq \mathcal{C}_X ,$$

and, if we consider the union of a fuzzy subset and its complement

$$E^*_\varphi \cup E^*_{\neg\varphi} = \{E_{\varphi\vee\neg\varphi} \, , \, \mu_{\varphi\vee\neg\varphi}\}$$

we obtain in general a *fuzzy subset* of (the universe) C_X.

4 Probability of "Fuzzy Events"

For simplicity we refer to variables X with a finite codomain. First of all we notice that in this context the concept of fuzzy event, as introduced by Zadeh, is nothing else than an ordinary event of the kind E_φ = "You claim that X has the property φ". We recall that for any probability distribution on X the global assessment $\{\mu_\varphi(x), P(x)\}_{x\in\mathcal{C}_X}$ is coherent (see Corollary 2) and so coherently extendible to E_φ. According to the disintegration formula, we get only one coherent value for the probability of E_φ, that is

$$P(E_\varphi) = \sum_{x\in\mathcal{C}_X} \mu_{\varphi_i}(x)P(x),$$

which coincides with the definition proposed by Zadeh in [25].

Now let $\mathcal{C}$ be a finite family of fuzzy subsets $E^*_{\varphi i} = (E_{\varphi_i}, \mu_{\varphi_i})$ of X (with the events $\{E_{\varphi_i}\}_{\varphi_i}$ logically independent with respect to X). We denote with $< \mathcal{C} >$ its closure with respect to intersection, unions and by $\mathcal{F}_{<\mathcal{C}>}$ the sets of events E_{φ_i} related to the elements of $< \mathcal{C} >$.

It is easy to prove (see also [6]) that for every t-norm $\odot$ and t-conorm of the class of Frank, the probability assessment $\{P_\odot(E_{\varphi_i}), P_\odot(E_{\varphi_i} \wedge E_{\varphi_j})\}$ with

$$P_\odot(E_{\varphi_i}) = \sum_{x\in\mathcal{C}_X} \mu_{\varphi_i}(x)P(x)$$

$$P_\odot(E_{\varphi_i} \wedge E_{\varphi_j}) = \sum_{x\in\mathcal{C}_X} (\mu_{\varphi_i} \odot \mu_{\varphi_j})(x)P(x)$$

is coherent.

So we can extend the assessment also to the events $E_{\varphi_i} \vee E_{\varphi_j}$ and this extension is univocally determined by coherence: it satisfies the following equation

$$P_\odot(E_{\varphi_i} \vee E\varphi_j) = \sum_{x\in\mathcal{C}_X} (\mu_{\varphi_i} \oplus \mu_{\varphi_j})(x)P(x) = P_\odot(E_{\varphi_i}) + P_\odot(E_{\varphi_j}) - P_\odot(E_{\varphi_i} \wedge E_{\varphi_i}),$$

where $\oplus$ is the dual t-conorm of $\odot$.

Remark 2. *The assumption of logical independence of events E_φ, E_ψ with respect to X is necessary for all the above assertions, however this condition is not strong in this context, in fact, for example also E_φ and $E_{\neg\varphi}$ are logically independent, even if they are seemingly linked.*

The above assessment $P_{\odot}$ is a coherent conditional probability, so by using Theorem 3 it can be furthermore extended to any conditional event $A|B$ where A, B are events of the algebra $\mathcal{B}$ generated by $\{E_{\varphi_j}|(X = x), E_{\varphi_i}|(X = x), (X = x) : x \in \mathcal{C}_X\}$, with $B \neq \emptyset$. This extension is not unique in general, but for the events $A = E_{\varphi_i}$ and $B = E_{\varphi_j}$ $(i \neq j)$, with $P_{\odot}(E_{\varphi_j}) > 0$ the only coherent extension is:

$$P_{\odot}(E_{\varphi_i}|E_{\varphi_j}) = \frac{\sum_{x \in \mathcal{C}_X} (\mu_{\varphi_i} \odot \mu_{\varphi_j})(x) P(x)}{\sum_{x \in \mathcal{C}_X} \mu_{\varphi_j}(x) P(x)}. \quad (4)$$

We call the above extension of P a coherent $\odot$-extension.

When $P_{\odot}(E_{\varphi_j}) = 0$, we obtain in general a not unique extension to the events $E_{\varphi_i}|E_{\varphi_j}$.

We note that we have $P_{\odot}(E_{\varphi_j}) = 0$ if and only if $P(x) = 0$ for every x such that $\mu_{\varphi}(x) > 0$. In this case to obtain a unique extension we need to have also the conditional probability $P(\cdot|B)$, where B is the logical sum of the events x such that $P(E_{\varphi_j}|x) = 0$ (see Remark 1).

In fact, since P_{β} of condition *b)* of Theorem 2 is related to $P(x|B)$ and there is at least one event x in B such that $P(x|B) > 0$ and so

$$P_{\odot}(E_{\varphi_i}|E_{\varphi_j}) = \frac{\sum_{x \in \mathcal{C}_X} (\mu_{\varphi_i} \odot \mu_{\varphi_j})(x) P(x|B)}{\sum_{x} \mu_{\varphi_j}(x) P(x|B)}. \quad (5)$$

Remark 3. *The values $P_{\odot}(E_{\varphi_i}|E_{\varphi_j})$ through equation (5) are coherent only when the events E_{φ_i} and E_{φ_j} are logically independent with respect to X. For instance, the same formula cannot be used for obtaining the coherent extension of $P_{\odot}$ to $E_{\varphi_j}|E_{\varphi_j} \wedge E_{\varphi_i}$ (or $E_{\varphi_j}|E_{\varphi_j}$) which is necessarily 1, independently of the Frank t-norm used for computing the coherent values of $P_{\odot}(E_{\varphi_i} \wedge E_{\varphi_j})$.*

Theorem 4. *Let $\odot$ be a strict t-norm. Given a strictly positive probability distribution on X, if $P(E_{\varphi_i}|E_{\varphi_j}) = 1$, then E_{φ_i} and E_{φ_j} are not logically independent with respect to X.*

Proof - If E_{φ_i} and E_{φ_j} were logically independent with respect to X, the relevant coherent conditional probability would be

$$P_{\odot}(E_{\varphi_i}|E_{\varphi_j}) = \frac{\sum_{x \in \mathcal{C}_X} (\mu_{\varphi_i} \odot \mu_{\varphi_j})(x) P(x)}{\sum_{x \in \mathcal{C}_X} \mu_{\varphi_j}(x) P(x)}.$$

Then $P_{\odot}(E_{\varphi_i}|E_{\varphi_j}) = 1$ implies, for any $x \in \mathcal{C}_X$, the equality $\mu_{\varphi_i} \odot \mu_{\varphi_j}(x) = \mu_{\varphi_j}(x)$ that contradicts the fact that $\odot$ is a strict t-norm.

Then the two events cannot be logically independent with respect to X.

Note that by construction of the family $< \mathcal{C} >$, any pair of events E_φ, E_ψ in $\mathcal{F}_{<\mathcal{C}>}$ not logically independent is such that $E_\varphi \subseteq E_\psi$ or $E_\psi \subseteq E_\varphi$. Thus, the coherent values of the relevant (conditional) events are univocally determined and they are computed through the min t-norm (as shown in the next result).

Theorem 5. *Let $\mathcal{C} = \{E^*_{\varphi_i}\}_i$ be a family of fuzzy sets related to a variable X, with E_{φ_i} logical independent, and consider the class $< \mathcal{C} >$ obtained as closure of $\mathcal{C}$ with respect to the union, intersection of fuzzy sets. The assessment*

$$\{P_\odot(A|(X = x)), P(X = x), P_\odot(A|B) \,:\, A, B \in \mathcal{F}_{<\mathcal{C}>} \text{ with } P_\odot(B) > 0\},$$

where $\odot$ is the t-norm of the minimum, is coherent.

Proof - By construction $< \mathcal{C} >$ involves events E_{φ_i} related to $\mathcal{C}$ that are logically independent and the union and intersection of them. Then, by the above results the assessment

$$\mathcal{P}_1 = \{P(E_{\varphi_i}|(X = x)), P(X = x)\}_{x \in \mathcal{C}_X},$$

is coherent since E_{φ_i} are logically independent.

Moreover, again from the above results also $P_\odot(B|(X = x)), P_\odot(B)$ are coherent with $\mathcal{P}_1$ since the coherent values on them can be obtained through any Frank t-norm and so also through minimum.

We need to prove that also $P_\odot(A|B)$ is coherent with the previous assessment. Actually B and A are logically dependent on $E_{\varphi_i} \in \mathcal{C}$ so their coherent values have been computed through minimum t-norm and t-conorm as well as $P_\odot(A \wedge B)$, so the following cases can occur:

- A and B logically independent, then $P_\odot(A|B)$ can be computed as ratio of $P_\odot(A \wedge B)$ and $P_\odot(B)$ since the latter is positive;
- $A \subseteq B$, then $P_\odot(A \wedge B)$ is obtained through the minimum of $P_\odot(A)$ and $P_\odot(B)$ and then by the ratio we get $P_\odot(A|B)$ since $P_\odot(B)$ is positive;
- $B \subseteq A$, then $P_\odot(A \wedge B)$ is obtained through the minimum of $P_\odot(A)$ and $P_\odot(B)$ and then by the ratio we get $P_\odot(A|B) = 1$.

5 Fuzzy Inclusion

By using the above concepts and results it is immediate to understand the concept of "degree of inclusion", introduced in [23]. It is in fact seen as a fuzzy relation on the family $< \mathcal{C} >$ of fuzzy subsets E^*_π of $\mathcal{C}_X$, which should measure "how much" a given element of $< \mathcal{C} >$ is "contained" into another element of $< \mathcal{C} >$. Fuzzy inclusion (with different approaches) has been studied by many authors (see, e.g., [1, 9, 13, 20, 22, 24]).

Consider a probability distribution P on the set $\mathcal{H}$ of the events $x = \{X = x\}$. Finally consider a Frank t-norm $\odot$ and dual t-conorm $\oplus$.

Thus, since the assessment $P(\cdot|\cdot)$ defined on the following set of conditional events

$$\mathcal{C} = \{E_\varphi|x \,,\; E_\psi|x \,,\; x \,:\, x \in \mathcal{C}_X\}$$

is coherent by Corollary 2, it can be extended (preserving coherence) to any set $\mathcal{D} \supset \mathcal{C}$.

We recall from [23] the following

Definition 4. *The degree* $I(E^*_{\varphi_i}, E^*_{\varphi_j})$ *of fuzzy inclusion of the fuzzy subset* $E^*_{\varphi_i} = (E_{\varphi_i}, \mu_{\varphi_i})$ *in the fuzzy subset* $E^*_{\varphi_j} = (E_{\varphi_j}, \mu_{\varphi_j})$ *is a function*

$$I : \mathcal{F}_{<\mathcal{C}>} \times \mathcal{F}_{<\mathcal{C}>} \to [0,1]$$

with

$$I(E^*_{\varphi_i}, E^*_{\varphi_j}) = P_{\odot}(E_{\varphi_j}|E_{\varphi_i}) \, ,$$

obtained as any coherent $\odot$*-extension of* $P(\cdot|\cdot)$ *from* $\mathcal{C} \cup \mathcal{H}$ *to the conditional event* $E_{\varphi_j}|E_{\varphi_i} \in \mathcal{F}_{<\mathcal{C}>}$.

The semantic behind the above definition is the following: "the more" $E^*_{\varphi_i}$ is included in $E^*_{\varphi_j}$, "the more" if we claim the property φ_i we are willing to claim also the property φ_j.

We say that a fuzzy set $E^*_{\varphi_i}$ is included in $E^*_{\varphi_j}$ if the degree $I(E^*_{\varphi_i}, E^*_{\varphi_j})$ is maximum, that is equal to 1. Clearly, any fuzzy set $E^*_{\varphi_j}$ is included in itself if $I(E^*_{\varphi_j}, E^*_{\varphi_j}) = 1$ for every coherent extension $P_{\odot}$ of P (see Remark 3). So fuzzy inclusion is reflexive (as in the crisp case).

Moreover, the degree of inclusion of two fuzzy sets has the lowest possible value 0 when they are "disjoint" (i.e., the corresponding membership functions have disjoint supports). In fact in this case $P_{\odot}(E_{\varphi_j}|E_{\varphi_i})$ is equal to zero, for every coherent extension $P_{\odot}$ of P.

We recall the following definition of inclusion due to Zadeh [25].

Definition 5. *A fuzzy subset* E^*_{φ} *of* $\mathcal{C}_X$ *is included in a fuzzy subset* E^*_{ψ} *(in symbol* $E^*_{\varphi} \sqsubseteq E^*_{\psi}$*) iff*

$$\mu_{\varphi}(x) \leq \mu_{\psi}(x) \text{ for any } x \in \mathcal{C}_X . \tag{6}$$

In [23] the following result has been proved:

Proposition 1. *Let* X *be a variable with finite range. If two fuzzy subsets* $E^*_{\varphi} = (E_{\varphi}, \mu_{\varphi})$, $E^*_{\psi} = (E_{\psi}, \mu_{\psi})$ *of* $\mathcal{C}_X$ *are such that* $I(E^*_{\varphi}, E^*_{\psi}) = 1$ *for any probability distribution on* $\mathcal{C}_X$*, then they satisfy Zadeh's definition of inclusion.*

Conversely, given two fuzzy subsets $E^*_{\varphi} = (E_{\varphi}, \mu_{\varphi})$, $E^*_{\psi} = (E_{\psi}, \mu_{\psi})$ *of* $\mathcal{C}_X$ *and their intersection* $E^*_{\varphi} \cap E^*_{\psi} = \{E_{\varphi \wedge \psi}\, , \, \mu_{\varphi \wedge \psi}\}$*, take*

$$\mu_{\varphi \wedge \psi}(x) = \min\{\mu_{\varphi}(x), \mu_{\psi}(x)\}$$

for any x*. If* $\mu_{\varphi}(x) \leq \mu_{\psi}(x)$ *for any* $x \in \mathcal{C}_X$*, then* $I(E^*_{\varphi}, E^*_{\psi}) = 1$ *for any probability distribution on* $\mathcal{C}_X$.

In [23] there is also a discussion of the connections between fuzzy inclusion and *similarity*, whose *degree* is defined as

$$S(E^*_{\varphi}, E^*_{\psi}) = P\big((E_{\varphi} \wedge E_{\psi})|(E_{\varphi} \vee E_{\psi})\big) \, ,$$

proving, in particular, the following equality

$$(I_1 + I_2 - I_1\, I_2)\, S = I_1\, I_2\,, \tag{7}$$

where

$$I_1 = I(E^*_\varphi, E^*_\psi)\ ,\ I_2 = I(E^*_\psi, E^*_\varphi)\ ,\ S = S(E^*_\varphi, E^*_\psi)\,.$$

6 Maximum Degree of Fuzzy Inclusion

Given a conditional event $E|H$, notice that $P(E) = 1$ does not imply $P(E|H) = 1$ (as in the usual framework where it is necessary to assume $P(H) > 0$). We can take instead $P(H) = 0$ (the *conditioning* event H – which *must* be a *possible* one – may in fact have *zero probability*, since in the assignment of $P(E|H)$ we are driven only by coherence [5]). Then a probability equal to 1 can be, in our framework, updated (see e.g. [5]).

Moreover, notice that $P(E|H) = 1$ does not imply the inclusion $H \subseteq E$ (corresponding to implication, in terms of events as propositions): take in fact, e.g., an event E with $P(E) > 0$ and an event $H \supset E$ such that $P(H) = P(E)$, that is $P(\neg E \wedge H) = 0$. (In particular, if $\neg H \vee E = \Omega$, then $H \subseteq E$, so we certainly have $P(E|H) = 1$).

In most real situations, a base of knowledge is given by an arbitrary set $\mathcal{C}$ of (conditional) events, with a function P (a coherent conditional probability) that summarizes the relevant state of information.

Now, as far as fuzzy inclusion is concerned, let us consider a set $\mathcal{D}$ of (ordered) pairs (E_φ, E_ψ) of fuzzy subsets of $\mathcal{F}_\mathcal{C}$ (such that $E_\varphi|E_\psi \in \mathcal{C}$) with degree $I(E_\psi, E_\varphi)$ of fuzzy inclusion (of E_ψ in E_φ) equal to 1, i.e. with $P(E_\varphi|E_\psi) = 1$. We call any such set a MDFI–set ("Maximum Degree of Fuzzy Inclusion").

Notice that, if $I(E_\varphi, E_\psi) = 1$, then by eq. (7) the reverse degree of inclusion $I(E_\psi, E_\varphi)$ *coincides with the similarity* $S(E_\varphi, E_\psi)$.

We wonder now whether, given $\mathcal{D}$, it is possible to find further pairs of fuzzy subsets in $\mathcal{F}_\mathcal{C}$ with maximum degree of fuzzy inclusion (MDFI–pairs).

This enlarging of the given assessment to new events (maintaining the rules required to the conditional probability P, i.e. coherence) is clearly a particular way of making inference.

Even if for any coherent assessment on $\mathcal{C}$ its enlargement to a family $\mathcal{K} \supseteq \mathcal{C}$ is not (in general) unique, nevertheless for some events we can have a unique coherent extension, so giving rise to the important concept of *entailment*.

Note that also in the case of coherent $\odot$-extension (see eq. (4)) we obtain a unique value, since we choose a t-norm $\odot$.

Definition 6. *The MDFI–set $\mathcal{D} \subseteq \mathcal{F}_\mathcal{C}$ entails the pair (E_φ, E_ψ) of fuzzy sets if the only coherent value for $P(E_\psi|E_\varphi)$ is 1.*

In other words, the pair (E_φ, E_ψ) is entailed by the MDFI–set $\mathcal{D}$ if every possible *extension* of the probability assessment P on $\mathcal{C}$ such that $P(E_{\psi_i}|E_{\varphi_i}) = 1$ (i.e. $(E_{\varphi_i}, E_{\psi_i}) \in \mathcal{D}$) assigns the value 1 also to $P(E_\psi|E_\varphi)$.

In the following we denote by $\mathcal{E}$ the set of entailed MDFI–pairs, that is $(E^*_\varphi, E^*_\psi) \in \mathcal{E}$ means that the only coherent value for $P(E_\varphi|E_\psi)$ is 1 (or equivalently the degree of inclusion $I(E^*_\psi, E^*_\varphi)$ of E^*_ψ in E^*_φ is 1).

Theorem 6. *Let $\mathcal{C}$ be a set of conditional events, P a coherent probability assessment on $\mathcal{C}$, $\mathcal{D}$ a MDFI–set and $\mathcal{E}$ the set of entailed MDFI–pairs. Then the following properties hold:*

(i) $\mathcal{D}$ *entails* (E^*_ψ, E^*_ψ), *i.e.* $(E^*_\psi, E^*_\psi) \in \mathcal{E}$, *for any* $E^*_\psi \neq (\emptyset, \mu_\emptyset)$,

(ii) *if* $(E^*_\psi \sqsubseteq E^*_\varphi\,,\ E^*_\varphi \sqsubseteq E^*_\psi)$ *and* $(E^*_\delta, E^*_\psi) \in \mathcal{D}$ *then* $(E^*_\delta, E^*_\varphi) \in \mathcal{E}$,

(iii) *if* $(E^*_\psi \sqsubseteq E^*_\varphi)$ *and* $(E^*_\psi, E^*_\delta) \in \mathcal{D}$ *then* $(E^*_\varphi, E^*_\delta) \in \mathcal{E}$,

(iv) *if* $(E^*_\delta, E^*_\psi \wedge E^*_\varphi)\,,\ (E^*_\varphi, E^*_\psi) \in \mathcal{D}$ *then* $(E^*_\delta, E^*_\psi) \in \mathcal{E}$,

(v) *if* $(E^*_\varphi, E^*_\psi)\,,\ (E^*_\delta, E^*_\psi) \in \mathcal{D}$ *then* $(E^*_\delta, E^*_\psi \wedge E^*_\varphi) \in \mathcal{E}$,

(vi) *if* $(E^*_\varphi, E^*_\psi)\,,\ (E^*_\psi, E^*_\varphi)\,,\ (E^*_\delta, E^*_\psi) \in \mathcal{D}$
then $(E^*_\delta, E^*_\varphi) \in \mathcal{E}$,

(vii) *if* $(E^*_\varphi, E^*_\psi)\,,\ (E^*_\delta, E^*_\psi) \in \mathcal{D}$ *then* $(E^*_\varphi \wedge E^*_\delta, E^*_\psi) \in \mathcal{E}$,

(viii) *if* $(E^*_\delta, E^*_\psi)\,,\ (E^*_\delta, E^*_\varphi) \in \mathcal{D}$ *then* $(E^*_\delta, E^*_\psi \vee E^*_\varphi) \in \mathcal{E}$.

Proof - *(i)*, *(ii)* trivially follow from the notion of Zadeh inclusion (eq. (6)) and elementary properties of conditional probability.

(iii) Since $(E^*_\psi \sqsubseteq E^*_\varphi)$, from eq. (6) it follows that $\mu_\psi(x) \leq \mu_\varphi$ for any x, (considering the same t-norm in order to build the intersection of two fuzzy sets) we have, for any E^*_δ,

$$\mu_{\psi\wedge\delta}(x) \leq \mu_{\varphi\wedge\delta}$$

for any x and so $P(E_\varphi \wedge E_\delta) \geq P(E_\psi \wedge E_\delta)$.

Moreover, $P(E_\psi|E_\delta) = 1$ implies

$$P(E_\varphi|E_\delta) = P(E_\varphi \wedge E_\delta|E_\delta) \geq P(E_\psi \wedge E_\delta|E_\delta) = 1.$$

(iv): from $P(E_\delta|E_\psi \wedge E_\varphi) = P(E_\varphi|E_\psi) = 1$ it follows, by the disintegration formula (1) for $P(\cdot|E_\psi)$,

$$P(E_\delta|E_\psi) = P(E_\delta|E_\psi \wedge E_\varphi)P(E_\varphi|E_\psi) + P(E_\delta|E_\psi \wedge \neg E_\varphi)P(\neg E_\varphi|E_\psi) =$$

$$= P(E_\delta|E_\psi \wedge E_\varphi)P(E_\varphi|E_\psi) + P(E_\delta|E_\psi \wedge \neg E_\varphi) \cdot 0 = 1\,.$$

(v): since $1 = P(E_\varphi|E_\psi) = P(E_\delta|E_\psi)$, we have that

$$1 = P(E_\delta|E_\psi) = P(E_\delta|E_\psi \wedge E_\varphi)P(E_\varphi|E_\psi) + P(E_\delta|E_\psi \wedge \neg E_\varphi)P(\neg E_\varphi|E_\psi) =$$

$$= P(E_\delta|E_\psi \wedge E_\varphi)\,.$$

(vi): since at least one conditioning event must have positive probability $P(\cdot|E_\psi \vee E_\varphi \vee E_\delta)$, it follows from the premises that $E_\psi, E_\varphi, E_\delta$ have all positive probability; then

$$P(E_\psi \wedge E_\delta) = P(E_\psi) = P(E_\psi \wedge E_\varphi) = P(E_\varphi)\,,$$

which implies $P(\neg E_\psi \wedge E_\varphi) = 0$.

Moreover,

$$P(E_\psi \wedge E_\varphi \wedge E_\delta) = P(E_\psi)P(E_\varphi \wedge E_\delta|E_\psi) = P(E_\varphi)P(E_\psi \wedge E_\delta|E_\varphi)\,,$$

so that

$$P(E_\psi \wedge E_\varphi \wedge E_\delta) = P(E_\psi)P(E_\varphi|E_\psi)P(E_\delta|E_\psi \wedge E_\varphi) =$$

$$= P(E_\varphi)P(E_\psi|E_\varphi)P(E_\delta|E_\psi \wedge E_\varphi) = P(E_\psi) = P(E_\varphi)\,.$$

It follows

$$P(E_\varphi \wedge E_\delta) = P(E_\psi \wedge E_\varphi \wedge E_\delta) + P(\neg E_\psi \wedge E_\varphi \wedge E_\delta) = P(E_\psi \wedge E_\varphi \wedge E_\delta) = P(E_\varphi)\,,$$

and so $P(E_\delta|E_\varphi) = 1$.

(vii): since

$$1 \geq P(E_\varphi \vee E_\delta|E_\psi) = P(E_\varphi|E_\psi) + P(E_\delta|E_\psi) - P(E_\varphi \wedge E_\delta|E_\psi) = 2 - P(E_\varphi \wedge E_\delta|E_\psi)\,,$$

it follows $P(E_\varphi \wedge E_\delta|E_\psi) = 1$.

(viii): since

$$P(E_\delta|E_\psi \vee E_\varphi) = P((E_\delta \wedge E_\psi) \vee (E_\delta \wedge E_\varphi)|(E_\psi \vee E_\varphi)) = P((E_\delta \vee E_\psi)|(E_\psi \vee E_\varphi)) +$$

$$+P((E_\delta \vee E_\varphi)|(E_\psi \vee E_\varphi)) - P((E_\psi \wedge E_\varphi \wedge E_\delta)|(E_\psi \vee E_\varphi)) = P(E_\delta|E_\psi)P(E_\psi|E_\psi \vee E_\varphi) +$$
$$+P(E_\delta|E_\varphi)P(E_\varphi|E_\psi \vee E_\varphi) - P(E_\delta|E_\psi \vee E_\varphi)P(E_\psi \wedge E_\varphi|E_\psi \vee E_\varphi) = P(E_\psi|E_\psi \vee E_\varphi) +$$
$$+P(E_\varphi|E_\psi \vee E_\varphi) - P(E_\delta|E_\psi \vee E_\varphi)P(E_\psi \wedge E_\varphi|E_\psi \vee E_\varphi) \geq 1\,,$$

we get $P(E_\delta|E_\psi \vee E_\varphi) = 1$.

7 Default Rules

In a previous paper [8], given a *coherent* conditional probability P on a family $\mathcal{C}$ of conditional events, a *default rule* (denoted by $H \longmapsto E$) between events (i.e. crisp sets), has been defined as any conditional event $E|H \in \mathcal{C}$ such that $P(E|H) = 1$.

By applying Theorem 6 to crisp sets, we get that any MDFI–pair of crisp sets can be interpreted as a default rule. In fact, notices that properties *(i)–(viii)* correspond to those that, in default logic (see, e.g., [17]), are called, respectively, Reflexivity, Left Logical Equivalence, Right Weakening, Cut, Cautious Monotonicity, Equivalence, And, Or.

Thus a MDFI-set can be looked on as a set of generalized set default rules between fuzzy sets.

In particular, taking into account eq.(7) relating similarity and fuzzy inclusion, it follows that if $S(E^*_\varphi, E^*_\psi) = 1$ (i.e. E^*_φ and E^*_ψ have the maximum degree of similarity) then either inclusions $I(E^*_\varphi, E^*_\psi)$ and $I(E^*_\psi, E^*_\varphi)$ are maximal or both $I(E^*_\varphi, E^*_\psi)$ and $I(E^*_\psi, E^*_\varphi)$ are not maximal. Then, under $S(E^*_\varphi, E^*_\psi) = 1$, if $I(E^*_\varphi, E^*_\psi)$ is maximal, then necessarily the converse inclusion $I(E^*_\psi, E^*_\varphi)$ is maximal and in this case the equivalence property *(vi)* says that, $E^*_\varphi \longmapsto E^*_\delta$ implies also $E^*_\psi \longmapsto E^*_\delta$.

References

1. Burillo, P., Frago, N., Fuentes, R.: Inclusion grade and fuzzy implication operators. Fuzzy Sets and Systems 114, 417–430 (2000)
2. Coletti, G.: Coherent numerical and ordinal probabilistic assessments. IEEE Trans. on Systems, Man, and Cybernetics 24, 1747–1754 (1994)
3. Coletti, G., Scozzafava, R.: Characterization of Coherent Conditional Probabilities as a Tool for their Assessment and Extension. International Journal of Uncertainty, Fuzziness and Knowledge-Based Systems 4, 103–127 (1996)
4. Coletti, G., Scozzafava, R.: From conditional events to conditional measures: a new axiomatic approach. Annals of Mathematics and Artificial Intelligence 32, 373–392 (2001)
5. Coletti, G., Scozzafava, R.: Probabilistic logic in a coherent setting (Trends in logic n.15). Kluwer, Dordrecht (2002)
6. Coletti, G., Scozzafava, R.: Conditional Probability, Fuzzy Sets, and Possibility: a Unifying View. Fuzzy Sets and Systems 144, 227–249 (2004)
7. Coletti, G., Scozzafava, R.: Conditional Probability and Fuzzy Information. Computational Statistics & Data Analysis 51, 115–132 (2006)
8. Coletti, G., Scozzafava, R., Vantaggi, B.: Coherent conditional probability as a tool for default reasoning. In: Bouchon-Meunier, B., Foulloy, L., Yager, R.R. (eds.) Intelligent Systems for Information Processing: from Representation to Applications, pp. 191–202. Elsevier (2003)
9. De Baets, B., De Meyer, H., Naessens, H.: On rational cardinality-based inclusion measures. Fuzzy Sets and Systems 128, 169–183 (2002)
10. de Finetti, B.: Sull'impostazione assiomatica del calcolo delle probabilitá. Annali Univ. Trieste, vol. 19, pp. 3–55 (1949); Engl. transl. In: Ch.5 of Probability, Induction, Statistics. Wiley, London (1972)
11. Dubins, L.E.: Finitely Additive Conditional Probabilities, Conglomerability and Disintegration. The Annals of Probability 3, 89–99 (1975)
12. Dubois, D., Moral, S., Prade, H.: A semantics for possibility theory based on likelihoods. J. Math. Anal. Appl. 205, 359–380 (1997)
13. Fan, J., Xie, W., Pei, J.: Subsethood measure: new definitions. Fuzzy Sets and Systems 106, 201–209 (1999)
14. Frank, M.J.: On the simultaneous associativity of $F(x,y)$ and $x+y-F(x,y)$. Aequationes Math. 19, 194–226 (1979)
15. Gilio, A.: Algorithms for precise and imprecise conditional probability assessments. In: Coletti, G., Dubois, D., Scozzafava, R. (eds.) Mathematical Models for Handling Partial Knowledge in Artificial Intelligence, pp. 231–254. Plenum Press, New York (1995)

16. Hisdal, E.: Are grades of membership probabilities. Fuzzy Sets and Systems 25, 325–348 (1988)
17. Lehmann, D., Magidor, M.: What does a conditional knowledge base entail? Artificial Intelligence 55, 1–60 (1992)
18. Lawry, J.: Appropriateness measures: An uncertainty model for vague concepts. Synthese 161, 255–269 (2008)
19. Loginov, V.I.: Probability treatment of Zadeh membership functions and theris use in patter recognition. Engineering Cybernetics, 68–69 (1966)
20. Kosko, B.: Fuzziness versus probability. Int. J. General Systems 17, 211–240 (1990)
21. Krauss, P.H.: Representation of Conditional Probability Measures on Boolean Algebras. Acta. Math. Acad. Scient. Hungar. 19, 229–241 (1968)
22. Kuncheva, L.: Fuzzy rough sets: application to feature selection. Fuzzy Sets and Systems 51, 147–153 (1992)
23. Scozzafava, R., Vantaggi, B.: Fuzzy Inclusion and Similarity through Coherent Conditional Probability. Fuzzy Sets and Systems 160, 292–305 (2009)
24. Young, V.: Fuzzy subsethood. Fuzzy Sets and Systems 77, 371–384 (1996)
25. Zadeh, L.: Fuzzy sets. Information and Control 8, 338–353 (1965)

Fuzzy Answer Set Programming: An Introduction

Marjon Blondeel[1], Steven Schockaert[2], Dirk Vermeir[1], and Martine De Cock[3]

[1] Vrije Universiteit Brussel, Department of Computer Science, Pleinlaan 2, 1050 Brussel, Belgium
mblondee@vub.ac.be, dvermeir@vub.ac.be

[2] Cardiff University, School of Computer Science and Informatics, 5 The Parade, Cardiff, CF24 3AA, UK
S.Schockaert@cs.cardiff.ac.uk

[3] Universiteit Gent, Department of Applied Mathematics and Computer Science, Krijgslaan 281 (S9), 9000 Gent, Belgium
Martine.DeCock@UGent.be

Abstract. In this chapter, we present a tutorial about fuzzy answer set programming (FASP); we give a gentle introduction to its basic ideas and definitions. FASP is a combination of answer set programming and fuzzy logics which has recently been proposed. From the answer set semantics, FASP inherits the declarative nonmonotonic reasoning capabilities, while fuzzy logic adds the power to model continuous problems. FASP can be tailored towards different applications since fuzzy logics gives a great flexibility, e.g. by the possibility to use different generalizations of the classical connectives. In this chapter, we consider a rather general form of FASP programs; the connectives can in principal be interpreted by arbitrary $[0,1]^n \rightarrow [0,1]$-mappings. Despite that very general connectives are allowed, the presented framework turns out to be an intuitive extension of answer set programming.

1 Introduction

In this chapter we will present and illustrate the basic definitions of fuzzy answer set programming (FASP). In recent years a variety of approaches to FASP have been proposed (e.g. [7], [13], [15], [25]). This framework is a generalization of answer set programming using fuzzy logics, a class of logics whose semantics are based on truth values taken from the unit interval $[0,1]$ [11]. Answer set programming (ASP) [1] is a tool for modeling combinatorial search problems in a declarative way. It has its roots in logic programming and nonmonotonic reasoning.

Nonmonotonicity enables human-like reasoning; humans constantly revise their knowledge when they obtain new information. In contrast, classical logic works monotonically; when new knowledge is added, the set of conclusions that can be inferred increases. To overcome this limitation of classical logic when imitating human reasoning, several nonmonotonic logics, e.g. autoepistemic logic [17] and default logic [20], and logic programming with negation-as-failure such as ASP [9] have been proposed. In logic programming, and ASP in particular, nonmonotonicty is obtained by the negation-as-failure operator "not". The difference with classical negation $\neg$ is that $\neg a$ is true if we can derive $\neg a$, whereas not$a$ is true is we fail to derive a. Note that this means that

R.R. Yager et al. (Eds.): *Soft Computing: State of the Art Theory*, STUDFUZZ 291, pp. 209–222.
DOI: 10.1007/978-3-642-34922-5_15

ASP can deal with incomplete information; one can draw conclusions if information is absent.

The basic idea of ASP is that a search problem is translated into an ASP program, i.e. a set of rules of the form $\alpha \leftarrow \beta$. Such a rule indicates that whenever the body β holds, the head α holds as well. The expression α is a disjunction of literals and β a conjunction of extended literals. A literal is an atom or an expression of the form $\neg a$ with a an atom. An extended literal is a literal or an expression of the form $\text{not}\, l$ with l a literal. Given an ASP program, the idea is to find a minimal set of literals that can be derived from the program. A program can have several of such "answer sets" or none at all. The answer sets then correspond to the solutions of the original search problem. Let us consider a concrete example. Suppose one wants to color the vertices of a graph in either black or white but adjacent nodes must be colored differently. This search problem can be modeled by the program P:

$$
\begin{aligned}
\text{black}(X) &\leftarrow \text{not}\,\text{white}(X)\\
\text{white}(X) &\leftarrow \text{not}\,\text{black}(X)\\
&\leftarrow \text{edge}(X,Y) \wedge \text{white}(X) \wedge \text{white}(Y)\\
&\leftarrow \text{edge}(X,Y) \wedge \text{black}(X) \wedge \text{black}(Y)
\end{aligned}
$$

The first two rules express that each node should have exactly one of the two colors. The last two rules are constraints expressing that two nodes connected by an edge should have a different color. The empty head of a constraint can be thought of as always being "false". Hence, a constraint rule only holds if its body is false as well. Note that we use variables X and Y; this is to allow a compact description of the problem. By grounding the program, i.e. replacing the variables in all meaningful ways, one gets all the rules. For instance, for a graph with nodes a and b, the first rule from the program P above, gives rise to the grounded rules "$\text{black}(a) \leftarrow \text{not}\,\text{white}(a)$" and "$\text{black}(b) \leftarrow \text{not}\,\text{white}(b)$". In addition, a number of facts, rules of the form "$\text{edge}(a,b) \leftarrow$" with a and b nodes, are added to the program. The empty body of a fact can be thought of as always being "true". Hence, such a rule implies that there is an edge between node a and b, since a rule with a body that is true can only hold if the head is true as well. After grounding the program, the answer set semantics defines the solutions to the program. For instance, if there are three nodes a, b and c such that there is an edge between a and b and one between b and c, then there are two answer sets. One of these will contain the atoms $\text{black}(a)$, $\text{white}(b)$ and $\text{black}(c)$ and the other the atoms $\text{white}(a)$, $\text{black}(b)$ and $\text{white}(c)$.

Unfortunately, ASP is not suitable for expressing continuous optimization problems since it is limited to expressing problems in boolean logic. For example, suppose one wants to travel by car from one city to another in Winter. The driving time that is needed to do this depends on several factors; for instance the amount of snow, the distance and the traffic. These concepts are a matter of degree rather than boolean properties, thus we cannot directly use ASP to model this problem. One solution to this problem is to allow propositions to be true to a certain degree in $[0,1]$ and to generalize the syntax and semantics of ASP using fuzzy logics. We can then write the rule

$$\text{driving time} \leftarrow f(\text{snow}, \text{distance}, \text{traffic})$$

where "driving time", "snow" and "traffic" now have to be seen as atoms that can be assigned a degree in $[0,1]$. The function f defines how these degrees have to be combined to give the driving time. Note that it is not realistic to assume that f is a linear function. For example, if it starts to snow, not even taking into account the other factors, then the driving time will increase very fast; after that the increase of driving time due to the snow will slow down. In practice, one can use statistical information to define f. Finally, remark that FASP is used to deal with partial truth and not with uncertainty or vagueness. See [8] for a discussion on the difference between these concepts. To deal with uncertainty, among others, ASP can be extended with possibility theory (e.g. [18]) or with probability theory (e.g. [3]).

The basic idea of FASP is to model search problems with continuous domains. A continuous search problem can then be translated into a FASP program, i.e. a set of rules of the form $\alpha \leftarrow \beta$ where α and β are built from atoms, expressions of the form not$a$ with a an atom, constants and connectives that can in principal be interpreted by arbitrary $[0,1]^n \to [0,1]$-mappings. Such a rule now intuitively means that the truth degree of α must be greater or equal to the truth degree of β. Reconsider our example about the driving time. The rule

$$\text{snow} \leftarrow 0.2$$

can be used to indicate that it snows to *at least* degree 0.2 which could mean, depending on how you define the degree of snow, that the snow melts immediately when it touches the ground. This rule is thus satisfied if it snows to degree 0.9. However this attaches a higher value to "snow" than the rule actually supports. If the degree of "snow" does not depend on other atoms, it is reasonable to attach the degree 0.2 to "snow". This is in line with the idea of ASP which attempts to make as few literals true as possible to satisfy the rules of a program. Hence here we are interested in finding the lowest truth degrees that we can assign to each of the atoms, such that the rules are still satisfied. Although α and β may be built from very general connectives, FASP can model search problems entirely similar as ASP does for search problems with discrete domains.

Although it has been studied by several authors, FASP is by far not as developed as ASP. For example, very little is known about its computational complexity and few techniques are known to compute the answer sets of FASP programs. Also, many extensions proposed for ASP have not yet been considered in FASP. With the exceptions of e.g. [15], [21] and [23], most work is even restricted to FASP programs with exactly one atom in the head.

In the following section, we will give the necessary background on fuzzy logics followed by an introduction to FASP in Section 3. We present a motivating real life example in Section 4 and some remarks and open problems about FASP in Section 5.

2 Background on Fuzzy Logics

Fuzzy logics [11] form a class of logics whose semantics are based on truth degrees taken from the unit interval $[0,1]$. We will consider general fuzzy logics whose formulas are built from a set of atoms A, the truth constants in $[0,1] \cap \mathbb{Q}$ and arbitrary n-ary connectives for each $n \in \mathbb{N}$. A *fuzzy interpretation* is a mapping $I : A \to [0,1]$, also called a *fuzzy set* on A. The set of all fuzzy sets on A will be written as $\mathscr{F}(A)$. We can extend a fuzzy interpretation I as follows. Each n-ary connective f is interpreted by a function $\mathtt{f} : [0,1]^n \to [0,1]$. For instance, the n-ary connective "average" can correspond to the function $[0,1]^n \to [0,1] : (x_1,\ldots,x_n) \mapsto \frac{1}{n}\sum_{i=1}^{n} x_i$. We define $[f(\alpha_1,\ldots,\alpha_n)]_I = \mathtt{f}([\alpha_1]_I,\ldots,[\alpha_n]_I)$ for formulas α_i $(1 \leq i \leq n)$. For $c \in [0,1] \cap \mathbb{Q}$ we have $[c]_I = c$. If C is a set of formulas we say that I is a *fuzzy model* of C iff $[\alpha]_I = 1$ for all $\alpha \in C$; we write this as $I \models C$. For fuzzy interpretations $I_1, I_2 \in \mathscr{F}(A)$ we write $I_1 \leq I_2$ iff $I_1(a) \leq I_2(a)$ for all $a \in A$. If $I_1 \leq I_2$ and $I_1 \neq I_2$, we write $I_1 < I_2$. A fuzzy model I is a *minimal fuzzy model* of a set of formulas C if there does not exist a fuzzy model J of C such that $J < I$.

We will now recall some generalizations of the classical connectives. Specifically, *triangular norms* (short t-norm) are used to generalize classical conjunction. These are mappings $\mathtt{T} : [0,1]^2 \to [0,1]$ that are commutative, associative, increasing and for which it holds that $\mathtt{T}(x,1) = x$ for each $x \in [0,1]$. Disjunction can be generalized by a *triangular conorm* (short t-conorm). These are mappings $\mathtt{S} : [0,1]^2 \to [0,1]$ that are commutative, associative, increasing and for which it holds that $\mathtt{S}(x,0) = x$ for each $x \in [0,1]$. Logical implication can be generalized by an implicator, i.e. a function $\mathtt{I} : [0,1]^2 \to [0,1]$ such that $\mathtt{I}(0,0) = \mathtt{I}(0,1) = \mathtt{I}(1,1) = 1$ and $\mathtt{I}(1,0) = 0$ and $\mathtt{I}$ is decreasing in the first component and increasing in the second. Given a t-norm $\mathtt{T}$, the *residual implicator* $\mathtt{I}_\mathtt{T}$ of $\mathtt{T}$, defined as

$$\mathtt{I}_\mathtt{T}(x,y) = \sup\{\lambda \mid \lambda \in [0,1] \text{ and } \mathtt{T}(x,\lambda) \leq y\}$$

satisfies all these conditions. If $\mathtt{T}$ is a left-continuous t-norm, then for all $x,y \in [0,1]$ it holds that $x \leq y$ iff $\mathtt{I}_\mathtt{T}(x,y) = 1$ ([11]). In general, residual implicators are a good choice to generalize classical implication since they satisfy a generalization of the modus ponens rule: $\mathtt{T}(x, \mathtt{I}_\mathtt{T}(x,y)) \leq \min(x,y)$. For continuous t-norms $\mathtt{T}$ there is an even stronger property: $\mathtt{T}(x, \mathtt{I}_\mathtt{T}(x,y)) = \min(x,y)$. Consider a residual implicator $\mathtt{I}$ and a t-norm $\mathtt{T}$. The *biresiduum* of $\mathtt{I}$ and $\mathtt{T}$ is defined as $\mathtt{E}_{\mathtt{T},\mathtt{I}}(x,y) = \mathtt{T}(\mathtt{I}(x,y), \mathtt{I}(y,x))$. This function is a generalization of the logical equivalence. Note that $\mathtt{I}$ does not need to be the residual implicator of $\mathtt{T}$; for an arbitrary residual implicator $\mathtt{I}$ it holds that either $\mathtt{I}(x,y) = 1$ or $\mathtt{I}(y,x) = 1$. Finally, negation can be generalized by a negator, i.e. a function $\mathtt{N} : [0,1] \to [0,1]$ such that $\mathtt{N}$ is decreasing, $\mathtt{N}(1) = 0$ and $\mathtt{N}(0) = 1$. Every implicator $\mathtt{I}$ induces a negator $\mathtt{N}_\mathtt{I}$ defined as $\mathtt{N}_\mathtt{I}(x) = \mathtt{I}(x,0)$.

Logics whose semantics are based on (left-)continuous triangular norms form an important subclass of fuzzy logics; they generalize the classical logical connectives in a natural way. In examples we will often use the fuzzy logic based on the Łukasiewicz t-norm. For the connectives conjunction $\otimes$, disjunction $\oplus$, implication $\to$ and negation $\neg$, and a fuzzy interpretation $I \in \mathscr{F}(A)$ we then have

- $[\alpha \otimes \beta]_I = \max([\alpha]_I + [\beta]_I - 1, 0)$
- $[\alpha \oplus \beta]_I = \min([\alpha]_I + [\beta]_I, 1)$
- $[\alpha \to \beta]_I = \min(1 - [\alpha]_I + [\beta]_I, 1)$
- $[\neg\alpha]_I = 1 - [\alpha]_I$

Łukasiewicz logic is often used in applications because it preserves many nice properties from classical logic. Moreover, among the t-norm based logics, Łukasiewicz logic is the only one with a continuous residual implicator. This means that a set of formulas in Łukasiewicz logic can be seen as a set of constraints on continuous functions. This logic is also closely related to mixed integer programming. McNaughton [16] showed this in a non-constructive way and Hähnle [10] gave a concrete translation from a set of formulas in Łukasiewicz logic into a mixed integer program. Finally, Łukasiewicz logic is also very close to linear logic, see e.g. [6].

More general, truth values do not need to be values in $[0,1]$. An arbitrary complete lattice will do the trick as well. We restrict to the unit interval because it is intuitive and convenient in practice.

3 Fuzzy Answer Set Programming (FASP)

Recall the example from the introduction of the chapter. Suppose you want to travel by car from one city to another and you want to have an idea about the time needed to do this. The driving time can depend on several factors, for instance the amount of snow, the distance and the traffic. These concepts are a matter of degree rather than boolean properties, thus we cannot directly use ASP to model this problem. One solution is to allow propositions to be true to a certain degree in $[0,1]$ and to generalize the syntax and semantics of ASP using fuzzy logics. We now see "driving time", "snow" and "traffic" as atoms that can be assigned a truth value in $[0,1]$. To do this, an appropriate rescaling is needed. For instance, "snow" will have truth value 0 if there is no snow at all and it will have a truth degree $x > 0$ if there is snow, but it will be given a different value depending on how much snow falls from the sky and if it melts or not. The degree of "driving time" then depends on $f(\text{snow}, \text{distance}, \text{traffic})$ with f corresponding to a $[0,1]^3 \to [0,1]$-mapping that is increasing in each of its arguments. In practice, this function f can be defined by using statistical information. We can then write the rule

$$\text{driving time} \leftarrow f(\text{snow}, \text{distance}, \text{traffic}).$$

The syntax and semantics of FASP, as we will define below, can deal with such general functions f. A lower bound on the driving time can then be found in an answer set that corresponds to a solution of the FASP program.

3.1 Programs and Fuzzy Models

Consider a set of atoms A. An atom corresponds to a property that may have a certain truth degree in $[0,1]$, not restricted to only 1 (true) or 0 (false).

Definition 1. *A* FASP program *is a finite set of rules of the form*

$$r : g(a_1, \ldots, a_n) \leftarrow f(b_1, \ldots, b_m, \text{not}_1\, c_1, \ldots, \text{not}_k\, c_k),$$

with $a_i, b_j, c_l \in A \cup ([0,1] \cap \mathbb{Q})$ *(*$i \in \{1, \ldots, n\}$*,* $j \in \{1, \ldots, m\}$ *and* $l \in \{1, \ldots, k\}$*),* f *and* g *resp.* $(m+k)$*-ary and* n*-ary connectives. We assume* f *resp.* g *is interpreted by a function* $\mathtt{f} : [0,1]^{m+k} \to [0,1]$ *resp.* $\mathtt{g} : [0,1]^n \to [0,1]$ *such that* $\mathtt{f}$ *and* $\mathtt{g}$ *are increasing in all their arguments. We also assume that* $\leftarrow$ *is interpreted by a residual implicator and each negation-as-failure operator* not_j *corresponds to a negator* $\mathtt{N_j}$*. We refer to the rule by its label* r*.*

Note that in line with the tradition in logic programming we write a rule as $\alpha \leftarrow \beta$ where $\leftarrow$ is actually an implication $\rightarrow$. Also remark that in the definition of rules, we restrict to rational numbers to ensure that the language remains recursively enumerable. The expression $g(a_1,\ldots,a_n)$ is called the *head* r_h of rule r and $f(b_1,\ldots,b_m,$ $\text{not}_1\, c_1,\ldots,\text{not}_k\, c_k)$ is called the *body* r_b. Typically the connectives correspond to the connectives from a given fuzzy logic (see Section 2), but other choices, e.g. averaging operators, can be useful as well. A rule of the form "$0 \leftarrow a$" is usually written as "$\leftarrow a$" and a rule of the form "$a \leftarrow 1$" as "$a \leftarrow$".

If the connectives in Definition 1 are restricted to compositions of the classical connectives, i.e. conjunctions in the body and disjunctions in the head, and the truth values are restricted to 0 and 1, we obtain the same syntax as classical ASP. Note that in classical ASP, it is for example not needed to consider disjunctions in the body of rules since a rule $a \leftarrow b \vee c$ can be expressed by the two rules $a \leftarrow b$ and $a \leftarrow c$. As will become clear, for FASP this is not the case.

By using fuzzy interpretations (see Section 2), one can assign truth degrees to atoms and rules. For instance, in a FASP program, a rule

$$\text{open} \leftarrow 0.5$$

is modeled by a fuzzy interpretation I iff $I(\text{open}) \geq 0.5$.

Definition 2. *A fuzzy interpretation I of a FASP program P is an element of $\mathscr{F}(\mathscr{B}_P)$, with $\mathscr{B}_P$ the set of atoms occurring in P. A fuzzy interpretation I is called a* fuzzy model *of P iff $[r]_I = 1$ for all $r \in P$.*

If we restrict the fuzzy interpretations in Definition 2 to mappings $\mathscr{B}_P \rightarrow \{0,1\}$, we get classical interpretations of ASP programs.

Recall that we are interested in the "minimal" knowledge that can be derived from a program: from a single rule "open $\leftarrow$ 0.5", we want to derive that the truth degree of "open" is 0.5. One can use minimal fuzzy models to deal with this.

Example 1. Consider the following program P:

$$\begin{aligned} r_1 &: \text{open} &\leftarrow \text{not}\,\text{closed} \\ r_2 &: \text{closed} &\leftarrow \text{not}\,\text{open} \end{aligned}$$

We assume that "$\leftarrow$" and "not" correspond to resp. the Łukasiewicz implicator and the Łukasiewicz negator. The properties "open" and "closed" can be given a value $[0,1]$ depending on the extent, e.g. the angle, to which a door is opened resp. closed. Each combination of degrees of "open" and "closed", not necessarily meaningful, is represented by a fuzzy interpretation. The rule r_1 intuitively means that the door is open to a degree greater or equal than the extent to which the door is not closed. Rule r_2 implies the opposite property. Specifically, a fuzzy interpretation I models the program P iff

$$\begin{aligned} I(\text{open}) &\geq 1 - I(\text{closed}) \\ I(\text{closed}) &\geq 1 - I(\text{open}). \end{aligned}$$

By considering for example the rule

$$r_3 : \text{open} \leftarrow 0.5$$

we add the information that the door must be open to at least degree 0.6. The minimal fuzzy model of the program only containing rule r_3 is the fuzzy interpretation I such that $I(\text{open}) = 0.6$. As will become clear later, the fuzzy interpretation I with $I(\text{open}) = 0.6$ and $I(\text{closed}) = 0.4$ is an answer set of the program consisting of rule r_1, r_2 and r_3.

One can consider different types of programs, depending on the rules they contain. Programs without negation-as-failure are called *positive*, programs with exactly one atom in the head are called *normal* and normal programs that are positive are called *simple*. Let us discuss these programs more in detail. We start with generalizing the idea of forward chaining from ASP to simple programs.

3.2 Simple Programs and Answer Sets

For simple programs, the minimal fuzzy model exists and is unique [24]. Similar to ASP, minimal fuzzy models of simple FASP programs can be characterized by forward chaining, as illustrated below and subsequently defined more formally.

Example 2. Consider the program P:

$$\begin{aligned} a &\leftarrow 0.1 \\ b &\leftarrow 0.8 \\ c &\leftarrow a \oplus b \\ a &\leftarrow b \otimes c \end{aligned}$$

First, consider the fuzzy interpretation $I_0 : \mathscr{B}_P \rightarrow [0,1] : x \mapsto 0$ by which every atom has truth degree 0. However, I_0 is not a fuzzy model: for example the first rule imposes that the truth degree of a must be greater or equal to 0.1. Let us increase the truth values by defining $I_1 : \mathscr{B}_P \rightarrow [0,1]$. To model $a \leftarrow 0.1$ and $b \leftarrow 0.8$, we put $I_1(a) = 0.1$ and $I_1(b) = 0.8$. To model the third rule, we put $I_1(c) = \max(I_1(a) \oplus I_1(b), 0) = 0.9$. We check if I_1 models the last rule. This is not the case since $I_1(a) = 0.1$ and $I_1(b) \otimes I_1(c) = 0.7$. We need to adjust I_1 once more: we define a new fuzzy interpretation I_2 with $I_2(a) = 0.7$, $I_2(b) = I_1(b)$ and $I_2(c) = I_1(c)$ The fuzzy interpretation I_2 is the unique minimal fuzzy model of P.

Definition 3. *Consider a simple FASP program P. A fuzzy interpretation I is the* answer set *of P if it is the minimal fuzzy model of P.*

For simple programs the minimal fuzzy model, i.e. the answer set, coincides with the least fixpoint of the immediate consequence operator Π_P [7]. This operator maps fuzzy interpretations to fuzzy interpretations and is defined as

$$\Pi_P(I)(a) = \sup\{[r_b]_I \mid (a \leftarrow r_b) \in P\},$$

for $a \in \mathscr{B}_P$ and a fuzzy interpretation I. Intuitively, the minimal fuzzy model of a simple FASP program corresponds to the maximal information one can derive by forward chaining until no new knowledge can be discovered anymore.

Note that, unlike ASP, it is not always possible to compute this fixpoint in a finite number of steps. Consider for instance the program containing the single rule "$a \leftarrow$

$\frac{a+1}{2}$". It will take infinitely many steps taken by the immediate consequence operator to find the least fixpoint $I(a) = 1$ [22].

If constraints, i.e. rules in which the head is a constant, are allowed in simple programs, the least fixpoint of the immediate consequence will exist, but there may be no fuzzy model at all.

Example 3. Consider the program P:

$$\begin{aligned} a &\leftarrow 1 \\ 0 &\leftarrow a \end{aligned}$$

The least fixpoint of Π_P is the fuzzy interpretation I with $I(a) = 1$. However, P has no fuzzy model since there cannot exist a fuzzy interpretation M such that $M(a) \geq 1$ and $0 \geq M(a)$. To deal with constraints one can use the fact that a fuzzy interpretation I is an answer set of $P \cup C$, with P a simple FASP program and C a set of constraints iff I is an answer set of P and a fuzzy model of C.

3.3 Positive Programs and Answer Sets

If the heads of the rules in a positive program P can be more general formulas than only single atoms, P can have several minimal fuzzy models, or none at all. In any case, they present the minimal knowledge that can be derived from P.

Example 4. Consider the program P:

$$\begin{aligned} a \oplus b &\leftarrow 0.3 \\ a &\leftarrow b \\ b &\leftarrow 0.1 \end{aligned}$$

A minimal fuzzy model of P is the fuzzy interpretation I such that $I(a) = 0.2$ and $I(b) = 0.1$. But, for example I' such that $I'(a) = 0.15$ and $I'(b) = 0.15$ is a minimal fuzzy model as well. On the other hand the program

$$\begin{aligned} \min(a, 0.5) &\leftarrow b \\ b &\leftarrow 0.1 \oplus a \end{aligned}$$

has no (minimal) fuzzy models. Indeed, for a fuzzy interpretation I to model this program it must hold that $\min(0.1 + I(a), 1) \leq \min(I(a), 0.5)$. If $0.1 + I(a) \leq 1$, this would imply that $0.1 + I(a) < I(a)$ and if $0.1 + I(a) \geq 1$, then it would follow that $1 < 0.5$.

Definition 4. *Consider a positive FASP program P. A fuzzy interpretation I is an* answer set *of P if it is a minimal fuzzy model of P.*

Note that the immediate consequence operator cannot be used for programs with more than one atom in the heads of rules; the truth value of the body of a rule does not necessarily have an equal impact on the truth values of each of the atoms in the head.

3.4 General Programs and Answer Sets

In this section, we will generalize the definitions of ASP to arbitrary FASP programs. For FASP programs that are not positive, answer sets can no longer be defined directly in terms of minimal fuzzy models.

Example 5. Consider for example the program

$$\begin{aligned} a &\leftarrow a \\ 0 &\leftarrow \text{not}\, a \end{aligned}$$

with "not" interpreted by the Łukasiewicz negator. The only minimal fuzzy model is I such that $I(a) = 1$. However, the justification for deriving a truth value for a only depends on itself, so this fuzzy model is not in line with the intuition of forward chaining. To solve this problem, we will reduce a general FASP program to a positive FASP program.

Intuitively, we "guess" an answer set I and replace all negation-as-failure literals $\text{not}\, c$ by their fuzzy interpretation $[\text{not}\, c]_I$. For the program from Example 1

$$\begin{aligned} \text{open} &\leftarrow \text{not}\,\text{closed} \\ \text{closed} &\leftarrow \text{not}\,\text{open} \end{aligned}$$

a suitable guess would be I with $I(\text{open}) = 0.6$ and $I(\text{closed}) = 0.4$; a door is closed to the degree 0.4 if it is opened to the degree $1 - 0.4 = 0.6$. Let us now consider the same program, but we replace "not closed" and "not open" by their fuzzy interpretations under I:

$$\begin{aligned} \text{closed} &\leftarrow 0.4 \\ \text{open} &\leftarrow 0.6 \end{aligned}$$

The minimal fuzzy model of this program is exactly I. Hence, I was a stable guess and we say that it is answer set of the program.

Note that I_x with $I_x(\text{open}) = x$ and $I_x(\text{closed}) = 1 - x$ with $x \in [0, 1]$ are stable guesses as well.

Definition 5. *Consider a FASP program P and a fuzzy interpretation I. The* reduct r^I *of a rule r in P w.r.t. I is obtained by replacing all expressions of the form* $\text{not}_j\, a$ *by the fuzzy interpretation* $[\text{not}_j\, a]_I$. *The* reduct P^I *of P w.r.t. I is the set of rules*

$$P^I = \{r^I \mid r \in P\}.$$

Note that this definition generalizes the well-known Gelfond-Lifschitz transformation [9], used to transform general ASP programs to positive ASP programs.

Formally, we have the following definition.

Definition 6. *Consider a FASP program P and a fuzzy interpretation I. I is called an* answer set *of P iff I is an answer set of* P^I.

A FASP program can have several answer sets as in Example 5, or none at all, as in Example 6.

Example 6. Consider the program P consisting of the one rule

$$p \leftarrow \text{not}\, p$$

with "not" interpreted by the negator $\mathtt{N} : [0,1] \to [0,1]$ with $\mathtt{N}(x) = 0$ if $x > 0$ and $\mathtt{N}(0) = 1$. For each fuzzy interpretation I with $I(p) > 0$ we have that P^I is is the positive program consisting of the rule

$$p \leftarrow 0.$$

The unique minimal model of P^I is J with $J(p) = 0$, hence our original guess I is clearly not a minimal model of P^I. If, on the other hand, we start with this fuzzy interpretation $J(p) = 0$, then we obtain for P^J the rule

$$p \leftarrow 1.$$

J is not a fuzzy model of P^J, let alone a minimal fuzzy model. We conclude that P has no answer sets.

However, if a different negator is used, this program can have an answer set for instance if not is interpreted by the Łukasiewicz negator. For a guess $M(p) = x$ with $x \in [0,1]$, we now obtain for P^M the rule

$$p \leftarrow 1 - x.$$

Hence, M is the minimal fuzzy model of P^M if $x = 1 - x$ or $x = 0.5$.

By Definition 6 it follows that an answer set of a FASP program P is also a fuzzy model of P. One can even prove that it must be a minimal fuzzy model of P [12]. The converse however does not hold.

Example 7. Recall the program from Example 5

$$\begin{aligned} a &\leftarrow a \\ 0 &\leftarrow \text{not}\, a \end{aligned}$$

with "not" interpreted by the Łukasiewicz negator. The only minimal fuzzy model is the fuzzy interpretation I such that $I(a) = 1$. However it is not an answer set since I is not a minimal fuzzy model of

$$\begin{aligned} a &\leftarrow a \\ 0 &\leftarrow 0 \end{aligned}$$

4 Motivating Example

Forest fires cause massive loss of vegetation and animal life. If a fire is detected on time, suppression units are able to reach the fire in its initial stages which is important to avoid huge losses. Moreover suppression costs will be considerably reduced. Wireless sensor networks can be effectively used for this purpose [26]. These networks consist of a number of devices that can sense their environment and communicate wirelessly. We will use FASP to determine, given measurements made by the sensors about the temperature, if there are sensors that are not working optimally and if so, within what range we can reasonably assume the temperature to be.

Suppose we have n sensors. By assuming an appropriate linear rescaling, we can see temperature as a value in $[0,1]$. Although we will not be able to derive an exact temperature, we will try to find a subinterval of $[0,1]$ in which we may assume the temperature to be. More specifically, for each sensor $i \in \{1,\ldots,n\}$, we denote the upper bound on the exact temperature at its location as t_i^{up} and the lower bound as t_i^{low}. The measured temperature is t_i'. The sensor network defines a weighted graph G as follows. The vertices are the sensors and there is an edge with weight $w_{ij} \in [0,1]$ between the vertices corresponding to sensor i and sensor j. The value w_{ij} is such that we can reasonably assume, based on the locations of sensors i and j that the temperature difference between these locations must be less than $1 - w_{ij}$. Finally, we suppose that b_i represents the error on the temperature measured by sensor i.

We can write the following program P. For $i, j \in \{1,\ldots,n\}$ we have the rules

1. $t_i^{low} \leftarrow t_i' \otimes \text{not}\, b_i$
2. $t_i^{up} \leftarrow t_i' \oplus b_i$
3. $b_i \oplus b_j \leftarrow (t_i' \nleftrightarrow t_j') \otimes w_{ij}$

where we use the connectives from Łukasiewicz logic and assume that the negation-as-failure operator is interpreted by the Łukasiewicz negation. The constants $t_i' \nleftrightarrow t_j'$, which define the degree to which t_i' and t_j' are different are defined as $1 - (t_i' \leftrightarrow t_j')$ with $\leftrightarrow$ the Łukasiewicz biresiduum.

Rules 1 and 2 define the relationship between the actual and the measured temperature. For a fuzzy interpretation I to model these rules, it most hold for each sensor i that $I(t_i') - I(b_i) \leq I(t_i^{low})$ and $I(t_i') + I(b_i) \leq I(t_i^{up})$. An answer set I is such that $I(t_i^{up})$ and $I(t_i^{low})$ will be chosen minimal. Rule 3 imposes that if the difference between t_i' and t_j' is too large with respect to $1 - w_{ij}$, then there must be something wrong with sensors i and/or j. The semantics of FASP makes sure that the "total error" is distributed over sensor i and j in a minimal way.

Consider as a concrete example a forest with three sensors. Suppose we have $t_1' = 0.4$, $t_2' = 0.9$ and $t_3' = 0.5$ and $w_{1,2} = 0.8$, $w_{1,3} = 0.8$ and $w_{2,3} = 0.8$. Hence, we have $I(t_1' \nleftrightarrow t_2') = 0.5$, $I(t_1' \nleftrightarrow t_3') = 0.1$ and $I(t_2' \nleftrightarrow t_3') = 0.4$.

For I to be a fuzzy model of rule 3, it most hold, for each sensor i that

$$\max(I(t_i' \nleftrightarrow t_j') + I(w_{ij}) - 1, 0) \leq I(b_i) + I(b_j),$$

or more specifically,

1. $0.3 \leq I(b_1) + I(b_2)$
2. $0 \leq I(b_1) + I(b_3)$
3. $0.2 \leq I(b_2) + I(b_3)$

A possible "guess" for an answer set I could be such that $I(b_1) = 0.29$, $I(b_2) = 0.01$ and $I(b_3) = 0.19$. From rules 1 and 2, we obtain that for I to be an answer set of P, i.e. a minimal fuzzy model of P^I, we must have $I(t_1^{low}) = 0.11$ and $I(t_1^{up}) = 0.69$, $I(t_2^{low}) = 0.89$ and $I(t_2^{up}) = 0.91$, $I(t_3^{low}) = 0.31$ and $I(t_3^{up}) = 0.69$.

Another possibility is a fuzzy interpretation J with $J(b_1) = 0.15$, $J(b_2) = 0.15$ and $J(b_3) = 0.05$.

The answer sets provide us with all possible ways in which the "total error" can be "divided" over the sensors. For each such setting, we also obtain an upper and lower bound on the actual temperature at the locations of the sensors.

5 Some Remarks about FASP

In ASP, there are two types of negation: negation-as-failure and strong negation. When an atom a and its negation $\neg a$ both appear in the head of rules, there is the possibility of inconsistency. In a fuzzy context, the classical definition of consistency must be modified since a literal a and its negation $\neg a$ can be both true in a consistent way. One solution could be to define fuzzy interpretations of a program P as elements in $\mathscr{L}_P$, i.e. the set of literals in P and to add the rules $1 \leftarrow a \oplus \neg a$ for each $a \in \mathscr{L}_P$. In [25], degrees of consistency of fuzzy interpretations are discussed.

Logic programming, which contains ASP as a special case, has had a significant impact on the development of nonmonotonic logics and vice versa [2]. It is closely related to e.g. autoepistemic, equilibrium and default logic. Equilibrium logic [19] is one of the most general approaches to ASP. Programs, seen as sets of formulas in equilibrium logic, can be arbitrary propositional theories without restrictions on where the two types of negation may occur. When restricting to the syntax of ASP, there is a one-to-one correspondence between the equilibrium models of a program and its answer sets. This result is generalized to FASP in [21] by introducing a fuzzy equilibrium logic. Even when very general constructs such as in this chapter are allowed, the answer sets of a FASP program correspond to its fuzzy equilibrium models. Furthermore, a reduction from the problem of finding fuzzy equilibrium models to the problem of solving a particular bilevel mixed integer program is proposed, allowing to implement reasoners already existing for these types of problems for FASP. By using the complexity of fuzzy equilibrium logic, it was also shown that computational complexity of FASP is equal to that of ASP in the general case. This means that in general, adding fuzziness to ASP, does not imply an increase in the computational complexity. In [4] some results about the computational complexity of FASP with Łukasiewicz semantics are presented. For simple programs a correspondence to an open problem was shown, which indicates that setting the complexity may not be easy. However, there is P-membership for several interesting subclasses. The correspondence between autoepistemic logic and ASP [14] was generalized in [5].

As was illustrated by examples, some FASP programs have no solutions. In practice however, it might be suitable to opt for an imperfect solution. One strategy is to add weights to the rules in a FASP program: rules do not have to be satisfied to degree 1. In [12], aggregated FASP is proposed. The idea is not to immediately state the extent to which each rule should be satisfied, but to let an aggregator function determine an overall score of suitability of a solution. Contrary to the case of classical ASP, there is not yet an efficient FASP solver, even without considering aggregators. A well-known technique for ASP consists of translating a program to a propositional theory whose models correspond to the answer sets of the program. In [12] the first steps towards an efficient FASP solver are taken by generalizing the ideas of translating an ASP program to a SAT instance. A FASP solver could then use existing techniques for solving fuzzy satisfiability problems such as mixed integer programming. The completion of a FASP program is introduced and it was shown that in case the program has no cyclic dependencies, which induce so called loops, the models of this completion are the answer sets of the original program. For programs that contain loops, a reduction to fuzzy SAT is proposed using a generalization of the notion of loop formulas in ASP to FASP [12]. One of the most important issues that still need to be tackled for building a good solver is optimizing the grounding of FASP programs.

6 Conclusions

We presented an introduction to fuzzy answer set programming, a recently developed framework that is suitable for modeling search problems with continuous domains. The syntax and semantics make FASP highly configurable and applicable in different domains. However, a lot of topics still need to be investigated, FASP is by far not as developed as ASP. Little is known about the computational complexity of specific fragments and there are almost no techniques available to actually calculate answer sets.

References

1. Baral, C.: Knowledge Representation, Reasoning and Declarative Problem Solving. Cambridge University Press (2003)
2. Baral, C., Gelfond, M.: Logic programming and knowledge representation. Journal of Logic Programming 19, 73–148 (1994)
3. Baral, C., Gelfond, M., Rushton, J.: Probabilistic reasoning in computer science. In: 7th International Conference on Logic Programming and Nonmonotonic Reasoning, pp. 21–33 (2004)
4. Blondeel, M., Schockaert, S., De Cock, M., Vermeir, D.: Complexity of fuzzy answer set programming under łukasiewicz semantics: First results. Poster Proceedings of the Fifth International Conference on Scalable Uncertainty Management, pp. 7–12 (2011)
5. Blondeel, M., Schockaert, S., De Cock, M., Vermeir, D.: Fuzzy Autoepistemic Logic: Reflecting about Knowledge of Truth Degrees. In: Liu, W. (ed.) ECSQARU 2011. LNCS, vol. 6717, pp. 616–627. Springer, Heidelberg (2011)
6. Ciabattoni, A., Luchi, D.: Two Connections Between Linear Logic and łukasiewicz Logics. In: Gottlob, G., Leitsch, A., Mundici, D. (eds.) KGC 1997. LNCS, vol. 1289, pp. 128–139. Springer, Heidelberg (1997)
7. Viegas Damásio, C., Moniz Pereira, L.: Antitonic Logic Programs. In: Eiter, T., Faber, W., Truszczyński, M. (eds.) LPNMR 2001. LNCS (LNAI), vol. 2173, pp. 379–392. Springer, Heidelberg (2001)
8. Dubois, D., Prade, H.: Possibility theory, probability theory and multiple-valued logics: a clarification. Annals of Mathematics and Artificial Intelligence 32(1-4), 35–66 (2001)
9. Gelfond, M., Lifschitz, V.: The stable model semantics for logic programming. In: Proceedings of the Fifth International Conference and Symposium on Logic Programming, pp. 1070–1080 (1988)
10. Hähnle, R.: Proof theory of many-valued logic - linear optimization - logic design: connections and interactions. Soft Computing 1, 107–119 (1997)

11. Hájek, P.: Metamathematics of Fuzzy Logic. Trends in Logic (1998)
12. Janssen, J.: Foundations of fuzzy answer set programming. PhD thesis, Ghent University and Vrije Universiteit Brussel (2011)
13. Janssen, J., Schockaert, S., Vermeir, D., De Cock, M.: General Fuzzy Answer Set Programs. In: Di Gesù, V., Pal, S.K., Petrosino, A. (eds.) WILF 2009. LNCS, vol. 5571, pp. 352–359. Springer, Heidelberg (2009)
14. Lifschitz, V., Schwarz, G.: Extended logic programs as autoepistemic theories. In: Proceedings of the Second International Workshop on Logic Programming and Nonmonotonic Reasoning, pp. 101–114 (1993)
15. Lukasiewicz, T., Straccia, U.: Tightly Integrated Fuzzy Description Logic Programs Under the Answer Set Semantics for the Semantic Web. In: Marchiori, M., Pan, J.Z., de Sainte Marie, C. (eds.) RR 2007. LNCS, vol. 4524, pp. 289–298. Springer, Heidelberg (2007)
16. McNaughton, R.: A theorem about infinite-valued sentential logic. The Journal of Symbolic Logic 16(1) (1951)
17. Moore, R.: Semantical considerations on nonmonotonic logic. In: Proceedings of the Eight International Joint Conference on Artificial Intelligence, pp. 272–279 (1983)
18. Nicolas, P., Garcia, L., Stéphan, I., Lefèvre, C.: Possibilistic uncertainty handling for answer set programming. Annals of Mathematics and Artificial Intelligence 47(1-2), 139–181 (2006)
19. Pearce, D.: A New Logical Characterisation of Stable Models and Answer Sets. In: Dix, J., Przymusinski, T.C., Moniz Pereira, L. (eds.) NMELP 1996. LNCS, vol. 1216, pp. 57–70. Springer, Heidelberg (1997)
20. Reiter, R.: A logic for default reasoning. Artificial Intelligence 13(1-2), 81–132 (1980)
21. Schockaert, S., Janssen, J., Vermeir, D., De Cock, M.: Answer Sets in a Fuzzy Equilibrium Logic. In: Polleres, A., Swift, T. (eds.) RR 2009. LNCS, vol. 5837, pp. 135–149. Springer, Heidelberg (2009)
22. Straccia, U.: Annotated answer set programming. In: Proceedings of the 11th International Conference on Information Processing and Management of Uncertainty in Knowledge-Based Systems (2006)
23. Straccia, U., Ojeda-Aciego, M., Damásio, C.V.: On fixed-points of multi-valued functions on complete lattices and their application to generalized logic programs. SIAM Journal on Computing 5, 1881–1911 (2009)
24. Van Nieuwenborgh, D.: Preferences in Answer Set Programming. PhD thesis, Vrije Universiteit Brussel (2005)
25. Van Nieuwenborgh, D., De Cock, M., Vermeir, D.: An introduction to fuzzy answer set programming. Annals of Mathematics and Artificial Intelligence 50(3-4), 363–388 (2007)
26. Yu, L., Wang, N., Meng, X.: Real-time forest fire detection with wireless sensor networks. In: Proceedings of the Wireless Communication Networking and Mobile Computing International Conference, pp. 1214–1217 (2005)

Applied Research in the Field of Automation of Learning and Knowledge Control

Shahnaz N. Shahbazova

Department of Information Technology and Programming
Azerbaijan Technical University
25 H. Cavid Ave., Baku, AZ1073 Azerbaijan
shahbazova@gmail.com

Abstract. This paper presents the results of research devoted to the implementation of an intelligent information system for learning and control of knowledge. The system is developed in order to create an effective environment capable of providing high-quality training functions with minimal involvement of the teacher, and to ensure adequate control of learning processes of individuals. The basic principles of the presented research are methods of analysis and algorithmic behavior of the teacher delivering the training and control of knowledge. The system is equipped with multiple solutions to a number of issues: organizing information material, formalizing the meaning of question-answer pairs in different circumstances, and accounting subjective opinions of experts.

Keywords: automated educational system, intelligent system, expert system, teacher's behavior, learning format, control knowledge, subjective expert's opinion, artificial neural networks, fuzzy logic, fuzzy rules, and linguistic variables.

1 Introduction

This paper presents the results of research activities related to the application of modern mathematical technologies, as well as methods of analytical and information modeling, to the development of an integrated learning system. The system provides students with the ability to study under the control of intelligent automated models and methods. Provided solutions, which are the fundamental elements presented here, have been evaluated to determine practical effectiveness of using the intelligent system in the learning process. Developed and implemented models and methods are capable of taking on intellectually complex tasks and carry out functions related to teaching and controlling knowledge procedures with minimal involvement of the teacher and the main educational institutions.

This article describes the solution to a knowledge control process, which allows for a whole new way of determining the volume of knowledge provided to and absorbed by a student. We study the algorithm of managing knowledge based on selection of questions corresponding to the current level of student's knowledge, as well as on determining the basic misconceptions and knowledge gaps.

R.R. Yager et al. (Eds.): *Soft Computing: State of the Art Theory*, STUDFUZZ 291, pp. 223–240.
DOI: 10.1007/978-3-642-34922-5_16

The developed decision-making algorithm is potentially able to determine the knowledge of students based on a minimum possible number of questions with accuracy comparable to traditional methods of examination carried out by teachers. The developed decision-making block uses fuzzy neural network technologies in combination with a knowledge base. It brings a new level of quality to decision-making processes. Its flexibility and transparency allows for easy modifications so it can work with practically with any educational material.

The proposed system is equipped with methods suitable for visualization of the outcome of the learning process, and building a "map of student" that allows a teacher to estimate, in a matter of seconds, the volume and quality of academic material learned by a student.

The methodology applied to the development of software package called Intelligent Information Systems Learning and Control Knowledge Student (IISLCKS) has been proposed. Multiple aspects of the system are analyzed and studied in [13, 14].

2 The Principles of Presentation Information Resources in System and the Functioning the Block of Control of the Educational Materials

At the initial stage of designing a complex system, one should plan for analysis of curriculum that ought to be split into separate educational materials (EMs). As the EMs, we understand the objects, phenomena, concepts and methods relevant to a specific domain of science and selected as topics that need to be covered in the process of education, Figure 1.

The set of EMs can be represented as a tree – a directed graph of the course content reflecting its hierarchical structure. Nodes (vertices) of a graph are single EMs, while the edges represent hierarchical relationships between them.

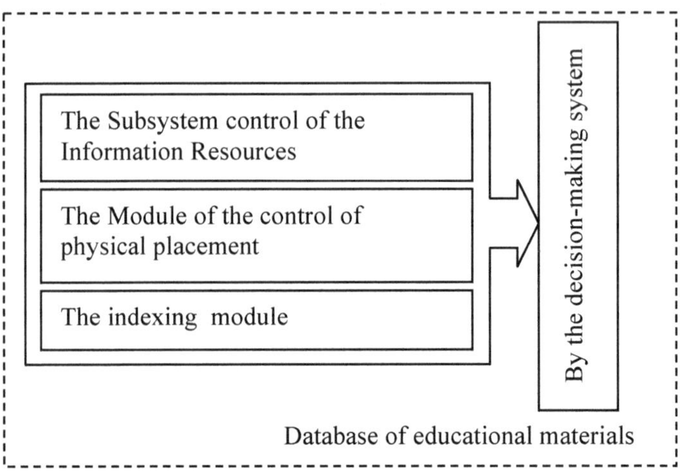

Fig. 1. The general model of a database of educational materials

The formation of a graph requires following the rules for constructing a hierarchical tree structure. For example, the presence of individual nodes that are not associated with higher-level nodes in the hierarchy of the EMs is unacceptable except for the root. At the same time, the group of EM nodes at any level should be implemented using a general principle of building trees.

In parallel with the construction of the graph, there is a need to compile a table of the EMs. The table, that requires the EMs' names, is similar to the table of content of the educational materials split (pre-processed) into chapters, sections, and paragraphs. However, the construction of the graph of educational material does not require, as opposed to chapters of contents, taking care of the sequence of EM statements. It is important to display a hierarchical structure of educational material.

After structuring and selecting the content of educational material, it is necessary to formulate the requirements regarding desired levels of presentation (visibility), easiness of learning, and automation (for example, for practical and laboratory classes). At the same time, for every EM it is necessary to generate two items: 1) the required level of initial knowledge that a student should have as a result of previous learning activities in other disciplines or subjects; and 2) the level that needs to be reached when the learning of a specific EM is successful. The educational material, represented a directed graph, is a set of interconnected various educational courses or individual topics in one academic discipline, Figure 2.

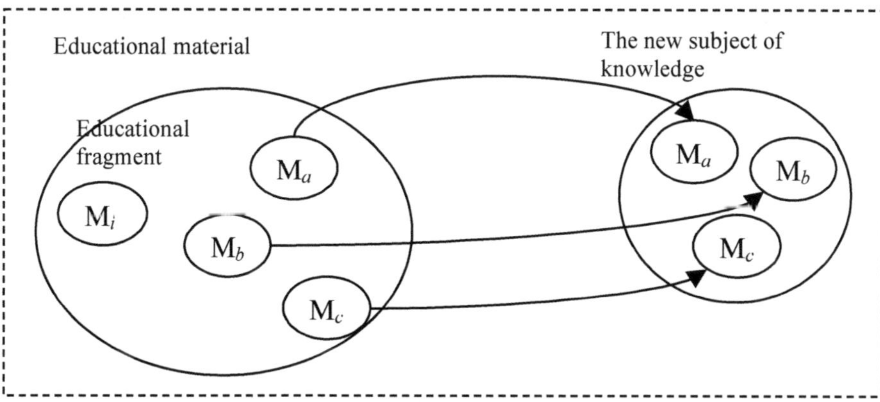

Fig. 2. Union of the set of educational fragments in the volume of knowledge eliminate some misconceptions

The proposed model of representing educational material allows to:

- define clearly the content of educational material and learning objectives;
- present content in a visual foreseeable manner;
- attract experts to discuss the completeness of the content and objectives at an early stage of design;
- provide a clear continuity of academic disciplines;
- determine the composition of educational complex;

- generate complete representation of the content of educational material, both for developers and users of the complex (teachers and students);
- generate requirements for the type, number and sequence of exercises for comprehension and consolidation of the theoretical material.

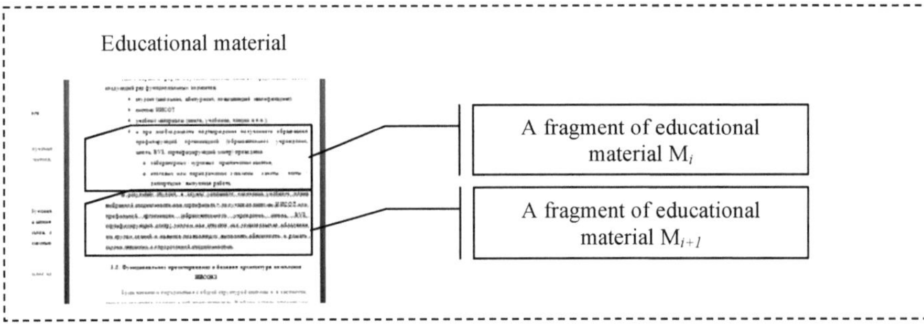

Fig. 3. The breakdown of educational materials to the minimum meaningful fragments

However, the model of the content of educational material does not contain answers to questions related to EMs, the sequence in which the EMs should be studied, as well as logical connection between them. These issues are considered during development of educational material.

The part of the system that controls the students' knowledge is based on the formal representation of the educational process. In some areas of science and technology, the educational material is divided into specific logically interconnected groups of courses, Figure 3. A set of educational materials (Ω) consists of subsets of educational courses (ω), and a complex network of intersected subsets, Figure 4.

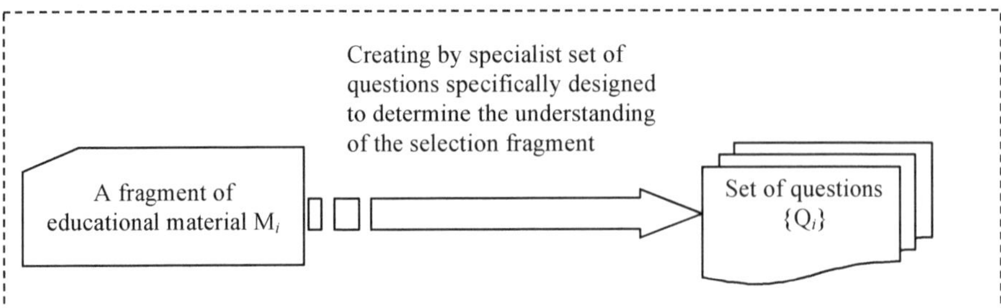

Fig. 4. Compilation of a set of questions determining understanding of a fragment of EM

An educational course (ω) consists of educational elements, which can be described as the head of the relevant courses, Figure 5.

The knowledge control system is based on question-answer objects representing links between the three components – information materials, the modulus of knowledge controls, and a student. The concept of question-answers (A) is illustrated in Figure 6. Even if questions are not so critical for the traditional learning processes, they play an important role in the proposed system.

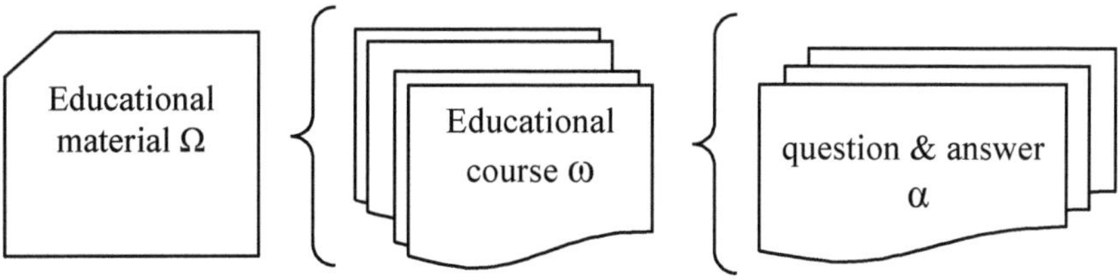

Fig. 5. The structure of educational resources

Wrong answers are pointers to the relevant sections of the knowledge. An important feature of the developed system for knowledge control is its flexibility in evaluation of incorrect answers. Incorrect answers are used to determine the student's level of error. For example, if the "value" of a correct answer is 100 points, then for incorrect answers give the student a number of points reflecting the error of a classification process – a distance from the correct answer.

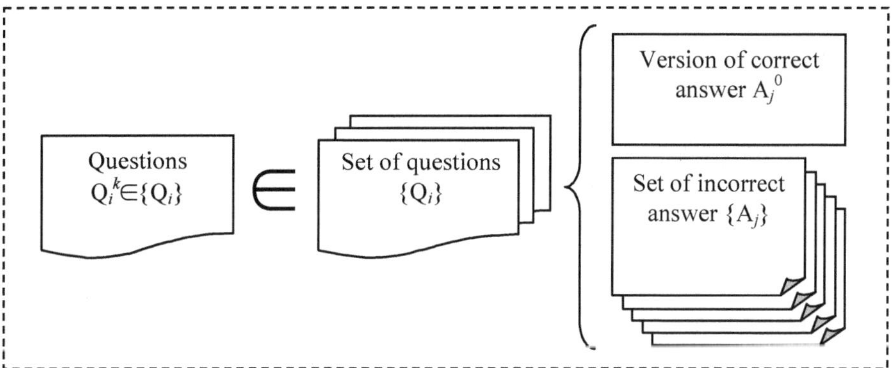

Fig. 6. Incorrect answers defining lack of understanding of educational material

Each question asked by the knowledge control system is associated with a group of answers, each of which is an interpretation of a certain level of knowledge or misconception. If the answer does not require a detailed analysis, the incorrect answers are some misconceptions in student substance. Each incorrect variant of the answer has been created with the aim of identifying gaps in the student's knowledge in a particular subject that is being studied, Figure 7.

Psychological characteristics of students do not play a significant role in the educational process. However, in the traditional learning process carried out by the teacher, the psychological factor is important and in most cases is taken into account. Implementations of modern techniques for distance learning allow for training on an individual basis and making psychological portraits of students. It is an important component of the educational process when a teacher is becoming increasingly

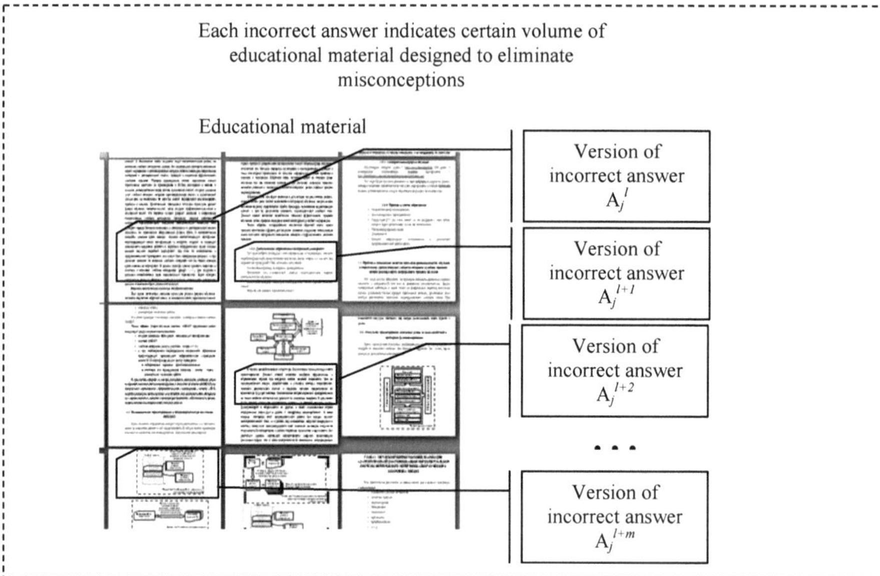

Fig. 7. Communication incorrect version of answers with relevant educational courses

remote to a student. Consideration of personal characteristics of students allows for simple indicators of successful testing, as well as recognizing person's advantages and disadvantages. All this leads to the construction of more accurate individual training plans.

3 The Model Mastering of Teaching Material and Methodology for Determining the Level of Knowledge

One of the important elements of the presented system is the automated task to create a fully individual educational environment. As is the case with other intellectual tasks let us take a real person – a teacher – and imitate her behavior via an algorithm. In this case, we look at a simple behavior, laconic and expedient enough, so it does not require complicated or too simplified implementation in an automated environment.

Let us suppose that during the course of teaching of a given student, her teacher notices a lack of knowledge of the previously covered material, or sees instability in the monitored results of knowledge absorption reflecting the lack of learning or understanding. It is logical that in such circumstances a teacher asks a student to return to the previous academic subjects in order to eliminate ambiguities and gaps in the student's knowledge.

In the streaming form of education, this approach is not acceptable. We see a lower score during evaluation of students' knowledge. In most cases, this is the consequence

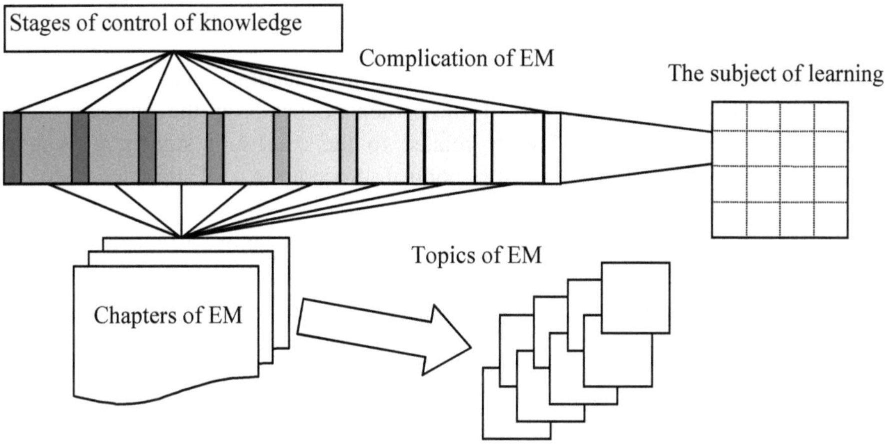

Fig. 8. Block diagram of the educational process

of not absorbing necessary amounts of educational material due to the weak capacity or skipping some classes. In order to correct this, the student is forced to improve independently. A set of educational topics that needs to be absorbed is determined. The absence or lack of knowledge of these topics will cause a significant deterioration of the ability to understand and assimilate all further material.

This procedure requires from students to possess a sufficiently high level of motivation and discipline that is not available to everyone. At the same time, if this problem is not property addressed, it leads to accumulation of the educational material required for reconsideration. Therefore, the occurrence of such situation in practice when we see a large group of students who consistently show poor results, i.e., students with the evaluations scores "bad" or "satisfactory", should be alarming. In educational institutions such students rarely graduate, and represent rather poorly educated citizens with nebulous prospects of finding jobs in their field and be useful to society.

The construction of the model means composing the matrix of regular and related EMs, the sequence of EMs to be studied, and the graph of logical connections between EMs. The process of creating of the model is carried out in four stages: 1) determining the priority of EMs; 2) relating priorities, constructing the sequences of studies in the form of lists of EMs; 3) forming logical relationships between EMs; and 4) constructing a graph of logical connections. The first and third stages are informal and are executed based on the analysis of educational materials provided by experts based on the experience gained during development of educational courses.

Thus, the EMs are based on the following elements: priorities for the EMs; logical relationships among EMs; and EM study sequences.

The process of completing of an EM plan should include analysis of texts from the study material for all the EMs, if such texts exist. The analysis of the content of educational material allows a more objective identification of prioirty relations and logical connections between the EMs.

The finalizing and improving of the model of educational materials include a sequence of manual steps. Many of these steps depend on the success of the educational process, and are directly related to the results of student knowledge evaluation processes obtained during control procedures. All this leads to the indiviually contrcuted curricula.

An importnat element of the learning system is its ability to make decisions regarding the level of difficulty of questions which shouldbe be asked to students. This should be preformed based on the result of answerings previous questions. The solution to this problem depends on many parameters, most of which are unknown to the system. A fairly accurate answer can be found with the help of mathematical apparatus of fuzzy logic.

The analysis of current situation depends on the following:

- questions correctly answered by a student;
- questions answered incorrectly;
- preliminary analysis of student's abilities;
- the number of correct answers to questions, coupled with their difficulty with respect to errors.

This list reflects the real computational tasks. A decision-making process is carried out in order to select questions, which according to the program, correspond to student's abilities. An incorrect answer triggers a re-evaluation process of the data about the student, and the next time she will be given less difficult questions. In the case of a correct answer, the program asks questions with progressive difficulty. This decision-making method allows an individual to make a progress during the learning process. Furthermore, it gives the most accurate evaluation of the student abilities.

At the end of the evaluation process, when a student and teacher want to sum up the results of the educational session, the number of correct answers and their complexity is available to the analysis program. The program updates the relevant database records of the student, and then begins the process of analysis that aims at providing updated and correct information about the student. This information can include: the current level of intelligence of the student; the comparison with the previous results of analysis of responses to which the student answered incorrectly, the visualization of the correct answers with commentary, as well as comments provided by the teacher while entering question into the database.

The importance of evaluation of the executed tests could be adjusted by the program and/or by the teacher. This approach allows for performing individual pre-tests and tests at different levels of difficulty.

As it has been previously stated due to the large number of external parameters a decision-making process is done with the help of the mathematical apparatus of fuzzy logic. The responsible subsystem also includes conducting tests that satisfy the following requirements:

1. protecting answers from unauthorized access;
2. preventing a student from modification of the number of correct answers to questions;
3. providing equal conditions for all tests.

During the process of testing, the next question is read from the database based on the inference results obtained from the Knowledge Base, located in the local network. The question is displayed in a form convenient for a student.

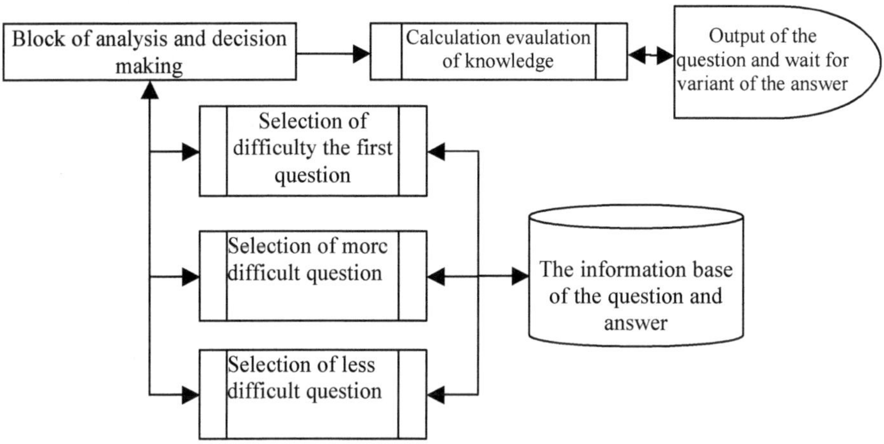

Fig. 9. A simplified diagram of the control system of knowledge

The Knowledge Base consists of fuzzy rules (1), which are detailed consideration of earlier logical constructions that interpret the data obtained by the fuzzy neural block [2, 4, 5].

$$\begin{aligned} &IF\ f_{input}(X_1)THEN\ f_{output}(Y_1) \\ &IF\ f_{input}(X_2)THEN\ f_{output}(Y_2) \\ &\quad\quad\quad\dots \\ &IF\ f_{input}(X_n)THEN\ f_{output}(Y_n) \end{aligned} \tag{1}$$

The selection of questions is done by execution of several fuzzy expressions, whose ultimate aim is to transfer its results to the decision-making block. The algorithm of sampling the first and subsequent questions is based on the results of the following tasks, Figure 9:

1. preliminary analysis of the student's knowledge – this task is to evaluate the student's knowledge, to select the first question: a lagging student will be asked a question from a group of simple questions, whereas the prepared student will be given more a difficult question, Figure 10;

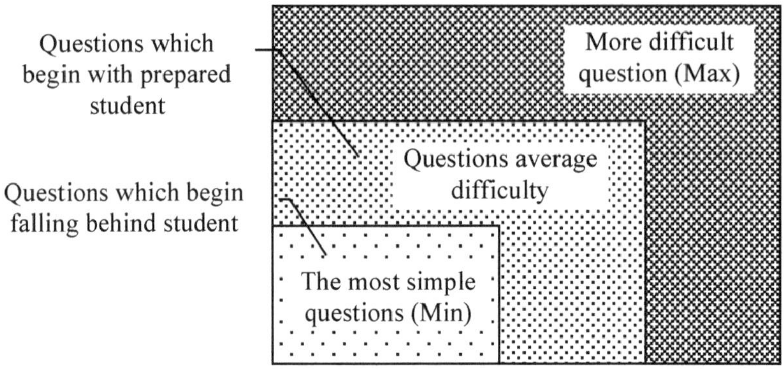

Fig. 10. The strategy of selecting the first question

Below, there are formulas that describe one of the functions of the decision-making block. They are related to the selection of the next question, which corresponds to the student's knowledge. With the correct answer program will select a more difficult question (2). An incorrect answer revaluation data about the student and the next will be given less difficult question (3).

2. selection of the next question if the previous question is answered correctly - in this case, a student is asked the question of increased complexity, Figure 11. The simplified formula used to select the next question would be:

$$\text{IF } (\text{Y}^{\text{A previous Answer}} = \text{True}) \text{ THEN } \text{f}^{\text{Increasing difficulty}}_{\text{The correct answer}}\left(\text{Q}^{\text{The diffuculty of the question}}\right)$$

$$Q = \frac{\left(Max\left(A^{+}\right) + Max\left(A^{-}\right)\right)}{2} \pm 2\% \qquad (2)$$

where, Q - the next question, (A+) - the level of the correct answer, Max (A+) - the maximum difficulty level of questions to which the student gave the correct answer, (A-) - value of the difficulty level of an incorrect response, Max (A-) - the maximum level of difficulty of questions to which a student gave a wrong answer, if there are

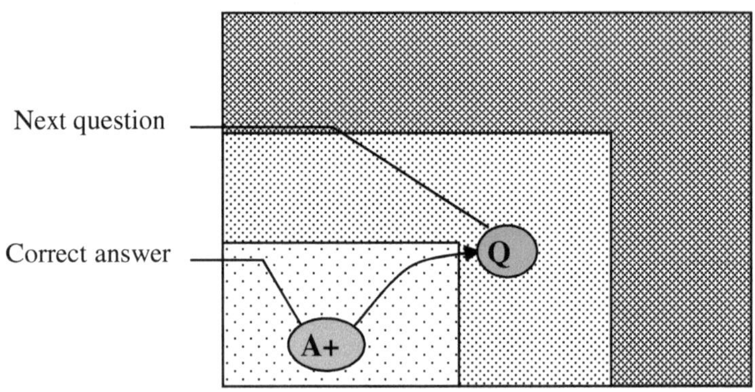

Fig. 11. Selection strategy question after correct answer

not wrong answers (the student answered all questions correctly) the default value of the maximum is taken, ± 2% - a maximum deviation in the range of which the next question is randomly selected.

3. selection of the next question if the previous question is answered incorrectly - in this case for complexity is reduced (Fig.12), the formula for selection of the next question would be:

IF $(Y^{\text{A previous Answer}} = False)$ THEN $f^{\text{Increasing difficulty}}_{\text{The incorrect answer}}(Q^{\text{The diffuculty of the question}})$

$$Q = \frac{(Max(A^-) + Min(A^+))}{2} \pm 2\% \tag{3}$$

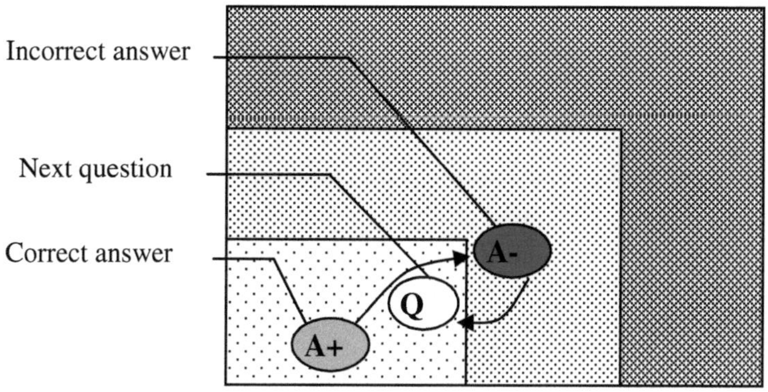

Fig. 12. Strategy selection the next question after incorrect answer

Maximum deviation creates a situation where for every group of students we do not have two students who will be asked the same questions even if the order of correct and incorrect answers is the same [7, 12, 14, 16].

4. processing the results and calculating a decision regarding continuation or termination of a testing/evaluation process – the number of correct answers is multiplied by their difficulty in respect to errors, multiple correct and incorrect answers are considered in order to performed evaluation, however if there is a high probability of uncertainty, a testing/evaluation process continues.

$$Z + jP = f\left(\frac{\sum_{i=1}^{N}(A_i^+)}{N}, \frac{\sum_{j=1}^{M}(A_j^-)}{M}, [A_1^+, A_2^+, ..., A_N^+], [A_1^-, A_2^-, ..., A_M^-] \right) \tag{4}$$

where, Z - evaluation of knowledge, P - uncertainty evaluation, f – input to the decision-making subprogram has the following characteristics: (Ai) - the set of values representing levels of difficulty of correct answers, (Aj-) - the set of difficulty levels of incorrect answers, N – the number of questions that received correct answers, M – the number of questions that received the incorrect answers.

The obtained complex number, formula (4), is a response of decision-making subprogram, which characterizes the evaluation of students' knowledge with some uncertainty [17, 18]. This uncertainty represents a level of confidence in the evaluation performed by decision-making block. Its value reflects the coverage of educational course, the higher the coverage, the less uncertainty, this in turn depends on the number of questions (Figure 13). For example, a given student can answer a small number of questions, and receive excellent evaluation, but the uncertainty in this case is very high. So, it is difficult to determine a proper number of questions that the student should be asked in order cover a meaningful amount of the educational course.

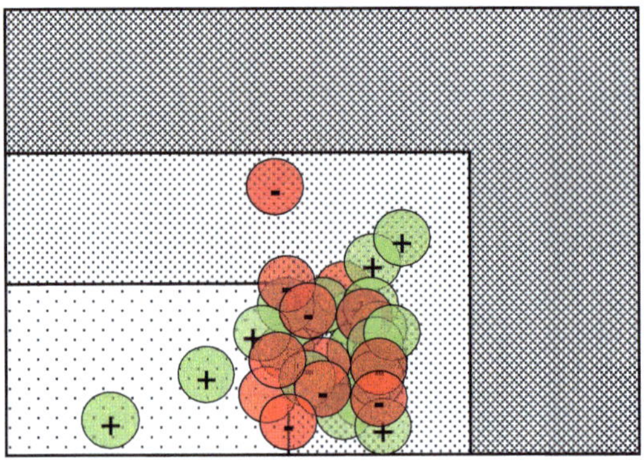

Fig. 13. An example of the distribution of answers at the end of testing

Thus, the system is equipped with a flexible algorithm for questioning. The teacher decides how much of the educational course needs to be included in questions, as well as the number of questions the student should be asked to determine the quality of students' knowledge. It could be for an interim test and a small volume of the educational course (10-20 questions - 20-30 minutes), or for a full-scale examination of the whole volume of the studied material (100-150 questions - 3-4 hours).

In the process, students can not only determine their knowledge, but also learn to analyze their own mistakes, which affect the generation of the curriculum. By achieving stable results in questions about a specific topic, students can go to the following sections of the course.

Aliasing in the training will give the necessary time to complete assimilation and retention of material, and then transition to a new, more difficult material. Each transition is accompanied by a small test on the previous analysis of the material with its absorption.

4 Definition of Membership Functions, the Principle of Defuzzification and Hybrid Method Correction Valuation of Errors

Using a threshold function of the logical structure constructed with application, the apparatus of fuzzy logic allowed us to avoid the problem of neural networks "black box" and use instead the scalar values of the weights and activation functions - a set of fuzzy expressions [10].

Initially, the decision-making bloc was developed based on two parallel technologies of fuzzy logic and neural networks. However, in many situations, the neural network was proved to be of a little use because of the impossibility to logically control and interpret the results [3, 9]. The transition to the use of a fuzzy neuron allowed qualitatively improves the mechanism of interim solutions, and simplifying the debugging and network training [6, 8].

The input fuzzy neurons use signal X_i to calculate the sum $\sum_{i=0}^{Max} f_{FN}(X_i)\Delta l \Lambda$ based on the size of fuzzy weights Δl and inputs $f_{FN}(X_i)\Delta l$, and taking into account the correction factor Λ.

Visually, the fuzzy neuron used in the system is shown in Figure 14 [11, 19].

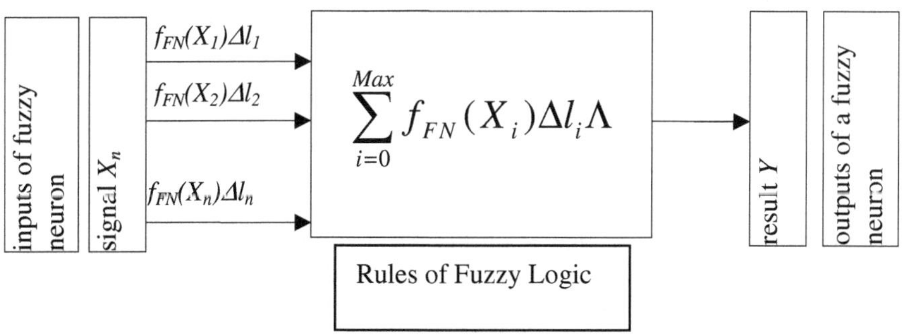

Fig. 14. The general scheme of the structure of a fuzzy neuron

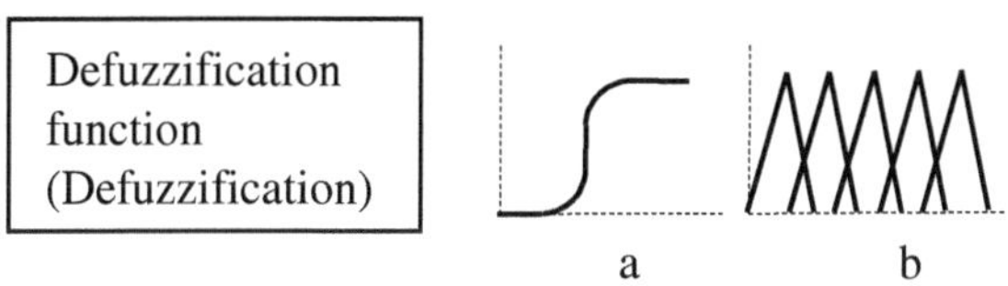

Fig. 15. The standard curve is a function of defuzzification: a) to evaluate the credited / not credited; b) to evaluate on a scale

Each signal, coming through the filter (configured to limit weak signals), is transformed into the domain of fuzzy sets associated with the available levels of activations. An output signal represents the aggregated value of system response signals passing through the defuzzification function, Figure 15. The output is a scalar representation of the fuzzy variable block included in the decision-making block.

The real scheme of the structure of fuzzy neural network used for the experiments is based on the well-proven technology ANFIS (Adaptive Neuro Fuzzy Inference System). Its realization for use in the educational system environment is shown in Figure 16 [1].

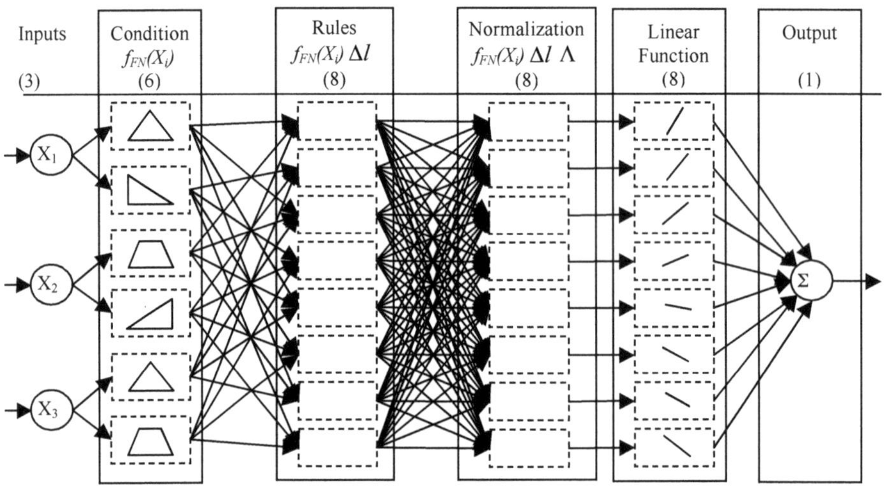

Fig. 16. Diagram of the structure of fuzzy neural network

In learning and testing phases of the fuzzy neural networks two evaluation techniques have been used to determine the quality of a decision-making: the traditional procedure of testing, and evaluation performed by the teacher.

During the experiments on various configurations of neural networks a single network has been selected. It has the mean square error equal to 0.227. Also, it was subjected to ten pilot phases involving groups of 5-6 subjects. The error was obtained by measuring the deviation between the evaluations of the student's knowledge provided by the teacher and the usual procedure of testing a specific curriculum. In calculating the evaluations, the confidence coefficients have been introduced: 3 for the teacher, and 1 for the evaluation obtained by the tests.

By the end of the learning procedures the obtained fuzzy neuron has been a system: {fuzzy signal} => {fuzzy activation function} => {fuzzy output} with the fuzzy output of a time variable representing the total representation of signals generated by the decision making block.

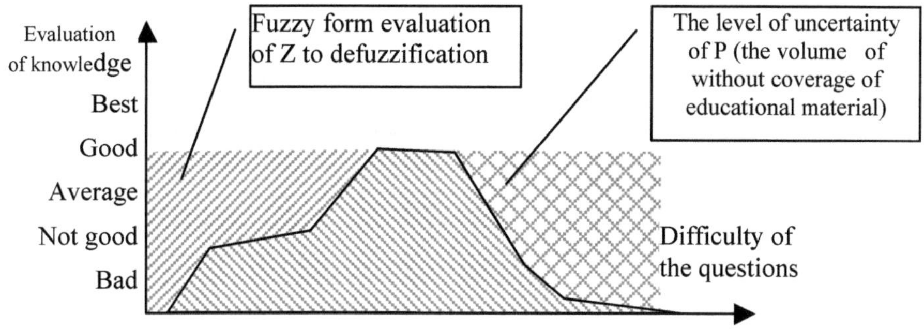

Fig. 17. An example of the resulting evaluation according to on the distribution and effectiveness of answers at the end of testing

Figure 17 shows the fuzzy function of knowledge evaluation that is convenient for processing in the computing environment. It represents the evaluation of student knowledge.

5 The General Model of Implementation of the Control Knowledge System

One of the most important factors influencing selection of models is related to simplicity and easiness of deploying the software package at virtually any educational institution. The only equipment needed is the hardware database server, which can be either a single server or a cluster with at least one server operating under a single management.

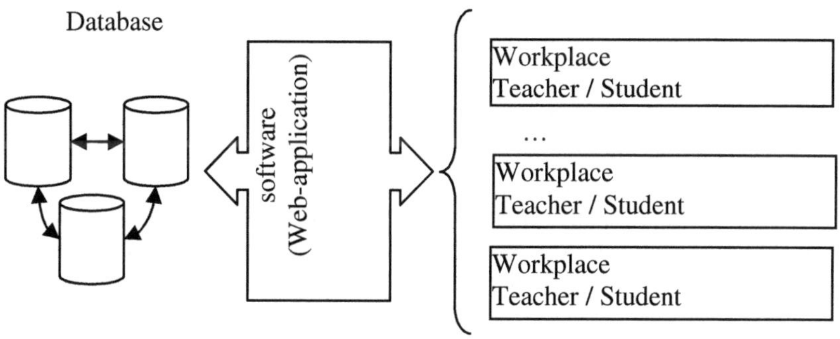

Fig. 18. The architecture of the system

All the logic of the system lies in the Web-application software. One of the most important parts of which is a knowledge control subsystem. During its development, the most perspective algorithms intellectualizing information processes have been used. As mentioned above, the basis of its power and flexibility is to use a combination of fuzzy logic and fuzzy neural network, Figure 19.

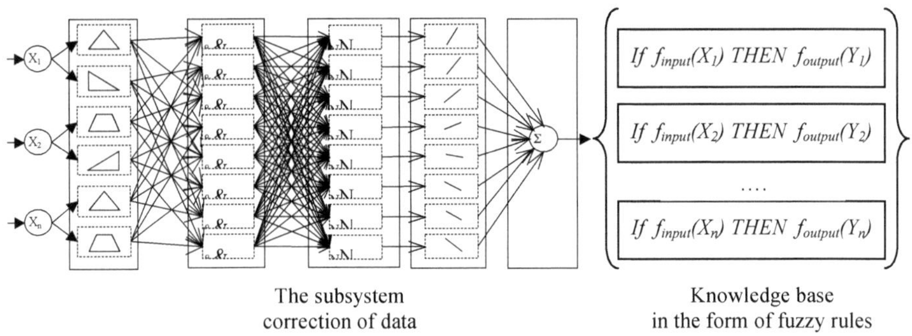

Fig. 19. The subsystem of decision-making

For implementing security policies in the system we have introduced several categories of users and define their roles and functional features in IISLCKS:

- Customers may be of two types: internal and external. An internal customer is anyone physically present at the university, and working on local computers. All queries of these customers are processed via the internal server. An external visitor can access the system through the Internet gateways.
- Authorized users are the users of information system, equipped with rights defined by the security policy:

  1) students – they are authorized users, registered in the system as consumers of services;
  2) corporate users – the authorized users, who are employees of educational institutions;
  3) content managers – corporate users who have rights to change data in a database information system.

Depending on the level of access we distinguish:

1) Teachers who have rights to add and edit their own training materials, and can be associated with a specific number of students.
2) Application administrators who are corporate users and have the rights to modify directories, shareware, information, lists of users. Application administrators shall assign permissions to users of information systems.
3) System administrators who ensure the normal functioning of information system, backing up and restoring data, carry out procedural maintenance work on the system, control system security logs, and carry out the monitoring of system resources.

The system's kernel allows teachers to create basis educational processes that satisfy the needs of a wide range of educational institutions, built educational materials which can be formalized, and build decisions making mechanisms expressed in the form of logical expressions. The structure of the access levels is simplistic, but at the same time providing the necessary protection and flexibility, Figure 20.

The access control mechanism consists of two groups of three-level privileges. This model is powerful enough and yet simple and functional, that fully satisfies the practical conditions and safety requirements.

	The developers of the system	Users of the system
The level of full access	System administrator DB Administrator	Rector Vice-Rector for Science
The level of monitoring and control	System programmers SQL Developers	Teaching staff Distance Methodists
The level of minimum access	Operators (Input educational materials, and testing)	Students (full-time, distance and distance learning)

Fig. 20. The hierarchy of levels of access control systems

6 Conclusion

This article presents the results of analytical work and modeling of information, which allows us to formulate and develop a set of models and methods required for building an autonomous learning system. The developed approaches deal with a complex set of functions required by teaching processes as well as the knowledge control procedures with minimum participation of teachers and the majors of educational institutions.

The presented solution for the control of knowledge brings the process of determining the volume of knowledge provided to each student to a whole new level. This algorithm is based on the original method of sample questions. Thanks to it, the system matches the current level of knowledge of students to the level of difficult of questions, as well as is able to determine the basic misconceptions and knowledge gaps.

Developed algorithm of decision-making is able to determine the level of knowledge of each student based on the smallest possible number of questions. This allows for quick evaluation of the knowledge with a high level of reliability comparable to the traditional method of survey, carried out by the teacher.

The decision-making block, built on fuzzy neural network technology coupled with a knowledge base containing a set of fuzzy rules, enables making well-balanced decisions. Flexibility and logic transparency of the approach allows its modifications so it can easily work with virtually any educational material.

The methodology required for development of the software package IISLCKS is presented. All aspects of the system are analyzed and investigated.

References

1. Jang, J.S.R.: ANFIS: Adaptive-Network-Based Fuzzy Inference System. IEEE Trans. Systems, Man, Cybernetics 23(5/6), 665–685 (1993)
2. Bonissone, Badami, Chiang, Khedkar, Marcelle, Schutten: Industrial Applications of Fuzzy Logic at General Electric. Proceedings of the IEEE 83(3), 450–465
3. Barsky, A.B.: Neural networks: recognition, management, decision-making. Finance and statistics, Moscow, pp. 30–63 (2004)
4. Bellman, R., Zadeh, L.A.: Decision-making in ambiguous circumstances, issues analysis and decision-making, pp. 180–199. Springer (1976)
5. Bernshteyn, L.S., Bojenyuk, A.V.: Fuzzy models of decision making: deduction, induction, analogy, pp. 78–99. Univ. Tsure, Taganrog (2001)
6. Bouchon-Meunier, B., Yager, R.R.: Fuzzy Logic and Soft Computing (Advances in Fuzzy Systems: Application and Theory), pp. 84–93, 103–119. World Scientific (1995)
7. Gorbunova, L.G.: On the realization of the rating system in pedagogical high schools. In: Proceedings of 2nd International Technical Conference "University Education", Part 1, Penza, pp. 105–106 (1998)
8. Hanss, M.: Applied Fuzzy Arithmetic: An Introduction with Engineering Applications, 1st edn., pp. 100–116, 139–147. Springer (2004)
9. Jang, Sun, C.-T.: Neuro-Fuzzy Modeling and Control. J.S.R. Proceedings of the IEEE 83(3), 378–406
10. Laurene, V.F.: Fundamentals of Neural Networks: Architectures, Algorithms and Applications, pp. 103–121. Prentice Hall, US edition (1993)
11. Nikravesh, M., Aminzadeh, F., Zadeh, L.A.: Soft Computing and Intelligent data analysis in oil exploration, pp. 273–287 (2003)
12. Nikravesh, M., Zadeh, L.A., Kacprzyk, J.: Soft Computing for Information Processing and Analysis, pp. 93–99 (2005)
13. Shahbazova, S., Freisleben, B.: A Network-Based Intellectual Information System for Learning and Testing. In: Fourth International Conference on Application of Fuzzy Systems and Soft Computing, Siegen, Germany, pp. 308–313 (2000)
14. Shahbazova, S., Zeynalova, S.: Decision-Making in Definition of Knowledge in the Conditions of Uncertainty of Educational Process. In: PCI 2010, Elm, vol. I, pp. 305–310 (2010)
15. Jang, J.S.R., Gulley, N.: The Fuzzy Logic Toolbox for use with MATLAB. The MathWorks Inc., Natick (1995)
16. Yager, R., Filev, D.: Essentials of fuzzy modeling and control. John Wiley and Sons, New York (1994)
17. Zadeh, L.A.: A new approach to the analysis of difficulty systems and decision processes. Mathematics Today, Knowledge, 23–37 (1974)
18. Zadeh, L.A., Kacprzyk, J.: Fuzzy Logic for the Management of Uncertainty, 1st Printing edn., pp. 75–84. Wiley-Interscience (1992)
19. Zadeh, L.A., Klir, G.J., Yuan, B.: Fuzzy Sets, Fuzzy Logic, and Fuzzy Systems: Selected Papers by Lotfi A. Zadeh, pp. 60–69 (1996)

Bio-inspired Optimization of Interval Type-2 Fuzzy Controllers

Oscar Castillo and Patricia Melin

Tijuana Institute of Technology, Division of Graduate Studies,
Calzada Tecnologico s/n, 22379, Tijuana, Mexico
{ocastillo,pmelin}@tectijuana.mx

Abstract. A review of the optimization methods used in the design of type-2 fuzzy systems, which are relatively novel models of imprecision, has been considered in this paper. The fundamental focus of the work has been based on the basic reasons of the need for optimizing type-2 fuzzy systems for different areas of application. Recently, bio-inspired methods have emerged as powerful optimization algorithms for solving complex problems. In the case of designing type-2 fuzzy systems for particular applications, the use of bio-inspired optimization methods have helped in the complex task of finding the appropriate parameter values and structure of the fuzzy systems. In this paper, we consider the application of genetic algorithms, particle swarm optimization and ant colony optimization as three different paradigms that help in the design of optimal type-2 fuzzy systems. We also provide a comparison of the different optimization methods for the case of designing type-2 fuzzy systems.

Keywords: Intelligent Control, Type-2 Fuzzy Logic, Interval Fuzzy Logic.

1 Introduction

Uncertainty affects decision-making and appears in a number of different forms. The concept of information is fully connected with the concept of uncertainty [17]. The most fundamental aspect of this connection is that the uncertainty involved in any problem-solving situation is a result of some information deficiency, which may be incomplete, imprecise, fragmentary, not fully reliable, vague, contradictory, or deficient in some other way. Uncertainty is an attribute of information [24]. The general framework of fuzzy reasoning allows handling much of this uncertainty and fuzzy systems that employ type-1 fuzzy sets represent uncertainty by numbers in the range [0, 1]. When something is uncertain, like a measurement, it is difficult to determine its exact value, and of course type-1 fuzzy sets make more sense than using crisp sets [14]. However, it is not reasonable to use an accurate membership function for something uncertain, so in this case what we need is higher order fuzzy sets, those which are able to handle these uncertainties, like the so called type-2 fuzzy sets [14]. So, the amount of uncertainty can be managed by using type-2 fuzzy logic because it offers

R.R. Yager et al. (Eds.): *Soft Computing: State of the Art Theory*, STUDFUZZ 291, pp. 241–254.
DOI: 10.1007/978-3-642-34922-5_17

better capabilities to handle linguistic uncertainties by modeling vagueness and unreliability of information [5] [23].

Recently, we have seen the use of type-2 fuzzy sets in Fuzzy Logic Systems (FLS) in different areas of application [1] [2] [6] [10] [12]. In this paper we deal with the application of interval type-2 fuzzy control to non-linear dynamic systems [3] [4] [5] [15] [19]. It is a well known fact, that in the control of real systems, the instrumentation elements (instrumentation amplifier, sensors, digital to analog, analog to digital converters, etc.) introduce some sort of unpredictable values in the information that has been collected [20]. So, the controllers designed under idealized conditions tend to behave in an inappropriate manner [11].

2 Fuzzy Logic Systems

In this section, a brief overview of type-1 and type-2 fuzzy systems is presented. This overview is considered to be necessary to understand the basic concepts needed to develop the methods and algorithms presented later in the paper.

2.1 Type-1 Fuzzy Logic Systems

Soft computing techniques have become an important research topic, which can be applied in the design of intelligent controllers, which utilize the human experience in a more natural form than the conventional mathematical approach [16, 18]. A FLS, described completely in terms of type-1 fuzzy sets is called a type-1 fuzzy logic system (type-1 FLS). In this paper, the fuzzy controller has two input variables, which are the error $e(t)$ and the error variation $\Delta e(t)$,

$$e(t) = r(t) - y(t) \tag{1}$$

$$\Delta e(t) = e(t) - e(t-1) \tag{2}$$

so the control system can be represented as in Figure 1.

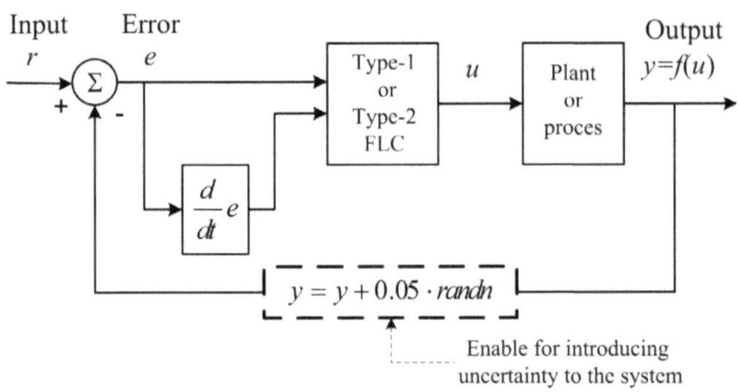

Fig. 1. System used for obtaining the experimental results

2.2 Type-2 Fuzzy Logic Systems

If for a type-1 membership function, as in Figure 2, we blur it to the left and to the right, as illustrated in Figure 3, then a type-2 membership function is obtained. In this case, for a specific value x', the membership function (u'), takes on different values, which are not all weighted the same, so we can assign an amplitude distribution to all of those points.

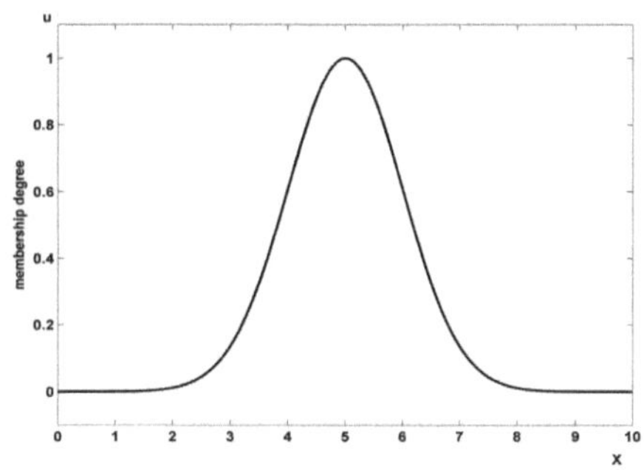

Fig. 2. Type-1 membership function

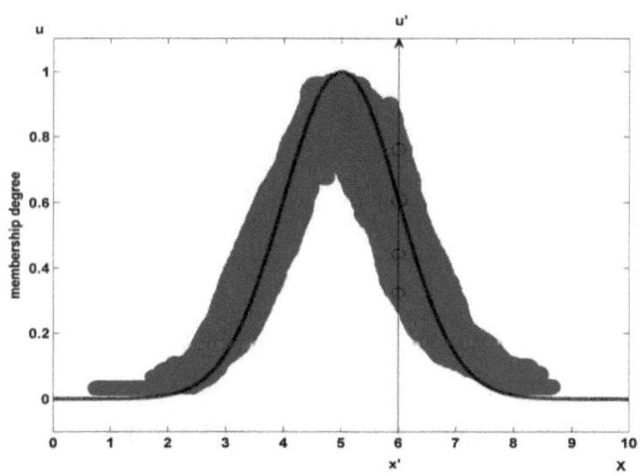

Fig. 3. Blurred type-1 membership function

A type-2 fuzzy set $\tilde{A}$, is characterized by the membership function [14, 17]:

$$\tilde{A} = \{((x,u), \mu_{\tilde{A}}(x,u)) \mid \forall x \in X, \forall u \in J_x \subseteq [0,1]\} \tag{3}$$

in which $0 \leq \mu_{\tilde{A}}(x,u) \leq 1$. Another expression for $\tilde{A}$ is,

$$\tilde{A} = \int_{x \in X} \int_{u \in J_x} \mu_{\tilde{A}}(x,u)/(x,u) \qquad J_x \subseteq [0,1] \tag{4}$$

Where $\int\int$ denotes the union over all admissible input variables x and u. For discrete universes of discourse $\int$ is replaced by $\sum$. In fact $J_x \subseteq [0,1]$ represents the primary membership of x, and $\mu_{\tilde{A}}(x,u)$ is a type-1 fuzzy set known as the secondary set. Hence, a type-2 membership grade can be any subset in [0,1], the primary membership, and corresponding to each primary membership, there is a secondary membership (which can also be in [0,1]) that defines the possibilities for the primary membership. Uncertainty is represented by a region, which is called the footprint of uncertainty (FOU). When $\mu_{\tilde{A}}(x,u)=1, \forall u \in J_x \subseteq [0,1]$ we have an interval type-2 membership function, as shown in Figure 4. The uniform shading for the FOU represents the entire interval type-2 fuzzy set and it can be described in terms of an upper membership function $\bar{\mu}_{\tilde{A}}(x)$ and a lower membership function $\underline{\mu}_{\tilde{A}}(x)$.

A FLS described using at least one type-2 fuzzy set is called a type-2 FLS. Type-1 FLSs are unable to directly handle rule uncertainties, because they use type-1 fuzzy sets that are certain [14]. On the other hand, type-2 FLSs, are very useful in circumstances where it is difficult to determine an exact membership function, and there are measurement uncertainties [7, 8, 15].

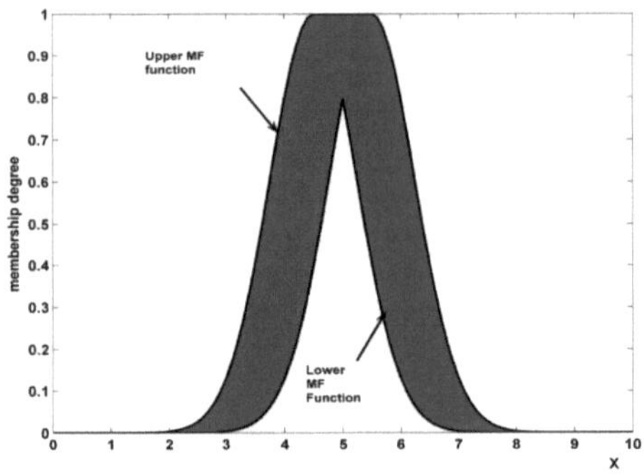

Fig. 4. Interval type-2 membership function

A type-2 FLS is again characterized by IF-THEN rules, but its antecedent or consequent sets are now of type-2. Similar to a type-1 FLS, a type-2 FLS includes a fuzzifier, a rule base, fuzzy inference engine, and an output processor, as we can see in Figure 5. The output processor includes type-reducer and defuzzifier; it generates a type-1 fuzzy set output (type-reducer) or a crisp number (defuzzifier).

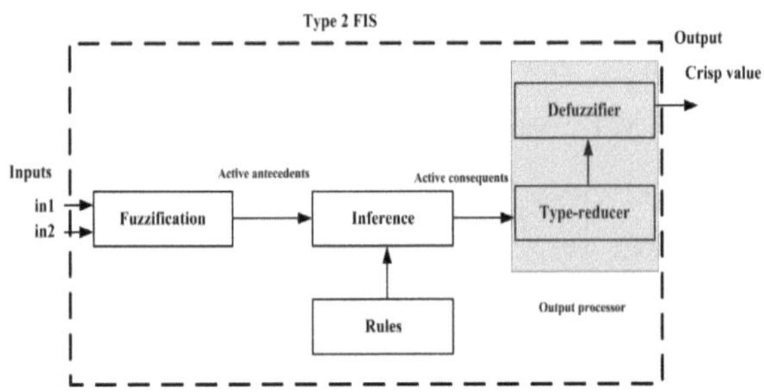

Fig. 5. Type-2 Fuzzy Logic System

2.2.1 Fuzzifier

The fuzzifier maps a crisp point $\mathbf{x}=(x_1,\ldots,x_p)^T \in X_1 x X_2 x \ldots x X_p \equiv \mathbf{X}$ into a type-2 fuzzy set $\tilde{A}_x$ in $\mathbf{X}$ [17], interval type-2 fuzzy sets in this case. We will use type-2 singleton fuzzifier, in a singleton fuzzification, the input fuzzy set has only a single point on nonzero membership [14]. $\tilde{A}_x$ is a type-2 fuzzy singleton if $\mu_{\tilde{A}_x}(x) = 1/1$ for $\mathbf{x}=\mathbf{x}'$ and $\mu_{\tilde{A}_x}(x) = 1/0$ for all other $\mathbf{x}\neq\mathbf{x}'$[17].

2.2.2 Rules

The structure of rules in a type-1 FLS and a type-2 FLS is the same, but in the latter the antecedents and the consequents will be represented by type-2 fuzzy sets. So for a type-2 FLS with p inputs $x_1 \in X_1,\ldots,x_p \in X_p$ and one output $y \in Y$, Multiple Input Single Output (MISO), if we assume there are M rules, the lth rule in the type-2 FLS can be written as follows [14]:

$$R^l\text{: IF } x_1 \text{ is } \tilde{F}_1^{\,l} \text{ and } \cdots \text{and } x_p \text{ is } \tilde{F}_p^{\,l}\text{, THEN } y \text{ is } \tilde{G}^l \qquad l=1,\ldots,M \tag{5}$$

2.2.3 Inference

In the type-2 FLS, the inference engine combines rules and gives a mapping from input type-2 fuzzy sets to output type-2 fuzzy sets. It is necessary to compute the join $\sqcup$, (unions) and the meet $\sqcap$ (intersections), as well as extended sup-star compositions (sup star compositions) of type-2 relations [14]. If $\tilde{F}^l{}_1 \times \cdots \times \tilde{F}^l{}_p = \tilde{A}^l$, equation (5) can be re-written as

$$R^l : \tilde{F}^l{}_1 \times \cdots \times \tilde{F}^l{}_p \rightarrow \tilde{G}^l = \tilde{A}^l \rightarrow \tilde{G}^l \qquad l=1,\ldots,M \tag{6}$$

R^l is described by the membership function $\mu_{R^l}(\mathbf{x}, y) = \mu_{R^l}(x_1,\ldots,x_p, y)$, where

$$\mu_{R^l}(\mathbf{x}, y) = \mu_{\tilde{A}^l \rightarrow \tilde{G}^l}(\mathbf{x}, y) \tag{7}$$

can be written as [14]:

$$\mu_{R^l}(\mathbf{x}, y) = \mu_{\tilde{A}^l \rightarrow \tilde{G}^l}(\mathbf{x}, y) = \mu_{\tilde{F}_1^l}(x_1) \sqcap \cdots \sqcap \mu_{\tilde{F}_p^l}(x_p) \sqcap \mu_{\tilde{G}^l}(y)$$

$$= [\sqcap_{i=1}^{p} \mu_{\tilde{F}^l{}_i}(x_i)] \sqcap \mu_{\tilde{G}^l}(y) \tag{8}$$

In general, the p-dimensional input to R^l is given by the type-2 fuzzy set $\tilde{A}_x$ whose membership function is

$$\mu_{\tilde{A}_x}(\mathbf{x}) = \mu_{\tilde{x}_1}(x_1) \sqcap \cdots \sqcap \mu_{\tilde{x}p}(x_p) = \sqcap_{i=1}^{p} \mu_{\tilde{x}i}(x_i) \tag{9}$$

where $\tilde{X}_i (i = 1,\ldots, p)$ are the labels of the fuzzy sets describing the inputs. Each rule R^l determines a type-2 fuzzy set $\tilde{B}^l = \tilde{A}_x \circ R^l$ such that [14]:

$$\mu_{\tilde{B}^l}(y) = \mu_{\tilde{A}_x \circ R^l} = \sqcup_{x \in \mathbf{X}} \left[\mu_{\tilde{A}_x}(\mathbf{x}) \sqcap \mu_{R^l}(\mathbf{x}, y) \right] \qquad y \in Y \;\; l=1,\ldots,M \tag{10}$$

This equation is the input/output relation in Figure 5 between the type-2 fuzzy set that activates one rule in the inference engine and the type-2 fuzzy set at the output of that engine [14]. In the FLS we used interval type-2 fuzzy sets and meet under product t-norm, so the result of the input and antecedent operations, which are contained in the firing set $\Pi_{i=1}^{p} \mu_{\tilde{F}_{i_i}}(x'_i \equiv F^l(\mathbf{x}')$, is an interval type-1 set [14],

$$F^l(\mathbf{x}') = \left[\underline{f}^l(\mathbf{x}'), \bar{f}^l(\mathbf{x}') \right] \equiv \left[\underline{f}^l, \bar{f}^l \right] \tag{11}$$

where

$$\underline{f}^l(\mathbf{x}') = \underline{\mu}_{\tilde{F}_1^l}(x_1') * \cdots * \underline{\mu}_{\tilde{F}_p^l}(x_p') \tag{12}$$

$$\bar{f}^l(\mathbf{x}') = \bar{\mu}_{\tilde{F}_1^l}(x_1') * \cdots * \bar{\mu}_{\tilde{F}_p^l}(x_p') \tag{13}$$

where * is the product operation.

2.2.4 Type Reducer

The type-reducer generates a type-1 fuzzy set output, which is then converted in a crisp output through the defuzzifier. This type-1 fuzzy set is also an interval set, for the case of our FLS we used center of sets (cos) type reduction, $Y_{\cos}$ which is expressed as [14]:

$$Y_{\cos}(\mathbf{x}) = [y_l, y_r] = \int_{y^1 \in [y_l^1, y_r^1]} \cdots \int_{y^M \in [y_l^M, y_r^M]} \int_{f^1 \in [\underline{f}^1, \bar{f}^1]} \cdots \int_{f^M \in [\underline{f}^M, \bar{f}^M]} 1 \Big/ \frac{\sum_{i=1}^{M} f^i y^i}{\sum_{i=1}^{M} f^i} \tag{14}$$

this interval set is determined by its two end points, y_l and y_r, which corresponds to the centroid of the type-2 interval consequent set $\tilde{G}^i$ [14],

$$C_{\tilde{G}^i} = \int_{\theta_1 \in J_{y1}} \cdots \int_{\theta_N \in J_{yN}} 1 \Big/ \frac{\sum_{i=1}^{N} y_i \theta_i}{\sum_{i=1}^{N} \theta_i} = [y_l^i, y_r^i] \tag{15}$$

before the computation of $Y_{\cos}(\mathbf{x})$, we must evaluate equation (15), and its two end points, y_l and y_r. If the values of f_i and y_i that are associated with y_l are denoted f_l^i and y_l^i, respectively, and the values of f_i and y_i that are associated with y_r are denoted f_r^i and y_r^i, respectively, from (14), we have [14]

$$y_l = \frac{\sum_{i=1}^{M} f_l^i y_l^i}{\sum_{i=1}^{M} f_l^i} \tag{16}$$

$$y_r = \frac{\sum_{i=1}^{M} f_r^i y_r^i}{\sum_{i=1}^{M} f_r^i} \tag{17}$$

2.2.5 Defuzzifier

From the type-reducer we obtain an interval set $Y_{\cos}$, to defuzzify it we use the average of y_l and y_r, so the defuzzified output of an interval singleton type-2 FLS is [14]

$$y(\mathbf{x}) = \frac{y_l + y_r}{2} \quad (18)$$

3 Bio-inspired Optimization Methods

In this section a brief overview of the basic concepts from bio-inspired optimization methods needed for this work is presented.

3.1 Particle Swarm Optimization

Particle swarm optimization is a population based stochastic optimization technique developed by Eberhart and Kennedy in 1995, inspired by social behavior of bird flocking or fish schooling [1]. PSO shares many similarities with evolutionary computation techniques such as the GA [9].

The system is initialized with a population of random solutions and searches for optima by updating generations. However, unlike the GA, the PSO has no evolution operators such as crossover and mutation. In the PSO, the potential solutions, called particles, fly through the problem space by following the current optimum particles [16]. Each particle keeps track of its coordinates in the problem space, which are associated with the best solution (fitness) it has achieved so far (The fitness value is also stored). This value is called *pbest*. Another "best" value that is tracked by the particle swarm optimizer is the best value, obtained so far by any particle in the neighbors of the particle. This location is called *lbest*. When a particle takes all the population as its topological neighbors, the best value is a global best and is called *gbest* [19].

The particle swarm optimization concept consists of, at each time step, changing the velocity of (accelerating) each particle toward its *pbest* and *lbest* locations (local version of PSO). Acceleration is weighted by a random term, with separate random numbers being generated for acceleration toward *pbest* and *lbest* locations [1]. In the past several years, PSO has been successfully applied in many research and application areas. It is demonstrated that PSO gets better results in a faster, cheaper way when compared with other methods [19]. Another reason that PSO is attractive is that there are few parameters to adjust. One version, with slight variations, works well in a wide variety of applications. Particle swarm optimization has been considered for approaches that can be used across a wide range of applications, as well as for specific applications focused on a specific requirement.

The basic algorithm of PSO has the following nomenclature:

x_z^i -Particle position

v_z^i -Particle velocity

w_{ij} -Inertia weight

p_z^i -Best "remembered" individual particle position

p_z^g -Best "remembered" swarm position

c_1, c_2 -Cognitive and Social parameters

r_1, r_2 -Random numbers between 0 and 1

The equation to calculate the velocity is:

$$v_{z+1}^i = w_{ij} v_z^i + c_1 r_1 \left(p_z^i - x_z^i \right) + c_2 r_2 \left(p_z^g - x_z^i \right) \tag{19}$$

and the position of the individual particles is updated as follows:

$$x_{z+1}^i = x_z^i + v_{z+1}^i \tag{20}$$

The basic PSO algorithm is defined as follows:

1) Initialize

a) *Set constants* $z_{\max}, c_1, c_2$

b) *Randomly initialize particle position* $x_0^i \in D$ *in* R^n *for* $i = 1, \ldots, p$

c) *Randomly initialize particle velocities* $0 \le v_0^i \le v_0^{\max}$ *for* $i = 1, \ldots, p$

d) *Set Z = 1*

2) Optimize

a) *Evaluate function value* f_k^i *using design space coordinates* x_k^i

b) *If* $f_z^i \le f_{best}^i$ *then* $f_{best}^i = f_z^i, p_z^i = x_z^i$.

c) *If* $f_z^i \le f_{best}^g$ *then* $f_{best}^g = f_z^i, p_z^g = x_z^i$.

d) *If stopping condition is satisfied then go to 3.*

e) *Update all particle velocities* v_z^i *for* $i = 1, \ldots, p$

f) *Update al particle positions* x_z^i *for* $i = 1, \ldots, p$

g) *Increment z.*

h) *Goto 2(a).*

3) Terminate

3.2 Genetic Algorithms

Genetic Algorithms (GAs) are adaptive heuristic search algorithms based on the evolutionary ideas of natural selection and genetic processes [8]. The basic principles of GAs were first proposed by John Holland in 1975, inspired by the mechanism of natural selection, where stronger individuals are likely the winners in a competing environment [9]. GA assumes that the potential solution of any problem is an individual and can be represented by a set of parameters. These parameters are regarded as the genes of a chromosome and can be structured by a string of values in binary form. A positive value, generally known as a fitness value, is used to reflect the degree of "goodness" of the chromosome for the problem, which would be highly related with its objective value. The pseudocode of a GA is as follows:

1. *Start with a randomly generated population of n chromosomes (candidate solutions to a problem).*
2. *Calculate the fitness of each chromosome in the population.*
3. *Repeat the following steps until n offspring have been created:*
    a. *Select a pair of parent chromosomes from the current population, the probability of selection being an increasing function of fitness. Selection is done with replacement, meaning that the same chromosome can be selected more than once to become a parent.*
    b. *With probability (crossover rate), perform crossover to the pair at a randomly chosen point to a form two offspring.*
    c. *Mutate the two offspring at each locus with probability (mutation rate), and place the resulting chromosomes in the new population.*
4. *Replace the current population with the new population.*
5. *Go to step 2.*

The simple procedure just described above is the basis for most applications of GAs found in the literature [21] [22].

3.3 Ant Colony Optimization

Ant Colony Optimization (ACO) is a probabilistic technique that can be used for solving problems that can be reduced to finding good paths along graphs. This method is inspired on the behavior presented by ants in finding paths from the nest or colony to the food source.

The S-ACO is an algorithmic implementation that adapts the behavior of real ants to solutions of minimum cost path problems on graphs [12]. A number of artificial ants build solutions for a certain optimization problem and exchange information about the quality of these solutions making allusion to the communication system of real ants [13].

Let us define the graph $G = (V, E)$, where V is the set of nodes and E is the matrix of the links between nodes. G has $n_G = |V|$ nodes. Let us define L^K as the number of hops in the path built by the ant k from the origin node to the destiny node. Therefore, it is necessary to find:

$$Q = \left\{ q_a, \ldots, q_f \middle| q_1 \in C \right\} \tag{21}$$

where Q is the set of nodes representing a continuous path with no obstacles; $q_a, \ldots, q_f$ are former nodes of the path and C is the set of possible configurations of the free space. If $x^k(t)$ denotes a Q solution in time t, $f(x^k(t))$ expresses the quality of the solution. The S-ACO algorithm is based on Equations (22), (23) and (24):

$$p_{ij}^k(t) = \begin{cases} \dfrac{\tau_{ij}^k}{\sum_{j \in N_{ij}^k} \tau_{ij}^{\alpha}(t)} & \text{if } j \in N_i^k \\ 0 & \text{if } j \notin N_i^k \end{cases} \tag{22}$$

$$\tau_{ij}(t) \leftarrow (1-\rho)\tau_{ij}(t) \tag{23}$$

$$\tau_{ij}(t+1) = \tau_{ij}(t) + \sum_{k=1}^{n_k} \tau_{ij}(t) \tag{24}$$

Equation (22) represents the probability for an ant k located on a node i selects the next node denoted by j, where, N_i^k is the set of feasible nodes (in a neighborhood) connected to node i with respect to ant k, τ_{ij} is the total pheromone concentration of link ij, and α is a positive constant used as a gain for the pheromone influence.

Equation (23) represents the evaporation pheromone update, where $\rho \in [0,1]$ is the evaporation rate value of the pheromone trail. The evaporation is added to the algorithm in order to force the exploration of the ants, and avoid premature convergence to sub-optimal solutions. For $\rho = 1$ the search becomes completely random.

Equation (24), represents the concentration pheromone update, where $\Delta\tau_{ij}^k$ is the amount of pheromone that an ant k deposits in a link ij in a time t.

The general steps of S-ACO are the following:

1. *Set a pheromone concentration τ_{ij} to each link (i,j).*
2. *Place a number k=1, 2,…, n_k in the nest.*
3. *Iteratively build a path to the food source (destiny node), using Equation (22) for every ant.*

- *Remove cycles and compute each route weight $f\left(x^k(t)\right)$. A cycle could be generated when there are no feasible candidates nodes, that is, for any i and any k, $N_i^k = \varnothing$; then the predecessor of that node is included as a former node of the path.*

4. *Apply evaporation using Equation (23).*
5. *Update of the pheromone concentration using Equation (24)*

6. *Finally, finish the algorithm in any of the three different ways:*
   - *When a maximum number of epochs has been reached.*
   - *When it has found an acceptable solution, with $f(x_k(t)) < \varepsilon$.*
   - *When all ants follow the same path.*

3.4 General Remarks about Optimization of Type-2 Fuzzy Systems

The problem of designing type-2 fuzzy systems can be solved with any of the above mentioned optimization methods. The main issue in any of these methods is deciding on the representation of the type-2 fuzzy system in the corresponding optimization paradigm. For example, in the case of GAs, the type-2 fuzzy systems must be represented in the chromosomes. On the other hand, in PSO the fuzzy system is represented as a particle in the optimization process. In the ACO method, the fuzzy system can be represented as one of the paths that the ants can follow in a graph. Also, the evaluation of the fuzzy system must be represented as an objective function in any of the methods.

4 General Overview of the Area and Future Trend

In this section a general overview of the area of type-2 fuzzy system optimization is presented. Also, possible future trends that we can envision based on the review of this area are presented. It has been well-known for a long time that designing fuzzy systems is a difficult task, and this is especially true in the case of type-2 fuzzy systems [5]. The use of GAs, ACO and PSO in designing type-1 fuzzy systems has become a standard practice for automatically designing this sort of systems [1] [2] [13] [21]. This trend has also continued to the type-2 fuzzy systems area, which has been accounted for with the review of papers presented in the previous sections. In the case of designing type-2 fuzzy systems the problem is more complicated due to the higher number of parameters to consider, making it of upmost importance the use of bio-inspired optimization techniques for achieving the optimal designs of this sort of systems. In this section a summary of the total number of papers published in the area of type-2 fuzzy system optimization is presented, so that the increasing trend occurring in this area can be better appreciated. Also, the distribution of papers according to the used optimization technique is presented, so that a general idea of how these different techniques are contributing to the automatic design of optimal type-2 fuzzy systems is obtained.

Figure 6 shows the distribution of the published papers in optimizing type-2 fuzzy systems according to the different bio-inspired optimization techniques previously mentioned. From Figure 6 it can be noted that the use of GAs have been decreasing recently, on the other hand the use of PSO, ACO and other methods have been increasing. The reason for the increase in use of PSO and ACO may be due to recent works in which either PSO or ACO have been able to outperform GAs for different applications. Regarding the question of which method would be the most appropriate for optimizing type-2 fuzzy systems, there is no easy answer. At the moment, what we can be sure of is that the techniques mentioned in this paper and probably newer ones that may appear in the future, would certainly be tested in the optimization of type-2 fuzzy systems because the problem of designing automatically these types of systems is complex enough to require their use.

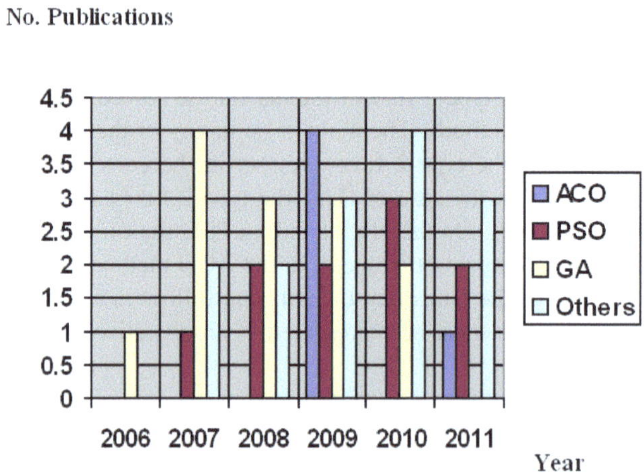

Fig. 6. Distribution of publications per area and year

There are other bio-inspired or nature-inspired techniques that at the moment have not been applied to the optimization of type-2 fuzzy systems that may be worth mentioning. For example, membrane computing, harmony computing, electromagnetism based computing, and other similar approaches have not been applied (to the moment) in the optimization of type-2 fuzzy systems. It is expected that these approaches and similar ones could be applied in the near future in the area of type-2 fuzzy system optimization. Of course, as new bio-inspired and nature-inspired optimization methods are being proposed at any time in this fruitful area of research, it is expected that newer optimization techniques would also be tried in the near future in the automatic design of optimal type-2 fuzzy systems.

5 Conclusions

In the previous sections we have presented a representative account of the different optimization methods that have been applied in the optimal design of type-2 fuzzy systems. To the moment, genetic algorithms have been used more frequently to optimize type-2 fuzzy systems. However, more recently PSO and ACO have attracted more attention and have also been applied with some degree of success to the problem of optimal design of type-2 fuzzy systems. There have been also other optimization methods applied to the optimization of type-2 fuzzy systems, like artificial immune systems and the chemical optimization paradigm. At this time, it would be very difficult to declare one of these optimization techniques as the best for optimizing type-2 fuzzy systems, as different techniques have had success for different applications of type-2 fuzzy logic. In any case, the need for bio-inspired optimization methods is justified due to the complexity of designing type-2 fuzzy systems.

References

[1] Bingül, Z., Karahan, O.: A Fuzzy Logic Controller tuned with PSO for 2 DOF robot trajectory control. Expert Systems with Applications 38(1), 1017–1031 (2011)

[2] Cao, J., Li, P., Liu, H., Brown, D.: Adaptive fuzzy controller for vehicle active suspensions with particle swarm optimization. In: Proceedings of SPIE-The International Society of Optical Engineering, vol. 7129 (2008)

[3] Castillo, O., Huesca, G., Valdez, F.: Evolutionary Computing for Topology Optimization of Type-2 Fuzzy Controllers. In: Castillo, O., Melin, P., Kacprzyk, J., Pedrycz, W. (eds.) Hybrid Intelligent Systems. STUD FUZZ, vol. 208, pp. 163–178. Springer, Heidelberg (2008)

[4] Castillo, O., Aguilar, L.T., Cazarez-Castro, N.R., Cardenas, S.: Systematic design of a stable type-2 fuzzy logic controller. Applied Soft Computing Journal 8, 1274–1279 (2008)

[5] Castillo, O., Melin, P., Alanis, A., Montiel, O., Sepulveda, R.: Optimization of interval type-2 fuzzy logic controllers using evolutionary algorithms. Journal of Soft Computing 15(6), 1145–1160 (2011)

[6] Castro, J.R., Castillo, O., Melin, P.: An Interval Type-2 Fuzzy Logic Toolbox for Control Applications. In: Proceedings of FUZZ-IEEE 2007, London, pp. 1–6 (2007)

[7] Castro, J.R., Castillo, O., Martinez, L.G.: Interval type-2 fuzzy logic toolbox. Engineering Letters 15(1), 14 (2007)

[8] Cordon, O., Gomide, F., Herrera, F., Hoffmann, F., Magdalena, L.: Ten years of genetic fuzzy systems: current framework and new trends. Fuzzy Sets and Systems 141, 5–31 (2004)

[9] Cordon, O., Herrera, F., Villar, P.: Analysis and guidelines to obtain a good uniform fuzzy partition granularity for fuzzy rule-based systems using simulated annealing. International Journal of Approximate Reasoning 25, 187–215 (2000)

[10] Dereli, T., Baykasoglu, A., Altun, K., Durmusoglu, A., Turksen, I.B.: Industrial applications of type-2 fuzzy sets and systems: A concise review. Computers in Industry 62, 125–137 (2011)

[11] Hagras, H.: Hierarchical type-2 fuzzy logic control architecture for autonomous mobile robots. IEEE Transactions on Fuzzy Systems 12, 524–539 (2004)

[12] Juang, C.-F., Hsu, C.-H.: Reinforcement ant optimized fuzzy controller for mobile-robot wall-following control. IEEE Transactions on Industrial Electronics 56(10), 3931–3940 (2009)

[13] Juang, C.-F., Hsu, C.-H.: Reinforcement interval type-2 fuzzy controller design by online rule generation and Q-value-aided ant colony optimization. IEEE Transactions on Systems, Man, and Cybernetics, Part B Cybernetics 39(6), 1528–1542 (2009)

[14] Karnik, N.N., Mendel, J.M.: An Introduction to Type-2 Fuzzy Logic Systems, Technical Report, University of Southern California (1998)

[15] Martinez, R., Castillo, O., Aguilar, L.T.: Optimization of interval type-2 fuzzy logic controllers for a perturbed autonomous wheeled mobile robot using genetic algorithms. Information Sciences 179(13), 2158–2174 (2009)

[16] Martinez, R., Rodriguez, A., Castillo, O., Aguilar, L.T.: Type-2 fuzzy logic controllers optimization using genetic algorithms and particle swarm optimization. In: Proceedings of the IEEE International Conference on Granular Computing, GrC 2010, pp. 724–727 (2010)
[17] Mendel, J.M.: Uncertainty, fuzzy logic, and signal processing. Signal Processing Journal 80, 913–933 (2000)
[18] Mohammadi, S.M.A., Gharaveisi, A.A., Mashinchi, M.: An evolutionary tuning technique for type-2 fuzzy logic controller in a non-linear system under uncertainty. In: Proceedings of the 18th Iranian Conference on Electrical Engineering, ICEE 2010, pp. 610–616 (2010)
[19] Oh, S.-K., Jang, H.-J., Pedrycz, W.: A comparative experimental study of type-1/type-2 fuzzy cascade controller based on genetic algorithms and particle swarm optimization. Expert Systems with Applications (2011) (article in press)
[20] Sepulveda, R., Castillo, O., Melin, P., Rodriguez-Diaz, A., Montiel, O.: Experimental study of intelligent controllers under uncertainty using type-1 and type-2 fuzzy logic. Information Sciences 177(10), 2023–2048 (2007)
[21] Wagner, C., Hagras, H.: A genetic algorithm based architecture for evolving type-2 fuzzy logic controllers for real world autonomous mobile robots. In: Proceedings of the IEEE Conference on Fuzzy Systems, London (2007)
[22] Wu, D., Tan, W.-W.: Genetic learning and performance evaluation of interval type-2 fuzzy logic controllers. Engineering Applications of Artificial Intelligence 19(8), 829–841 (2006)
[23] Yager, R.R.: Fuzzy subsets of type II in decisions. J. Cybernetics 10, 137–159 (1980)
[24] Zadeh, L.A.: The concept of a linguistic variable and its application to approximate reasoning. Information Sciences 8, 43–80 (1975)

System Modeling and Forecasting with Evolving Fuzzy Algorithms

Andre Lemos[1], Rosangela Ballini[2], Walmir Caminhas[1], and Fernando Gomide[3]

[1] Department of Electronic Engineering, Federal University of Minas Gerais, Av. Antônio Carlos, 6627, 31270-901 Belo Horizonte, MG, Brazil
{andrepl,caminhas}@cpdee.ufmg.br

[2] Institute of Economics, University of Campinas, 13083-857 Campinas, SP, Brazil
ballini@eco.unicamp.br

[3] School of Electrical and Computer Engineering, University of Campinas, 13083-852 Campinas, SP, Brazil
gomide@dca.fee.unicamp.br

Abstract. Many of the activities associated with the systems planning and operation require forecasts of future events. For instance, thermal models of distribution transformers with core immersed in oil are of utmost importance for power systems operation and safety. Its hot spot temperature determines the degradation speed of the insulation material and parts. High temperatures cause loss of mechanical stiffness, generating failures. Insulation degradation determines the lifetime limits of power transformers. Thermal models are needed to generate reliable data for lifetime forecasting methodologies. One of the greatest difficulties in thermal modeling is the non stationary nature of the transformers due to aging, parts replacement, and operational overloads. In this paper we use an evolving fuzzy model to build adaptive thermal models of distribution transformers. The model is an evolving fuzzy linear regression tree. The tree grows adaptively by replacing leaves with subtrees whenever they improve the model quality. The performance of the evolving regression is evaluated using actual data from an experimental transformer. The results suggest that the evolving fuzzy tree approach outperforms current state of the art models.

1 Introduction

The last decades have witnessed a steadily increase in energy consumption in major urban and rural areas worldwide. In particular, the raise in electric energy consumption has caused the number of distribution networks, and consequently the number of operating transformers, to increase. Distribution transformers with core oil-immersed are one of the most common equipments in electric energy distribution networks. Using reliable methods to predict transformer lifetime, the power companies can find the ideal value of the power of the transformer to be located in each place because models provide estimates the loading curve with which the transformer will work. With a suitable load forecast and methodology to predict transformers lifetime, companies can determine the electrical power that minimizes cost and maximizes transformers lifetime. This improves the cost-benefit of electricity distribution networks. Lifetime prediction is also

R.R. Yager et al. (Eds.): *Soft Computing: State of the Art Theory*, STUDFUZZ 291, pp. 255–268.
DOI: 10.1007/978-3-642-34922-5_18

important because it allows to evaluate the optimal replacement and maintenance period of power transformers. This prevents failures, unexpectedly interruptions of power supply, and improves distribution system safety [35].

Lifetime forecasting methodologies of power distribution transformers need thermal models [1]. The purpose of a thermal model is to estimate the temperature of the transformer insulation. For instance, it is known that the insulating paper is the part which degrades faster since, when subjected to high temperatures, it undergoes a polymerization process, causing the breakdown of the cellulose fibers. When the average fiber size is reduced, the paper loses its mechanical strength. By becoming brittle, the paper breaks due to vibration which occurs during the transformer operation, resulting in failure of electrical insulation. High temperatures reduce the lifetime of power transformers [1].

The IEEE Loading Guidelines [1] suggest a deterministic thermal model based on transient heating equations. This model was evaluated in [20] where discrepancies exceeding 10°C have been verified between model outputs and actual values. Since the model is used to estimate the temperature, which in turn is used to predict transformers lifetime, inaccurate thermal models mean inaccurate lifetime estimation.

Given the complexity to develop high accuracy deterministic models, several recent works develop thermal models using black box models from data. Neural networks and fuzzy systems are examples of data driven methods which have been shown to be efficient in thermal modeling [15, 16] because they are able to learn complex nonlinear relations and process imprecise data.

However, most if not all current forecasting models lack an important requirement to be useful in real world circumstances. They should be adaptive, but they are not. As any physical device, power transformer behavior changes through time due to variations in the environment, structural parts changes, maintenance, aging, etc. In general, a model may not be valid after a period of time if the structure and parameters of the model, or both, do not modify to accommodate changes.

The interest in adaptive system modeling has motivated the development of highly adaptive and intelligent systems denominated evolving intelligent systems (EIS) [3]. EIS models are self-developed from a stream of data. EIS are able to address problems of modeling, control, prediction, classification and data processing in a non-stationary, dynamic changing environments. Such systems embody online learning methods and one-pass incremental algorithms that evolve or gradually change individual models to guarantee life-long learning and self-organization of the system structure [5].

Pioneering work in this area were addressed in the realm of neural networks [2, 12, 37]. In the turn of the centuries the area has expanded to encompass fuzzy rule-based systems (evolving fuzzy systems) [4, 30] and neuro-fuzzy hybrids [22, 26]. During the last years, statistical modeling [17] and granular computing mechanisms [23] became important components of EIS.

Recently a new evolving fuzzy modeling technique called evolving fuzzy linear regression tree (eFT) [25] has been developed. Evolving fuzzy linear regression tree is a fuzzy regression tree with linear models in its leaves. The eFT can be built using data streams in an incremental and scalable manner. In general, an eFT encodes a nonlinear regression model whose output is defined as a weighted sum of linear local models.

The evolving approach for regression trees requires only features of the current sample plus a small amount of aggregated information to perform classification, function approximation and prediction tasks.

This chapter addresses the use of eFT to perform thermal modeling of power transformers. The aim is to predict the hot-spot temperature based on historical data. The main advantage of the eFT approach is its ability to adapt to changes, what makes it a good candidate in electrical energy distribution systems in which power transformers are key time-varying devices.

The remainder of the chapter is organized as follows. Section 2 briefly reviews the state of the art of evolving fuzzy systems area. Section 3 details the eFT model suggested by the authors recently. Section 4 uses eFT for adaptive thermal modeling of distribution power transformers and compares the results with advanced non-evolving and evolving methods. Finally, the conclusions and issues for further investigation are summarized in section 5.

2 Evolving Fuzzy Systems

Evolving fuzzy systems (eFS) can be seen as a synergy between fuzzy systems, as a mechanism for evolvable information compactation and representation, and recursive methods of machine learning [21]. The first eFS reported in the literature are dated from the beginning of this century [7, 22]. These models were developed to address a growing need for flexible, adaptive and interpretable models to develop intelligent sensors, autonomous navigation and control systems, classification and regression systems. Evolving fuzzy systems have advantages over other evolving black-box models such as evolving neural networks because they are linguistically interpretable. It is possible to extract information from the model structure as information granules (linguistic terms represented by fuzzy sets) [3].

Several forms of eFS can be found in literature, among them evolvable functional fuzzy rule-based systems in which the model structure (number of rules and rule antecedents) continuously evolves based on clusters created and/or excluded by recursive clustering algorithms [32]. The parameters of the consequents are updated using recursive least squares or its variations [28, 39].

One of the pioneering approaches in this area was introduced in [4] in the form of a functional fuzzy model in the form of Takagi-Sugeno. The evolving Takagi-Sugeno model (eTS) uses an incremental version of the subtractive clustering algorithm [9] with recursive evaluation of the information potential of new data samples to create new clusters or revise the existing ones. The rule consequent parameters are updated with the recursive least squares algorithm.

In [6] an extended version of eTS (xTS) was developed to include a mechanism to update the radius of influence of each rule recursively using data samples only. The xTS model also includes a new measure of the cluster quality to update the rule base.

The DENFIS (Dynamic Evolving Neural-Fuzzy System) [22] is another evolving TS type ofsystem derived from a distance-based recursive evolving clustering method (ECM) to adapt the rule base, and a weighted recursive least squares with forgetting factor algorithm to update rules consequent parameters.

FLEXFIS (Flexible Fuzzy Inference System), detailed in [30], uses a recursive clustering algorithm derived from a modification of the vector quantization technique [14] called eVQ (Evolving Vector Quantization) [29]. Consequent parameters estimation uses the weighted recursive least squares algorithm.

The ePL (evolving Participatory Learning) [27] is an evolving rule-based model based on the idea of participatory learning. Participatory learning [38] is a learning paradigm which assumes that learning and beliefs about the system to be modeled depend on what the learning process has already learned. The participatory learning mechanism is a potential candidate to develop evolving systems because it balances learning and model update, yet accounting for the knowledge accumulated.

The ePL model has been extended recently as a fuzzy rule-based system with multivariable membership functions, namely evolving Multivariable Gaussian (eMG) [24]. This model assumes that input variables may interact, avoids the curse of dimensionality when handling clusters formation, and introduces a sound and systematic approach for learning that results in an algorithm with few parameters.

Recently a new evolving fuzzy modeling approach called evolving Fuzzy linear regression Tree [25] has been developed. Linear regression trees [36] are generalizations of regression trees [8] in which the mean value of the output variable in the tree leaves are replaced by linear models. The internal nodes of these types of trees perform split tests that first divide the input space in non-overlapping regions and then associate a linear model to each resulting partition. Fuzzy linear regression trees replace the binary splitting decision in each internal node by a pair of membership functions, similarly as in fuzzy decision trees [40]-[19]. As a consequence, input space partitions may overlap. Evolving fuzzy linear regression trees are built from data streams in an incremental and scalable manner.

Next section focusses on the details of the structure and on the incremental learning algorithm developed for the evolving fuzzy tree model.

3 Evolving Fuzzy Linear Regression Tree

Linear regression trees generalize regression trees by assigning to leaves a linear model of the input variables instead of a zero-order model (e.g., the mean value of the output). Each region of the underlying input space comes with a linear model of the form $y_i = a_0 + a_1x_1 + ... + a_mx_m = \sum_{i=0}^{m} a_ix_i$, where $x_0 = 1$ and m is the number of input variables. In other words, linear regression trees are recursive structures capable to perform piecewise linear regression.

Figure 1 gives an example of a linear regression tree.

To estimate the output associated with a given input, start from the root (top) node and apply splitting decision tests until a leaf is reached. The output is computed using the input values and the linear model of the leaf. For example, for the tree shown in Figure 1, the model outputs -2.6 when inputs are $x_1 = 4$ and $x_2 = 3$.

Fuzzy linear regression trees replace the split tests in their internal nodes by two membership functions describing the concepts *less than* and *greater than*. The sharp partition boundaries of the original tree vanish when fuzzy sets are considered in the internal nodes of the tree. Given an input sample $x = (x_1, ..., x_m)$, all branches and leaves

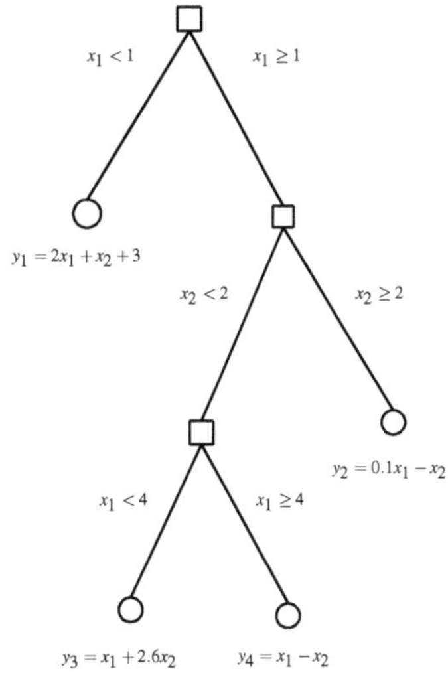

Fig. 1. Example of a linear regression tree

of a fuzzy linear regression tree fire to some degree. The weighted sum of the local linear models defines the tree output.

Reference [25] suggests sigmoid membership functions in each internal node of the tree as a way to represent the very nature of the concepts *less than* and *greater than*. Sigmoids (1) have two parameters, the center c and the spread σ. Depending on the sign of the parameter σ, the sigmoid is inherently open to the right or to the left. Figure 2 shows examples of membership functions describing the concepts *less than 5* and *greater than 5* choosing $\sigma = -0.5$ and $\sigma = 0.5$, respectively.

$$\mu(x) = \frac{1}{1 + \exp -\frac{1}{\sigma}(x - c)} \tag{1}$$

To compute the output for a given input, start at the root node and find the membership values for each pair of membership functions in the internal nodes considering all paths. Next, the membership values are combined using an aggregation operator, typically a *t-norm*. This results in a membership value linked to the model of each tree leaf. These values are used to compute the output as the weighted average of all local model outputs as follows:

$$\hat{y} = \frac{\sum_{i=1}^{l} y_i w_i}{\sum_{i=1}^{l} w_i} \tag{2}$$

where l is the number of leaves, and w_i is found using:

$$w_i = \mu_1(x)\ t\ \mu_2(x)\ t\ \dots\ t\ \mu_o(x) = T_{j=1}^{o} \mu_j(x) \tag{3}$$

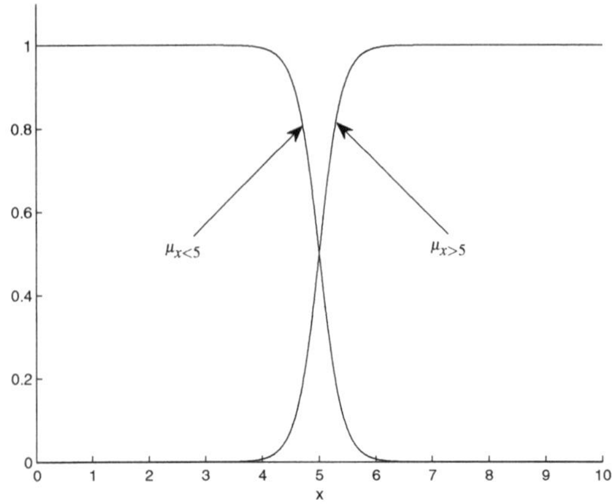

Fig. 2. Membership functions for *less than* and *greater than* 5

where t is a *t-norm* [33], o is the number of internal nodes reached from the root to leaf i, and μ_j is one of the sigmoid membership functions associated with node j.

3.1 Incremental Learning Algorithm

The data driven learning algorithm used to grow the tree starts with a single global linear model. The algorithm evolves the tree replacing leaves by subtrees using a statistical model selection test and the newest input data. All previously sampled values are discarded, and decisions are made using statistics computed recursively.

We assume that the tree leaves have λ candidate splits for each of the m input variables, totaling $\lambda \times m$ possible splits. Any of these can be used to replace an existing leaf. Every candidate split has a subtree comprising an internal node with two sigmoid membership functions (*less than* and *greater than*) centered on the split point, and two leaves (the left and the right one) containing linear models. Figure 3 illustrates a general candidate split.

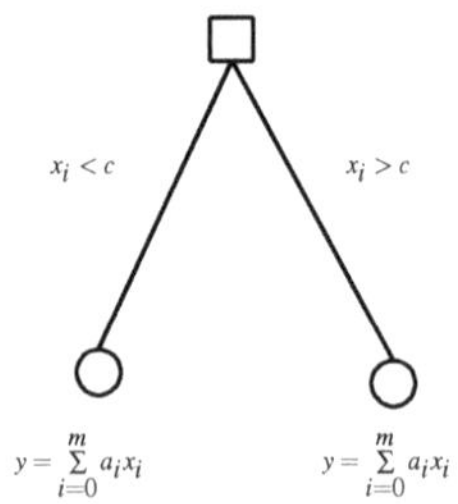

Fig. 3. General candidate split

As discussed above, leaves represent regions of the input space. Split points for different input variables are defined depending on the range of values in which samples fall within. They are chosen to uniformly divide the range of possible values into $\lambda+1$ intervals. Although each leaf defines a fuzzy region of the input space, splits are sharp and found using the centers of the membership functions of internal nodes.

Given an input, the model output can be estimated using equation (2) and the linear models of existing leaves can be updated using the weighted recursive least squares (WRLS) algorithm, similar as in [28, 39]. In this case, the updating formulas for parameters of the leaf i model at the k-th step are:

$$\begin{aligned} \gamma_i^{k+1} &= \gamma_i^k + Q_i^{k+1} x^k \Psi_i(x^k)\left[y_i^k - ((x^k)^T \gamma_i^k)\right] \\ Q_i^{k+1} &= Q_i^k - \frac{\Psi_i(x^k) Q_i^k x^k (x^k)^T Q_i^k}{1+(x^k)^T Q_i^k x^k} \end{aligned} \tag{4}$$

where $\Psi_i(x^k)$ for $i = 1, \ldots,$ is the normalized firing degree associated with the i-th local linear model:

$$\Psi_i(x^k) = \frac{w_i}{\sum_{j=1}^{l^k} w_j} \tag{5}$$

where l^k is the number of leaves at step k.

Next, membership function spreads σ_i lying in the path from the root to the highest active leaf are revised using the steepest descent method. For example, consider a fuzzy linear regression tree whose structure is as shown in Figure 1. Assume that for a given input, the leaf with the highest firing degree is y_2. In this case, the membership functions describing $x_1 \geq 1$ and $x_2 \geq 2$ should have their spreads updated.

Membership function spread updating aims to minimize at each step the squared error, in other words, the squared difference between the model output and the actual output:

$$e^k = \frac{1}{2}\left(\hat{y}^k - y^k\right)^2 \tag{6}$$

The recursive expression to update the spread of a membership function is:

$$\sigma_i = \sigma_i - \beta \frac{\partial e^k}{\partial \hat{y}^k} \frac{\partial \hat{y}^k}{\partial \mu_i} \frac{\partial \mu_i}{\partial \sigma_i} \tag{7}$$

where β is the basic learning rate. The partial derivatives of the error (6) and sigmoidal functions (1) are:

$$\frac{\partial e^k}{\partial \hat{y}^k} = \hat{y}^k - y^k \tag{8}$$

$$\frac{\partial \mu_i}{\partial \sigma_i} = \frac{\left(\exp -\frac{1}{\sigma_i}(x^k - c_i)\right)(x^k - c_i)}{1+\exp -\frac{1}{\sigma_i}(x^k - c_i)^2 \sigma_i^2} \tag{9}$$

The partial derivative $\partial \hat{y}^k / \partial \mu_i$ in (7) is obtained combining equations (2) and (3). To compute the derivative, we must first find paths from the root node to the leaves that pass through the membership function μ_i. Thus,

$$\frac{\partial \hat{y}^k}{\partial \mu_i} = \frac{\sum_{r=1}^{np^k} \frac{w_r}{\mu_i} y_r - \hat{y}^k \sum_{r=1}^{np^k} \frac{w_r}{\mu_i}}{\sum_{j=1}^{l^k} w_j} \tag{10}$$

where w_r results from *t-norm* aggregation of the membership functions in a path crossing μ_i; np^k is the number of paths satisfying this condition at step k. Note that (10) is valid only if the *t-norm* is the algebraic product. The product *t-norm* is adopted in all computational experiments reported in this chapter.

The linear models of all candidate splits associated with the selected leaf are also updated using (4). For each candidate split, the weights of all leaves (5), including those weights of the leaves of the candidate split, are revised . Next, the output of the resulting tree (with the selected leaf replaced by the candidate split) is computed. Finally, the linear models in the leaves of the selected candidate split are updated using (4).

Once a leaf is updated, tests are performed to evaluate whether or not the new tree is better than the previous. The test is a goodness of fit test that considers the accuracy and the number of parameters of the tree, with and without the subtree. Fundamentally, the test compares the quality of two models, the simpler original tree, and the more complex tree composed by the original tree with the subtree added. The assumption is that the simpler model can be nested in the complex model, and that the complex model should be more accurate. The test essentially tries to answer the following question: does the gain in accuracy worth the cost of having a more complex tree?

The test assumes that the parameters of the two tree models are estimated using the same samples of a data stream. It computes the following statistics [34]:

$$F_{inc} = \frac{(RSS_1 - RSS_2) \times (n_2 - p_2)}{RSS_2 \times (n_1 - n_2 + p_1)} \tag{11}$$

where RSS_1 and RSS_2 is the sum of residual squares of the simpler and complex model, respectively; p_1 and p_2 is the number of free parameters of each model; and n_1 and n_2 are the number of samples used to build the tree and set the candidate split parameters, respectively.

F_{inc} is distributed according to a Fisher's F distribution with $(n_2 - n_1 + p_2 - p_1, n_2 - p_2)$ degrees of freedom. The use of this statistics requires computation of the p-values (probability in the tail of the distribution) for all candidate splits of the last modified leaf. The candidate split associated with the smallest p-value is selected. The subtree of the selected candidate split replaces the corresponding leaf if its p-value is smaller than a confidence level α. It is necessary to introduce a multiple-comparison statistical correction because the same hypothesis is tested $\lambda \times m$ times using the same data set [34]. The Bonferroni correction [31] is used by dividing the desired significance level by the number of tests. Therefore, the selected subtree is added into the tree if

$$p\text{-value} < \frac{\alpha}{\lambda \times m} \tag{12}$$

Notice that the model selection test described requires the sum of squared residuals, the amount of candidate splits and samples processed to be updated at each step.

The evolving fuzzy linear regression trees algorithm is summarized in Algorithm 1. Additional details about the learning algorithm can be found in [25].

Algorithm 1. Algorithm to evolve fuzzy linear regression tree models

Compute the output and membership value of all leaves
Update the linear local models
Select the leaf with the highest membership value
for all inputs (m) **do**
    for all candidate splits (λ) **do**
        Estimate the output replacing the selected leaf with the candidate split
        Compute the p-value of the model selection test for the candidate split
    end for
end for
Select the candidate split with the minimum p-value
if p-value $< \frac{\alpha}{\lambda \times m}$ **then**
    Replace the selected leaf by the candidate split
end if

Four parameters need to be chosen, respectively:

- significance level α adopted for model selection tests;
- number of candidate splits for each variable, λ;
- initial spread of sigmoid membership functions, σ_0; and
- basic learning rate β to update the spreads.

As a rule of thumb, typical values of the significance level are 0.05 or 0.01. These are values that have shown to work well for a range of problems.

The balance between modeling accuracy and computational cost dictates how the number of candidate splits is chosen. Low values of λ may decrease accuracy of the model because fewer candidate splits will be created for each leaf. High values increase accuracy, but also increase the number of local models to be adapted.

Initial values for the spread of the membership functions are chosen from *a priori* information about the scale of the data. Appropriate adjustment of σ_0 speeds up the convergence of parameters to potentially optimal values.

The basic learning rate β usually is small values, normally between 10^{-5} and 10^{-1}.

4 Computational Experiments

In this section the eFT model is used to estimate the hot-spot temperature of an power transformer. Data used in the experiments were obtained from [13]. They were collected from measurements performed in a experimental power transformer. Data correspond to three 24-hour load curves, totalling 72 hours of data acquisition. The data was sampled from each sensor at 5-min intervals and different load current profiles. Two load curves exceed the transformer nominal power. Detailed information about the measurement system can be obtained from [13].

The model is a one-step ahead forecaster whose purpose is to predict the next (5 min) hot-spot temperature value using the actual and lagged values of hot-spot temperature (T) and load (K). Previous work [15] suggest as inputs the current and one step delayed values of hot-spot temperature and load:

$$\hat{T}^{k+1} = f(T^k, K^k, T^{k-1}, K^{k-1}) \tag{13}$$

The forecasting performance was evaluated using the root mean squared error (RMSE), the non-dimensional error (NDEI) and the mean absolute error (MAE). The NDEI is the ratio of the root mean squared error by the standard deviation of the target data. The error measures are computed as follows:

$$RMSE = \left(\frac{1}{n}\sum_{k=1}^{n}(y^k - \hat{y}^k)\right)^{\frac{1}{2}} \tag{14}$$

$$NDEI = \frac{RMSE}{std(y^k)} \tag{15}$$

$$MAE = \frac{1}{n}\sum_{k=1}^{n}\left|y^k - \hat{y}^k\right| \tag{16}$$

where n is the size of the test data set, y^k is the target output, and $std()$ is the standard deviation.

The experiment was conducted as follows. Data corresponding to 2 operation days were input to the eFT algorithm (576 observations) and the evolving model performance evaluated using data of the last day (288 observations), keeping the model structure and parameters fixed at the values found after evolving during the period of 2 days. The eFT parameters are $\alpha = 0.05$, $\sigma_0 = 0.01$, $\beta = 0.01$ and $\lambda = 25$.

Fig. 4 shows the eFT forecasts against the actual data for the last day. The resulting tree has 3 leaves and 2 internal nodes associated with the current temperature (T^k) and the first delayed load value (K^{k-1}).

Table 1 shows how eFT performs against the deterministic IEEE [1] model and alternative evolving and fixed structure data-driven modeling methods using the error measures (14,15,16). The multilayer perceptron (MLP) has one hidden layer with four neurons trained with backpropagation algorithm, the artificial neural network fuzzy inference system (ANFIS) has four fuzzy sets for each input variable and four fuzzy rules generated by means of the fuzzy c-means clustering procedure. The MLP used the following scheme for initialization phase: small weights values randomly assigned; $\alpha = 0.9$ as momentum parameter; 500 as the maximum number of epochs, and a adaptive learning-rate starting from $\eta = 0.01$ as initial step size. The ANFIS has 100 as maximum number of epochs, $\eta_i = 0.01$ as initial step size, $s_d = 0.9$ as step size decrease rate and $s_i = 1.1$ as step size increase rate. The parameters of the eTS model

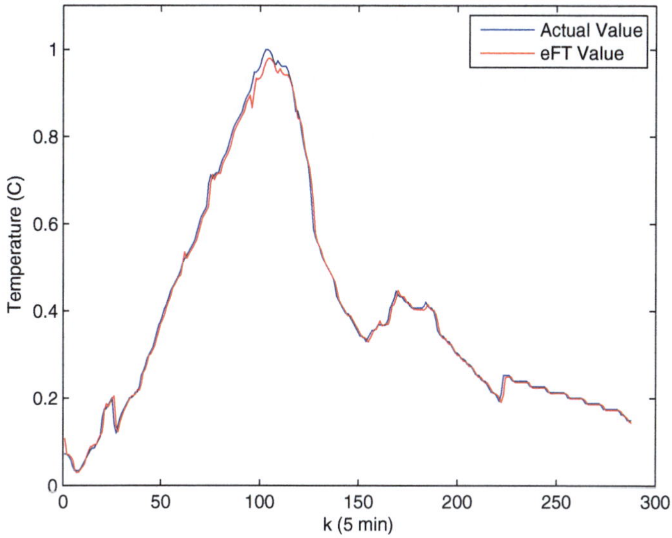

Fig. 4. Actual transformer temperature and eFT temperature forecast

Table 1. Performance of thermal modeling methods

Model	Number of rules (or nodes)	RMSE	NDEI	MAE
IEEE	-	0.4005	1.5671	0.3846
MLP [11]	4	0.0327	0.1278	0.0158
xTS [6]	8	0.0259	0.1015	0.0157
eTS [4]	8	0.0171	0.0670	0.0105
ANFIS [18]	4	0.0154	0.0604	0.0092
eFT	3	0.0123	0.0481	0.0079

were set to $r = 0.3$ and $\Omega = 750$. The xTS [6] has a $\Omega = 750$. The parameters of the models were those which produced their best results.

Table 1 suggests that the eFT model performs best among all models.

The error indices used in Table 1 are good to measure model accuracy. However, they do not reveal whether the results from one model is statistically superior to any other model. Therefore, is desired to employ some statistical test to help comparisons between any two models in terms of accuracy.

For instance, the MGN test [10] is a parametric test to compare the accuracy of two forecasting models. The statistics to perform the MGN test is computed as follows:

$$MGN = \frac{\hat{\rho}_{sd}}{\sqrt{\frac{1-\hat{\rho}_{sd}^2}{n-1}}} \tag{17}$$

where $\hat{\rho}_{sd}$ is the estimated correlation coefficient between $s = r_1 + r_2$, and $d = r_1 - r_2$, with r_1 and r_2 the residuals of the two models considered. In this case, the statistic is distributed as Student'T distribution with $n-1$ degrees of freedom. In MGN testing, the correlation between s and d will be zero if the forecasts are equally accurate.

Table 2 shows the pairwise comparisons between eFT and the alternative non-evolving and evolving approaches using the MGN test (17). The table shows the MGN statistics and the corresponding p-value (tail of the distribution). Table 2 confirms the superior accuracy of the eFT model for thermal modeling.

Table 2. MGN test evaluation for thermal modeling

Models	MGN	p-value
eFT vs IEEE	80.7356	0.0000
eFT vs MLP	36.2298	0.0000
eFT vs xTS	14.6436	0.0000
eFT vs eTS	8.9733	0.0000
eFT vs ANFIS	5.9374	0.0000

5 Conclusion

This paper has addressed evolving fuzzy modeling approaches emphasizing the fuzzy linear regression tree, a newly developed approach to build nonlinear fuzzy regression models from a set of local linear models in a tree structure. Evolving fuzzy tree models are build incrementally from a data stream. The usefulness and efficacy of the approach has been shown through adaptive thermal modeling of distribution transformers.

The evolving fuzzy tree modeling approach was evaluated using actual data from an experimental transformer. The experiments performed and the results obtained suggest that the method introduced in this chapter is a promising alternative to build adaptive thermal models to estimate transformer lifetime limits in actual electric energy distribution system.

Acknowledgement. The authors thank the Brazilian National Research Council, CNPq, for grants 141323/2009-4, 306343/2011-8, 309666/2007-4 and 304596/2009-4, respectively. The second author also acknowledges the support of FAPESP, the Research Foundation of the State of Sao Paulo, for grant 2011/13851-3. The third author is grateful for the support of FAPEMIG, the Research Foundation of the State of Minas Gerais, grant PPM-00252-09.

References

1. IEEE guide for loading mineral-oil-immersed transformers. IEEE Std. C57.91-1995 p. i (1996)
2. Alahakoon, D., Halgamuge, S., Srinivasan, B.: Dynamic self-organizing maps with controlled growth for knowledge discovery. IEEE Transactions on Neural Networks 11(3), 601–614 (2000)

3. Angelov, P.: Evolving takagi-sugeno fuzzy systems from streaming data, eTS+. In: Angelov, P., Filev, D., Kasabov, N. (eds.) Evolving Intelligent Systems: Methodology and Applications. Wiley-Interscience/IEEE Press (2010)
4. Angelov, P., Filev, D.: An approach to Online identification of Takagi-Suigeno fuzzy models. IEEE Transactions on Systems Man and Cybernetics Part B-Cybernetics 34(1), 484–498 (2004)
5. Angelov, P., Filev, D., Kasabov, N.: Guest editorial evolving fuzzy systems - preface to the special section. IEEE Transactions on Fuzzy Systems 16(6), 1390–1392 (2008)
6. Angelov, P., Zhou, X.: Evolving fuzzy systems from data streams in real-time. In: 2006 International Symposium on Evolving Fuzzy Systems, pp. 29–35 (2006)
7. Angelov, P.P.: Evolving rule-based models: a tool for design of flexible adaptive systems. Springer, London (2002)
8. Breiman, L., Friedman, J., Olshen, R., Stone, C.: Classification and regression trees, San Diego, California. Wadsworth Mathematics Series (1984)
9. Chiu, S.L.: Fuzzy model identification based on cluster estimation. Journal of Intelligent and Fuzzy Systems 2(3) (1994)
10. Diebold, F.X., Mariano, R.S.: Comparing predictive accuracy. Journal of Business and Economics Statistics 13, 253–263 (1995)
11. Duda, R.O., Hart, P.E., Stork, D.G.: Pattern Classification, 2nd edn. Wiley-Interscience (2000)
12. Fritzke, B.: Growing cell structures: a self-organizing network for unsupervised and supervised learning. Neural Networks 7, 1441–1460 (1994)
13. Galdi, V., Ippolito, L., Piccolo, A., Vaccaro, A.: Neural diagnostic system for transformer thermal overload protection. IEE Proceedings Electric Power Applications 147(5), 415–421 (2000)
14. Gray, R.: Vector quantization. IEEE ASSP Magazine 1(2), 4–29 (1984)
15. Hell, M., Costa, P., Gomide, F.: Participatory learning in power transformers thermal modeling. IEEE Transactions on Power Delivery 23(4), 2058–2067 (2008)
16. Hell, M., Costa Jr., P., Gomide, F.: Recurrent neurofuzzy network in thermal modeling of power transformers. IEEE Transactions on Power Delivery 22(2), 904–910 (2007)
17. Hisada, M., Ozawa, S., Zhang, K., Kasabov, N.: Incremental linear discriminant analysis for evolving feature spaces in multitask pattern recognition problems. Evolving Systems 1, 17–27 (2010)
18. Jang, J.: ANFIS - Adaptive-Network-Based Fuzzy Inference System. IEEE Transactions on Systems Man and Cybernetics 23(3) (1993)
19. Janikow, C.: Fuzzy decision trees: Issues and methods. IEEE Transactions on Systems Man and Cybernetics Part B-Cybernetics 28(1), 1–14 (1998)
20. Jardini, J.A., Brittes, J.L.P., Magrini, L.C., Bini, M.A., Yasuoka, J.: Power transformer temperature evaluation for overloading conditions. IEEE Transactions on Power Delivery 20(1), 179–184 (2005)
21. Kasabov, N., Filev, D.: Evolving intelligent systems: Methods, learning, & applications. In: 2006 International Symposium on Evolving Fuzzy Systems, pp. 8–18 (2006)
22. Kasabov, N., Song, Q.: DENFIS: Dynamic evolving neural-fuzzy inference system and its application for time-series prediction. IEEE Transactions on Fuzzy Systems 10(2), 144–154 (2002)
23. Leite, D., Costa Jr., P., Gomide, F.: Granular Approach for Evolving System Modeling. In: Hüllermeier, E., Kruse, R., Hoffmann, F. (eds.) IPMU 2010. LNCS (LNAI), vol. 6178, pp. 340–349. Springer, Heidelberg (2010)
24. Lemos, A., Gomide, F., Caminhas, W.: Multivariable gaussian evolving fuzzy modeling system. IEEE Transactions on Fuzzy Systems 19(1), 91–104 (2011)

25. Lemos, A., Gomide, F., Caminhas, W.: Fuzzy evolving linear regression trees. Evolving Systems 2(1), 1–14 (2011)
26. Leng, G., McGinnity, T., Prasad, G.: An approach for on-line extraction of fuzzy rules using a self-organising fuzzy neural network. Fuzzy Sets and Systems 150(2), 211–243 (2005)
27. Lima, E., Hell, M., Ballini, R., Gomide, F.: Evolving fuzzy modeling using participatory learning. In: Angelov, P., Filev, D., Kasabov, N. (eds.) Evolving Intelligent Systems: Methodology and Applications. Wiley-Interscience/IEEE Press (2009)
28. Ljung, L.: System Identification. Prentice-Hall (1999)
29. Lughofer, E.D.: Extensions of Vector Quantization for Incremental Clustering. Pattern Recognition 41(3), 995–1011 (2008)
30. Lughofer, E.D.: FLEXFIS: A Robust Incremental Learning Approach for Evolving Takagi-Sugeno Fuzzy Models. IEEE Transactions on Fuzzy Systems 16(6), 1393–1410 (2008)
31. Miller, R.: Simultaneous statistical inference. McGraw-Hill, Inc., New York (1966)
32. de Oliveira, J.V., Pedrycz, W.: Advances in Fuzzy Clustering and its Applications. John Wiley & Sons, Inc., New York (2007)
33. Pedrycz, W., Gomide, F.: Fuzzy Systems Engineering: Toward Human-Centric Computing. Wiley Interscience, NJ (2007)
34. Potts, D.: Incremental learning of linear model trees. In: ICML 2004: Proceedings of the Twenty-First International Conference on Machine Learning, p. 84. ACM, New York (2004)
35. Pylvanainen, J.K., Nousiainen, K., Verho, P.: Studies to utilize loading guides and ann for oil-immersed distribution transformer condition monitoring. IEEE Transactions on Power Delivery 22(1), 201–207 (2007)
36. Quinlan, R.: Learning with continuous classes. In: 5th Australian Joint Conference on Artificial Intelligence, pp. 236–243 (1992)
37. Williamson, J.R.: Gaussian ARTMAP: A neural network for past incremental learning of noisy multidimensional maps. Neural Networks 9(5), 881–897 (1996)
38. Yager, R.: A Model of Participatory Learning. IEEE Transactions on Systems Man and Cybernetics 20(5), 1229–1234 (1990)
39. Young, P.: Recursive estimation and time-series analysis: an introduction. Springer-Verlag New York, Inc., New York (1984)
40. Yuan, Y., Shaw, M.: Induction of fuzzy decision trees. Fuzzy Sets and Systems 69(2), 125–139 (1995)

Log-Domain Arithmetic for High-Speed Fuzzy Control on a Field-Programmable Gate Array

Ali Razib[1], Scott Dick[1], and Vincent Gaudet[2], Senior Member, IEEE

[1] Electrical and Computer Engineering
University of Alberta
Edmonton, Alberta, Canada
{arazib,sdick}@ualberta.ca
[2] Electrical and Computer Engineering
University of Waterloo
Waterloo, Ontario, Canada
vcgaudet@uwaterloo.ca

Abstract. Defuzzification has long been a bottleneck for fast implementations of fuzzy logic controllers, due to the large number of computationally expensive multiplication and division operations that are required. In this paper, we report a high-speed fuzzy inferential system based on log-domain arithmetic, which only requires addition, subtraction and multiplexing operations. The system is implemented on a Xilinx Virtex-II FPGA with a processing speed of 67.6 MFLIPS and a maximum combinational path delay of 4.2 ns. A pipelined version of the controller is also implemented, which achieves a speed of 248.7 MFLIPS. Although a small approximation error is introduced, software simulation and hardware implementation on FPGA confirm high similarity of the outputs for typical and log-domain control surfaces and a number of second-order plants.

1 Introduction

Fuzzy logic controllers, first proposed by Zadeh [1] and implemented in automatic control by Mamdani [2], are a particular class of intelligent controllers built on fuzzy expert systems that show substantial performance improvement over standard algorithms for these problems. However, the implementation of a fuzzy controller is not straightforward. Software simulations on general-purpose microprocessors are typically slow, implying the need for hardware implementations. Research on fuzzy hardware began with a voltage-mode analog circuit in the late 1960s. Although analog implementations [3] provide lower power consumption and smaller chip size [4], they are slower, less accurate, less flexible, and less scalable to larger problem sizes as compared to their digital counterparts. Thus, research into digital hardware implementations is ongoing, with the reported performance of ASIC implementations ranging from 20 to 50 million fuzzy logic inferences per second (MFLIPS) [5]. However, digital fuzzy hardware implementations have a performance bottleneck due to the large number of costly multiplication and division operations required in the defuzzifier, the last module in the fuzzy controller [6].

R.R. Yager et al. (Eds.): *Soft Computing: State of the Art Theory*, STUDFUZZ 291, pp. 269–287.
DOI: 10.1007/978-3-642-34922-5_19

For many applications involving the control of physical systems, current implementations of fuzzy hardware are adequate; a sampling rate of 50 MHz will be sufficient for a very large class of control problems. However, some control problems require a much higher sampling rate than this. Data communication networks are a good example; a number of network optimization goals (e.g. policing, active queue management) can be formulated as control problems. Linear control approaches are not very effective in these problems, and simulations often show that fuzzy controllers would be superior. However, even the fastest current fuzzy controllers are far too slow for communication network applications [26][27][28].

In this paper, we propose and implement a high-speed hardware-based fuzzy controller using log-domain arithmetic. The basic principle of our approach is to represent all fuzzy quantities by their logarithmic values, therefore transforming multiplication and division functions into much simpler addition and subtraction operations. Logarithmic values can be stored in lookup tables (LUTs) to facilitate rapid conversion between a log value and its exponential value. Summation of log-domain values is approximated as the maximum logarithmic value to simplify calculations. Although this introduces an approximation error, simulation and hardware implementation results indicate that it is minor. This paper refines our initial design from [29], conducts a much more thorough evaluation, and reports on an FPGA implementation of the log-domain fuzzy controller.

The rest of the paper is organized as follows. Basic concepts for fuzzy controllers are reviewed in Section 2. A survey of published literature on fast defuzzification is also presented. Section 3 reviews log-domain arithmetic principles and describes our log-domain controller. Simulation results for a number of second-order plants are reported in Section 4. Section 5 presents a hardware implementation with speed and resource usage comparisons between pipelined and non-pipelined versions of the log-domain controller. Section 6 concludes the paper.

2 Background Concepts

2.1 Fuzzy Sets and Fuzzy Control

The underlying mathematical construct of fuzzy logic is a fuzzy set, which is a generalization of the classical set theory [8]. It is a set having a characteristic function with a co-domain consisting of the unit interval [0,1] rather than the usual discrete set {0,1}, and the characteristic function is known as a *membership function*. This allows for a gradual transition from non-membership to full membership in the set [9]. In a fuzzy system, each input is associated with a group of fuzzy sets on that measurement dimension.

Figure 1 shows the general structure of a fuzzy controller. A preprocessor deals with any necessary formatting of input data. A fuzzifier module calculates the membership grade for each input in each fuzzy set defined on that dimension. A lookup table [11] usually contains the membership values for all possible (discrete) numerical inputs. This approach, albeit requiring more memory space, can be much faster than calculating the membership values in real-time [12]. The input-output mapping of the controller is controlled by fuzzy rules stored in the rulebase. In the current paper, we focus on TSK controllers, whose rules are of the form: IF x is A and y is B THEN $z = f(x, y)$ where A and B are fuzzy set antecedents and $z = f(x, y)$ is a crisp polynomial function or a constant (usually termed a *singleton*) forming the consequent [9]. Some of the benefits of using singleton consequents are simpler calculations and the possibility of setting extremal values for the control signal [10].

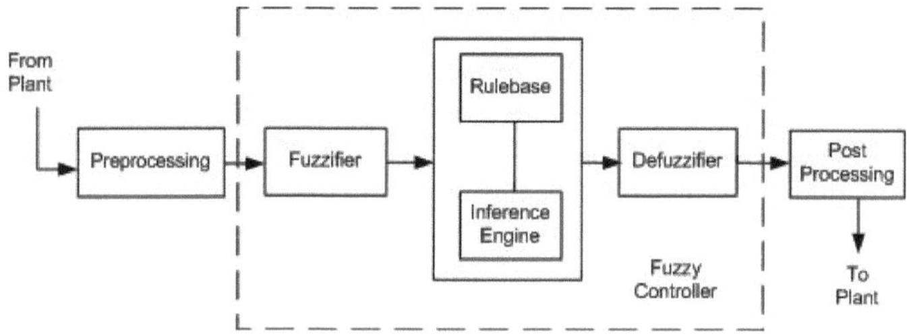

Fig. 1. Fuzzy controller block diagram (based on [10])

For each rule, the inference engine determines the *firing strength* α_i of a rule i, defined as the degree of fulfilment of the rule premise. The memberships for each antecedent predicate are aggregated using a fuzzy conjunction (a t-norm function, usually either the algebraic product or the minimum operator) to form the *activation* of the rule. For TSK controllers, the activation of a rule is used to weight the consequent function. The consequent functions are then *accumulated* using fuzzy disjunction, most commonly the maximum operator. This is the fuzzy output of the controller.

The fuzzy output from the inference engine has to be converted in the *defuzzifier* to a single real number to form a control signal to the plant. There are dozens of defuzzification schemes, such as Center of Gravity (COG), Mean of Maxima (MOM), and Bisector of Area (BOA) [13], with COG being the most common choice in the literature. The crisp control value u_{COG} is the abscissa of the center of gravity of the fuzzy set. For singleton consequents we have:

$$u_{COGS} = \frac{\sum_i y_i w_i}{\sum_i y_i}, \tag{1}$$

where y_i is the firing strength of each rule, and w_i is the consequent output. This is a widely used method, but it is also computationally expensive due to the numerous multiplication and division operations.

2.2 Fast Defuzzification

Defuzzification has always been a bottleneck for fast implementation of fuzzy systems. Therefore, a number of techniques have been proposed to speed up this processing stage. In [14], a heuristic approach based on adapting any fuzzy output shape into one single triangle and estimating the centroid position is presented. The processing time for this approach is reported to be 23 times less than that of COG defuzzification. Runkler and Glesner [15] propose a centroid-approximation algorithm (DECADE) that avoids multiplication and division operations, and the maximum error is reported to be about 7%.

Eisele *et al.* [16] present an approach that reduces the number of computations by skipping all the regions for which the output possibility distribution is zero. One drawback of this method is that the implementation becomes complicated because of the extra circuitry needed to determine the relevance of the regions. Three different implementations of COG defuzzification are proposed in [17, 18]. These methods are specific to the case of trapezoidal output membership functions that form a fuzzy partition (and thus, specific to Mamdani-type controllers).

A COG method with only integer additions and one integer division is presented in [19]. The algorithm maps the real values of fuzzy membership functions onto an integer grid. It is reported to be 12.75 times faster than conventional COG for the truck-backer problem. Introduction of quantization error is one disadvantage. Although multiplication is eliminated, one division operation is still required.

3 Log-Domain Arithmetic and Controller

3.1 Logarithmic Arithmetic

Logarithmic-domain arithmetic has been used to accelerate hardware for many applications where there are a large number of multiplication and/or division operations, e.g. [20]. The fundamental principle of a log-domain system is to take the logarithmic transform of all quantities. For instance, instead of performing a computation $x = f(q_1, q_2)$, the quantity $log(x)$ is computed using an equivalent function $g(log(q_1), log(q_2))$. For example, if $x = q_1/q_2$, then $log(x) = log(q_1) - log(q_2)$, and hence a divider is replaced by a much simpler subtractor. Similarly, multiplication is transformed into addition. If all quantities fall in the range (0, 1], all the logarithmic values are either 0 or a negative number, and the negative signs can be safely ignored, further simplifying calculations. As fuzzy membership values are always within (0, 1], (assuming we ignore membership values equal to 0 in our computations, as is commonly done) fuzzy logic controllers are amenable to this type of implementation. In particular, since the COG method requires multiplication, summation and division, the defuzzifier may see the greatest potential improvement in performance.

A summation operation is complicated in the log-domain. If $x = q_1 + q_2$, with $q_1 > 0$ and $q_2 > 0$, then it can be shown [21]:

$$\begin{aligned} \log(x) &= \log(q_1 + q_2) \\ &= \max(\log(q_1), \log(q_2)) + \log(1 + e^{-|\log(q_1) - \log(q_2)|}) \end{aligned} \tag{2}$$

The second logarithmic term in equation (2) becomes almost zero when q_1 and q_2 are not close in value. This term is often ignored in log-domain arithmetic at a small loss in precision [20], but can also be crudely approximated by a correction factor for some applications [21]. For fuzzy logic, we can use absolute values of all logarithmic quantities, so minimum operations can be changed to maximum and vice versa in the inference engine.

3.2 Approximate Correction Factor

In this section we describe how a sequence of numbers is added in the log-domain controller. Say we need to calculate the sum x of a sequence of numbers $q_1, q_2, \ldots, q_n$, while the only information available to us is the set of their logarithmic values.

$$\text{Now, } \log(x) = \log(q_1 + q_2 + \ldots + q_n)$$

$$= \log(e^{\log(q_1)} + e^{\log(q_2)} + \ldots + e^{\log(q_n)}) \tag{3}$$

If q_1 is the maximum among all the q_i's, $log(q_1)$ is also the maximum among all the log values. Now, from (3),

$$\log(x) = \log(e^{\log(q_1)}(1 + e^{\log(q_2)-\log(q_1)} + \ldots + e^{\log(q_n)-\log(q_1)}))$$

$$= \log(q_1) + \log(1 + e^{\log(q_2)-\log(q_1)} + \ldots + e^{\log(q_n)-\log(q_1)}) \tag{4}$$

If q_2 is the second largest value, the impact of $e^{\log(q_2)-\log(q_1)}$ is the highest among all the other terms. So in our approximation, we treat them as follows:

$$\log(1 + e^{\log(q_2)-\log(q_1)} + \ldots + e^{\log(q_n)-\log(q_1)}) \approx e^{\log(q_2)-\log(q_1)} \tag{5}$$

This is the correction factor applied to our system, instead of the one given in (2). It involves determining two maximum (or, minimum) values instead of one. As will be shown in Section 4, the advantage of using this correction factor is that the control surface becomes better in terms of root-mean-square (RMS) difference with respect to some typical fuzzy controllers. However, step response curves for log-domain controllers with and without a correction factor show no significant difference. The drawback of the correction factor is the added computational cost. Four addition/subtraction operations and two lookup tables are required.

3.3 Design of the Controller

Figure 2 shows a dataflow block diagram of the log-domain controller. The inputs are first passed to the fuzzifier module. It consists of a lookup table (LUT) with $m+1$ columns, where m is the number of membership functions. The first column of each row contains the possible values of the inputs within the input universe and the other columns have the logarithmic membership values. As all the log values are negative, we store the values without the signs. We note that storing logarithmic values in the fuzzifier LUT does not require any extra hardware compared to a conventional LUT fuzzifier.

The inference engine module usually determines the minimum value among the antecedents for each rule. As discussed, all of our values are negative numbers, and so we use the maximum function instead of minimum. The outputs (A_i) are passed to the subtractor block, SUB1, inside defuzzifier. The other input for SUB1, the log of all consequent outputs (S_i), comes from the rulebase. The subtracted outputs D_i go to the COMP1 block, which calculates the first two maximum values, D_1 and D_2. The second maximum value is retained for later use in the CORRECTION block. At the same time, two minimum values, $A1$ and $A2$, are determined from A_i.

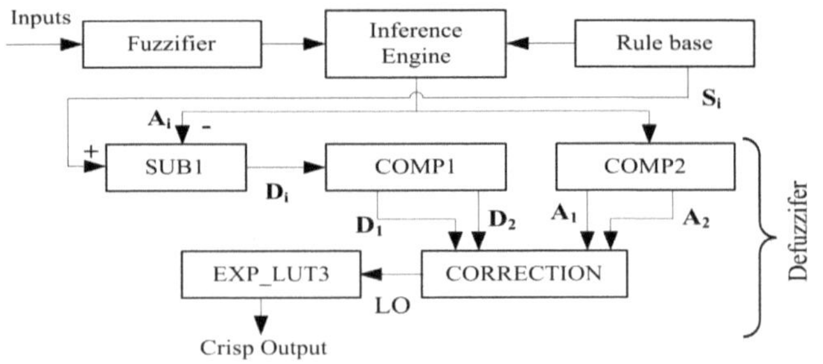

Fig. 2. Block diagram of the log-domain controller

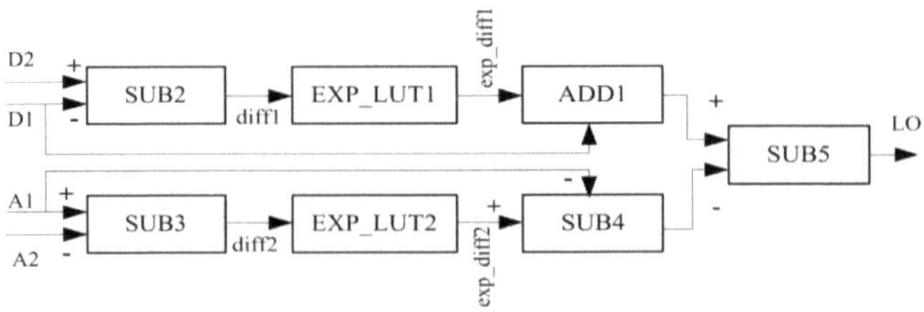

Fig. 3. Block diagram of CORRECTION block

Inside the CORRECTION block (see Fig. 3), a subtractor block, SUB2, computes the difference between D_2 and D_1, and passes this to EXP_LUT1 to get the corresponding exponential value, *exp_diff1*. An adder block, ADD1, computes *term1* as D_1 plus exp_diff1. Similarly, a parallel subtractor block, SUB2, is used for A_1 and A_2, and the output is sent to EXP_LUT2 to determine *exp_diff2*. Then A_1 is subtracted from *exp_diff2* to obtain *term2*. This subtraction functions as an addition with the exception that A_1 is changed back to its original negative value. The final subtractor block, SUB5, calculates the difference, *LO*, between *term1* and *term2*, and passes it back to defuzzifier module for exponential calculation, which results in the final crisp output. If the approximate correction factor is ignored, the CORRECTION block only contains an adder block that sums D_1 with A_1. This is a significant simplification with a negligible performance penalty.

4 Simulation Results

Three controllers: (1) a typical fuzzy controller, (2) a log-domain controller without a correction factor, and (3) one with the correction factor described in Section 3, are implemented in a MATLAB simulation.

Experiment 1

We compare and analyze the performance of the log-domain controllers (with and without correction factor) with a typical fuzzy controller. The benchmark controllers and the plant are taken from [22]. The transfer function of the second-order plant

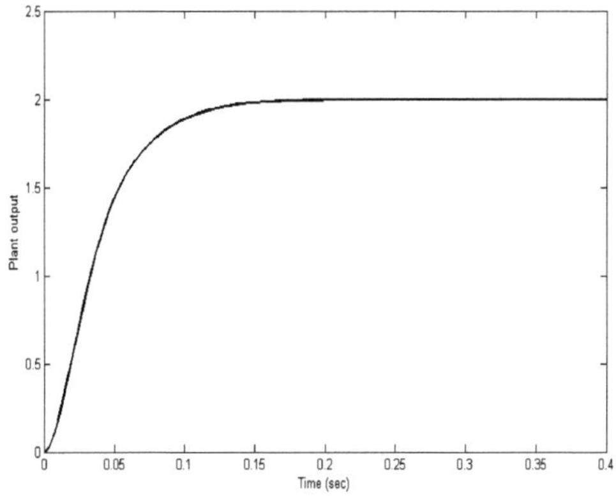

(a)

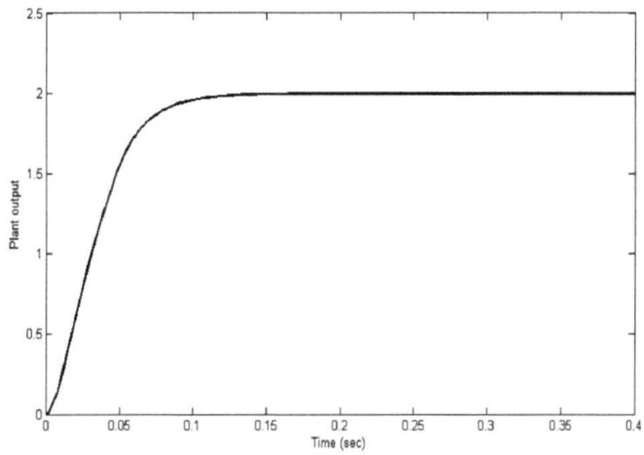

(b)

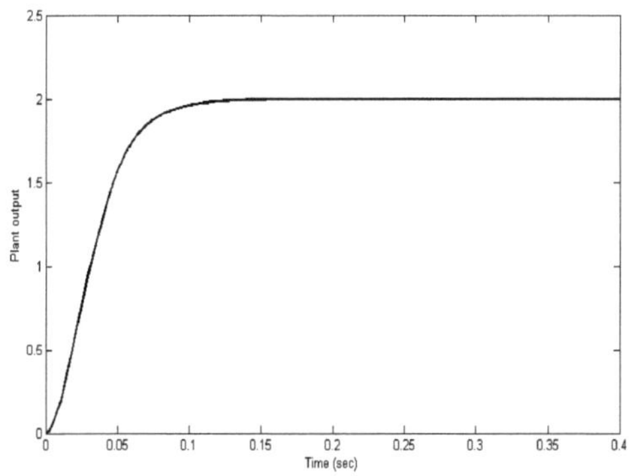

(c)

Fig. 4. Step responses for: (a) typical fuzzy control, (b) log-domain without correction, and (c) log-domain with correction

(representing a DC servomotor) used in our experiments is $\frac{1}{0.02s^2 + s}$. We measure the performance of our controllers by examining the step response curves and control surface plots. Rise time, settling time and overshoot of the step response and the RMS difference between control surfaces are compared.

Table 1. Rise time, settling time, and overshoot for a sampling period of 0.01 second

Controllers	Rise Time (s)	Settling Time (s)	Overshoot (%)
Typical fuzzy	0.0719	0.1298	0
Log-domain (uncorrected)	0.0567	0.1	0.0009
Log-domain (corrected)	0.0548	0.0997	0.0025

Results - A comparison of the step response of the different controllers is presented in Fig. 4. Table 1 illustrates the rise time, settling time, and overshoot for different controllers. *Rise time* is defined as the time the plant outputs take to get to 90% of the step size from a value of 10%. *Settling time* is the time for outputs to settle to within 2.5% of the steady state from control start. We express *overshoot* (output exceeds the steady state value) as a percentage relative to the final value of the plant output.

Table 2. RMS difference between control surfaces

Controllers	Log-domain (uncorrected)	Log-domain (corrected)
Typical	11.7040	5.2965
Log-domain (uncorrected)		16.5292

Figure 5 shows the control surface outputs generated for a typical fuzzy-logic controller, and log-domain controllers (with and without a correction factor). Table 2 reports the RMS difference between control surfaces of the controllers.

Analysis - The log-domain controller without correction uses only the maximum value to approximate the sum of a set of logarithmic values. While this approach relieves us from using computationally expensive multiplication and division operations, the RMS difference between the typical and log-domain controller (uncorrected) is higher (see also Fig. 5(b)). However, with a small correction factor involving the second-highest value in a sequence, the approximate output becomes much more similar to that of typical fuzzy-logic controller (see Fig. 5(c)). The corresponding RMS difference confirms the effectiveness of this approach through a greater than 50% reduction in value of the difference. The impact of having a correction factor is also clear from the RMS difference value between the log-domain controllers. There is not much obvious difference among the step responses of the different controllers, as shown in Fig. 4. Table 1, however, presents some very interesting insights. For each log-domain controller, the rise time is lower than that of the typical fuzzy controller. Log-domain controllers also have lower settling times, although a small overshoot is introduced.

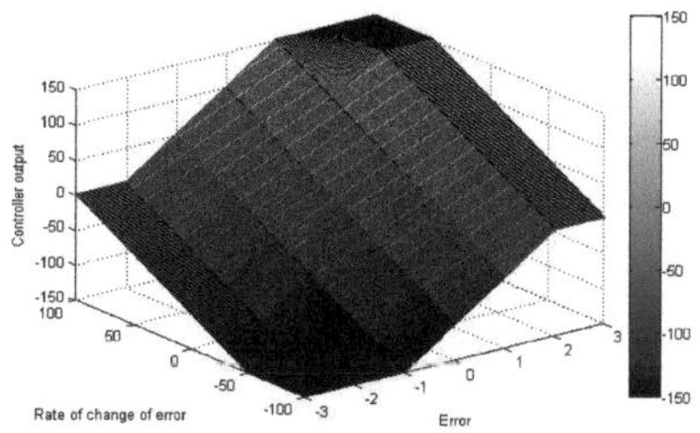

(a)

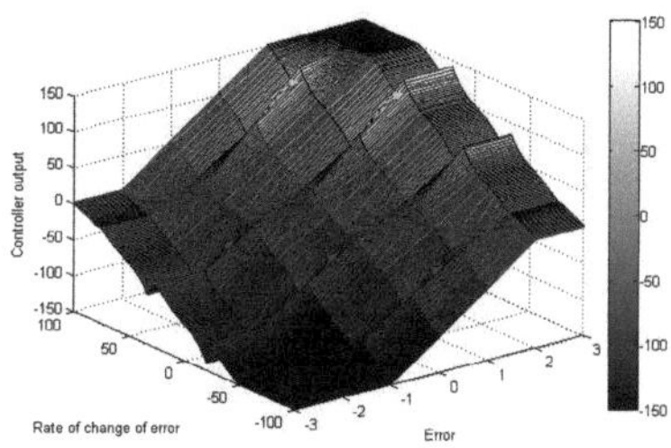

(b)

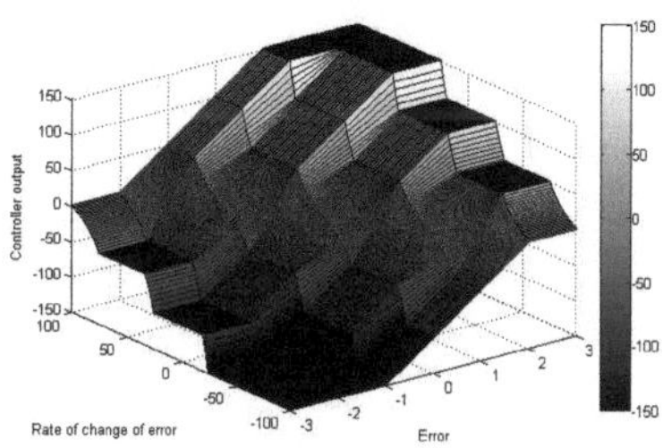

(c)

Fig. 5. Control surfaces of: (a) typical fuzzy control, (b) log-domain without correction, and (c) log-domain with correction

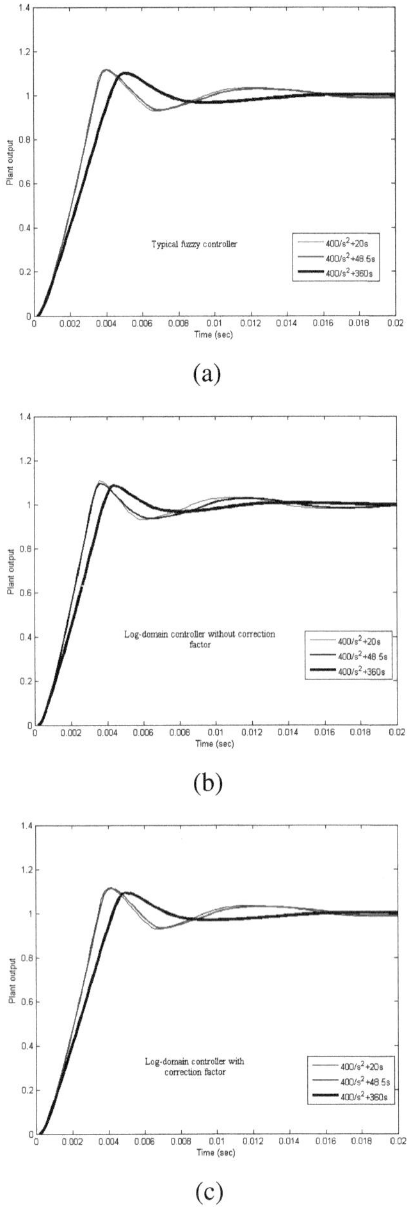

Fig. 6. Step responses of: (a) typical fuzzy control, (b) log-domain without correction, and (c) log-domain with correction

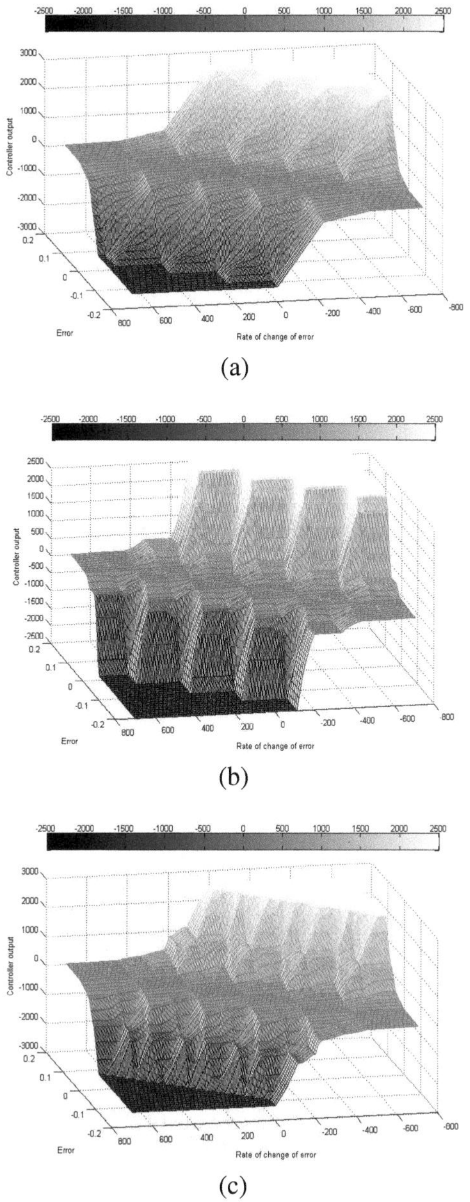

Fig. 7. Control surfaces of: (a) typical fuzzy, (b) log-domain without correction, and (c) log-domain with correction

Experiment 2

We have used three more second-order plants from the literature to verify and analyze the performance of our controller. The plant transfer functions have the general form of $400/(s^2+\sigma s)$, where σ is given three different values: 20, 48.5 and 360, respectively [24]. The membership functions and rule base for the fuzzy controller are also taken from [24].

Results - Figure 6 compares the step response of our log-domain controllers (both with and without correction) with a typical fuzzy controller for the plants mentioned above. Table 3 presents a comparison of rise time, settling time, and overshoot for different controllers and different plants.

Control surface outputs generated for a typical fuzzy controller, and log-domain controllers are shown in Fig. 7. Table 4 provides the root-mean-square (RMS) difference between control surfaces.

Analysis - The step responses confirm the effectiveness of the log-domain controllers. The plots from Fig. 6 are qualitatively similar, demonstrating that log-domain controllers perform as well as a typical fuzzy controller.

Table 3. Rise time, settling time and overshoot for different plants

Plant		Typical Fuzzy	Log-domain (uncorrected)	Log-domain (corrected)
$400/(s^2+20s)$	Rise Time	0.0024	0.0022	0.0024
	Settling Tine	0.0145	0.0126	0.0145
	Overshoot (%)	11.7373	10.8925	11.5596
$400/(s^2+48.5s)$	Rise Time	0.0024	0.0022	0.0024
	Settling Tine	0.0145	0.0132	0.0148
	Overshoot (%)	11.5669	9.6103	11.4486
$400/(s^2+360s)$	Rise Time	0.0030	0.0027	0.0030
	Settling Tine	0.0122	0.0099	0.0123
	Overshoot (%)	10.1534	8.7054	9.4037

Table 4. RMS difference between control surfaces

	Log-domain (uncorrected)	Log-domain (corrected)
Typical	355.0986	120.1319
Log-domain (uncorrected)		332.6803

Table 3 presents some interesting insights. The log-domain controller without correction performs better than both the typical fuzzy controller and its peer with correction for all cases of rise time, settling time and overshoot. The log-domain controller with correction factor has the same rise time and better overshoot compared to the typical controller for all the three plants, although the settling time is slightly higher for a couple of plants. This is a very promising result considering the fact that our log-domain controllers are much faster in processing than the typical fuzzy controller.

The control surface plots are also quite similar in shape for the log-domain and typical fuzzy controllers. Obviously, the surface of log-domain controller with correction factor better resembles the typical fuzzy controller, because of the approximate correction factor. Without this factor, the output is crudely approximated based on only the maximum value in a sequence. That is why there are some extra staircase outputs in the surface of log-domain controller without the correction factor. The RMS difference table also confirms this. A 67% reduction in output difference between the log-domain and typical fuzzy controller is obtained because of the correction factor. The impact of having a correction factor is also clear from the difference between the log-domain controllers.

5 Hardware Implementation

5.1 Design Considerations

We have implemented the log-domain controller without the correction factor on a Xilinx Virtex II FPGA device. The plant and the rulebase are taken from [22]. The transfer function of the second-order plant is $\frac{1}{0.02s^2+s}$. By using the substitution $s=\frac{z-1}{Tz}$, with T = 0.01 s being the sampling period, as in the backward rule approach [25], the continuous-time transfer function is converted to the discrete-time transfer function:

$$\frac{0.0033}{1-1.667\ z^{-1}+0.667\ z^{-2}} \tag{6}$$

This corresponds to the standard form of a second order digital filter:

$$\frac{a_0+a_1z^{-1}+a_2z^{-2}}{1+b_1z^{-1}+b_2z^{-2}}, \tag{7}$$

where a_0 = 0.0033, $a_1 = a_2 = 0$, b_1 = -1.667, and b_2 = 0.667. This type of digital filter is implemented in the following form [23]:

$$\begin{aligned} y(k) &= a_0x(k)+a_1x(k-1)+a_2x(k-2) \\ &-b_1y(k-1)-b_2y(k-2) \end{aligned}, \tag{8}$$

where k = current sample in time, $x(k)$ = input to the plant at kth sample, and $y(k)$ = output from the plant at kth sample.

The fuzzifier module of the log-domain controller accepts two input values, namely the error and rate of change of error, both being 10 bits wide. The error value uses 2 bits for the integer part and 8 bits for the fractional part. Simulation results show the possible error values ranging from 0.0 to 2.0. The rate of change values, however, can be anything between -39.0 and 0.0. Therefore, the 10 bits for rate input actually represent 6 bits for integer and 4 bits for the fractional part. As the rate values are negative, we only consider the absolute quantities without the sign. The lower number of bits for the fractional part does not have any visible impact because of the rate values being sparse. This bit width also determines the size of the LUTs used in the fuzzifier module. The error and rate LUTs have 1024 rows and five columns representing the logarithm of membership function values for NB, NM, ZR, PM and PB. The log values stored in the LUTs are all negative; therefore the sign is ignored. Each value is 16 bits wide with 4 bits for integer part and 12 bits for the fractional part, since the maximum log value can be 9.0. Another LUT contains 4096 rows each having an exponential value for an output value. Since the logarithm of the output can take any value within the range -11.0 to +5.0, we use 1 sign bit, 3 bits for the integer part and 8 bits for the fractional part. The use of 3 integer bits can only represent a negative number as low as -7; but it is justified by the fact that any lower number results in the exponential value being essentially zero. The output from the defuzzifier is represented by 20 bits, where 1 bit is used as sign, 7 integer bits and 12 fractional bits. Since the fuzzy controller output can go as high as 100.0, we have to use at least 7 integer bits to represent the value. Note that for a generalized hardware log-domain controller, the bit widths may be different, making the rows and columns in the LUTs bigger or smaller. The performance, however, should be the same except for a different amount of resource utilization.

The step signal is implemented on the same FPGA as the plant model and the fuzzy inference engine as a square wave to enable the plant outputs to be shown on the oscilloscope. On the positive cycle, the step value goes to 2 from 0; therefore, the first error value becomes 2.0 while the first rate value is zero. These values are sent to the log-domain controller, which produces a defuzzified output to be passed to the plant. The plant output is compared to 2.0 and the difference becomes the next error value. The next rate value is calculated based on the current and the previous error values. Eventually both the error and rate values become zero.

5.2 Log-Domain Controller

The block diagram of our log-domain controller with the plant implemented on FPGA is shown in Fig. 8.

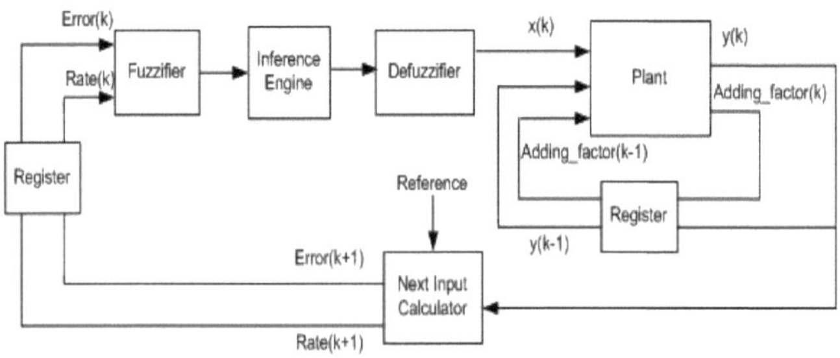

Fig. 8. Block diagram of log-domain controller with plant as implemented on Xilinx Virtex II FPGA

The fuzzifier module passes the corresponding row from both of the error and rate LUTs to the inference engine. The inference engine calculates the maximum value (log of firing strength for a rule) for each of the 25 possible pairs. Although the conventional task of inference engine is to calculate the minimum values, we use maximum instead, since all the negative log values are stored as positive. The six maximum values are passed to the defuzzifier module.

The defuzzifier module subtracts the log of firing strength values (log_fs) from the log of consequent values (log_c), and stores the results in an array named log_d. The maximum value among log_d (log_d_{max}) is calculated, while the minimum for log_fs is also determined. These two values are added together and the summation result (log_output) is used to find the corresponding exponential value from a LUT. This exponential value is assigned to the output, when the particular value of log_c (that contributes to the calculation of the value log_d_{max}) comes from a positive consequent value. Otherwise the exponential value is made negative before assigning it to the final output.

The output from the log-domain controller goes to the plant. It also accepts the plant output and an adding factor from the previous sample. The previous adding factor is used to calculate the current plant output, whereas the previous plant output contributes to generate the new adding factor. The fuzzy controller output (*x(k)*) gets multiplied by a_0 (0.0033) and added to Adding_factor(k-1) to generate the new plant output (*y(k)*). Adding_factor(k) is calculated as $Adding\_factor(k) = -b_1 * y(k) - b_2 * y(k-1)$

The plant output (*y(k)*) is passed to Next Input Calculator which generates the next error and rate values. Error(k+1) is calculated as step minus plant output at k^{th} sample, whereas Rate(k+1) is calculated in two steps. First, a new error value is subtracted from the previous error value, and then the result is multiplied by 1/T (100.0). All the multiplication operations are basically a group of additions because one of the operands is always a constant value. Figure 9 illustrates the operations that take place in the control system.

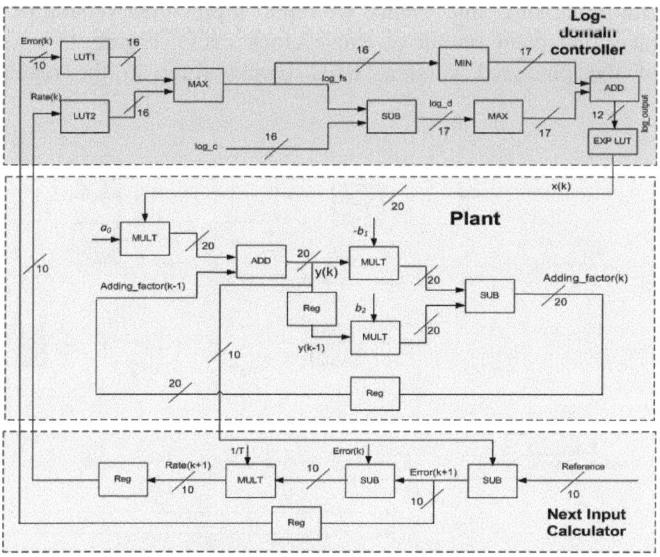

Fig. 9. Operations inside the control system

Results - Figure 10 displays the plant outputs on an oscilloscope for the log-domain controller. Here the positive cycle refers to the plant outputs for a step signal. The negative cycle simply mirrors the plant outputs along the horizontal axis. Note that in Figure 10, one waveform represents the reference and the other waveform represents the step response, i.e. the outputs of the plant.

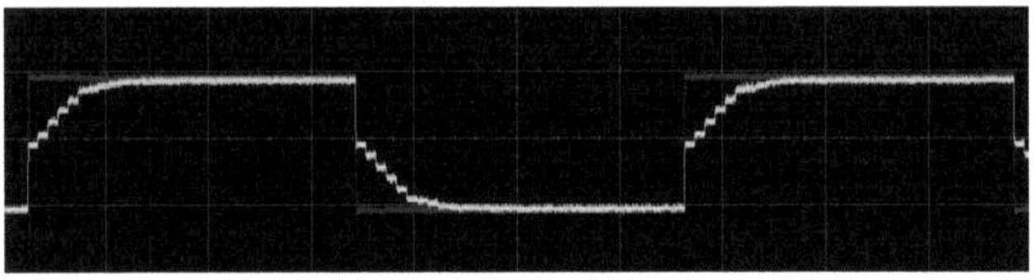

Fig. 10. Plant outputs for log-domain controller implemented on FPGA and displayed on an oscilloscope for both positive and negative cycles

Throughput - From the synthesis report, the maximum clock frequency of our log-domain controller is 67.6 MHz; since our controller produces one plant output per clock cycle, this implies a throughput of 67.6 MFLIPS. This is an improvement over the reported fastest fuzzy controller implementation [7] at 50 MFLIPS.

5.3 Pipelined Log-Domain Controller

One drawback of feedback systems is that the inputs at any sample depend on the outputs from previous samples. This results in a substantial waste of resources for systems where there is no dependency among the inputs. However, we believe our controller is more suitable for networking applications (e.g. per-packet inspection) where inputs would be independent of one another and the output ports cannot be idle for the entire processing time. Hence we report a pipelined version of the controller that generates one plant output at every clock cycle. Figure 11 shows the block diagram of the pipelined version. REG blocks refer to the registers used for pipelining.

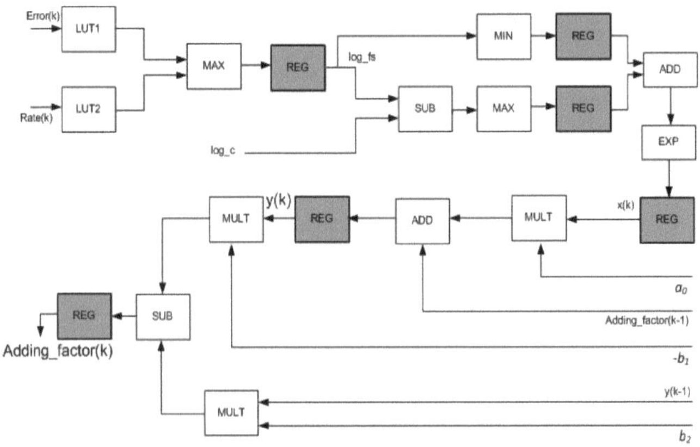

Fig. 11. Pipeline stages of log-domain controller

Results - Figure 12 shows the oscilloscope plot for pipelined version of the controller. Naturally, pipelining in a closed-loop control system results in significant chattering, due to the dependence between successive sample instants. As discussed, we believe that this would nto be a problem in other applications.

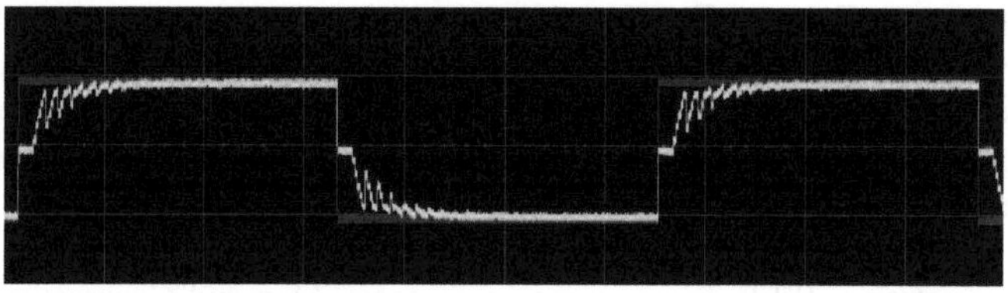

Fig. 12. Plant outputs for pipelined log-domain controller implemented on FPGA and displayed on an oscilloscope for both positive and negative cycles

Throughput - Pipelining results in a processing speed of 248.7 MFLIPS, a speedup of ~370%.

6 Conclusions

In this paper, a logarithmic-arithmetic-based fuzzy-logic controller is described, which results in a very high-speed hardware implementation. This approach removes computationally expensive multiplication and division operations that have been the bottlenecks for fuzzy control systems. As the simulation results illustrate, log-domain implementations – with and without the approximate correction factor – perform better than the typical fuzzy controller in terms of rise time and settling time of the step response curves. The hardware implementation on FPGA without the correction factor also achieves similar response to the step signal. The processing speed of the hardware version is 67.6 MFLIPS, which exceeds the fastest fuzzy controller implementation in the literature that we are aware of by 33%. Note, however, that that implementation [7] was performed with two 8-bit inputs, one 8-bit output, 5 membership functions for each variable and 25 rules. If we had implemented this controller on the same Xilinx Virtex II FPGA that we have used for our log-domain controller, it would have resulted in a different speed. Nevertheless, in this paper, we have been able to develop a very fast fuzzy controller using logarithmic arithmetic, which is a new approach for fuzzy logic control systems. The experiments show the ability of the controllers to produce outputs that are close to the expected ones. A further speedup to 248.7 MFLIPS is achieved by a pipelined version of the log-domain controller. We believe that this is a very promising result making the log-domain controllers potentially suitable for high-speed and large networks.

References

[1] Zadeh, L.: Fuzzy Sets. Information and Control 8(3), 338–353 (1965)

[2] Mamdani, E.H.: Application of Fuzzy Logic to Approximate Reasoning. IEEE Trans. Computers (26), 1182–1191 (1977)

[3] Han, I.S.: Membership Function Circuit for Neural/Fuzzy Hardware of Analog-Mixed Operation Based on the Programmable Conductance. In: IEEE Int. Fuzzy Systems Conf., pp. 1–4 (2007)

[4] Tigaeru, L.: Programmable analogue membership function circuit for hybrid-mode fuzzy systems. Electronics Letters 39(8), 642–644 (2003)

[5] Kalaykov, I., Tolt, G.: Fast fuzzy signal and image processing hardware. In: Annual Meeting of the North American Fuzzy Information Processing Society, pp. 7–12 (2002)

[6] Dick, S., Gaudet, V., Bai, H.: Bit-serial arithmetic: A novel approach to fuzzy hardware implementation. In: Annual Meeting of the North American Fuzzy Information Processing Society, pp. 1–6 (2008)

[7] Salcic, Z.: High-speed customizable fuzzy-logic processor: architecture and implementation. IEEE Transactions on Systems, Man and Cybernetics, Part A 31(6), 731–737 (2001)

[8] Stoll, R.R.: Set Theory and Logic. Dover Publications, New York (1979)

[9] Filev, D.P., Yager, R.R.: Essentials of Fuzzy Modeling and Control. Wiley, New York (1994)

[10] Jantzen, J.: Foundations of Fuzzy Control. Wiley, New York (2007)

[11] Ibrahim, A.: Fuzzy Logic for Embedded Systems Applications. Elsevier (2004)

[12] Kim, Y.D.: High Speed Flexible Fuzzy Hardware for Fuzzy Information Processing. IEEE Trans. on Systems, Man, and Cybernetics – Part A 27(1), 45–56 (1997)

[13] Lee, C.S., Lin, C.: Neural Fuzzy Systems: A Neuro-Fuzzy Synergism to Intelligent Systems. Prentice Hall, Alexandria (1996)

[14] Ginart, A., Sanchez, G.: Fast Defuzzification Method Based on Centroid Estimation. Applied Modelling and Simulation (2002)

[15] Runkler, T., Glesner, M.: DECADE—fast centroid approximation defuzzification for real time fuzzy control applications. In: Proceedings of the 1994 ACM Symposium on Applied Computing, pp. 161–165 (1994)

[16] Eisele, M., Hentschel, K., Kunemund, T.: Hardware realization of fast defuzzification by adaptive integration. In: Proceedings of the Fourth International Conference on Microelectronics for Neural Networks and Fuzzy Systems, pp. 318–323 (1994)

[17] Broekhoven, E., Baets, B.: Fast and accurate center of gravity defuzzification of fuzzy system outputs defined on trapezoidal fuzzy partitions. Fuzzy Sets and Systems 157(7), 904–918 (2006)

[18] Patel, A.: Transformation functions for trapezoidal membership functions. International Journal of Computational Cognition 2(3), 115–135 (2004)

[19] Lee, S.-G., Miyazaki, M., Kim, J.-I.: Design of Very High-Speed Integer Fuzzy Controller Without Multiplications by Using VHDL. In: Apolloni, B., Howlett, R.J., Jain, L. (eds.) KES 2007, Part I. LNCS (LNAI), vol. 4692, pp. 93–100. Springer, Heidelberg (2007)

[20] Robertson, P., Villebrun, E., Hoeher, P.: A Comparison of Optimal and Sub-Optimal MAP Decoding Algorithms Operating in the Log Domain. In: IEEE Int. Conf. Communications, Seattle, WA, pp. 1009–1013 (1995)

[21] Erfanian, J., Pasupathy, S., Gulak, G.: Reduced Complexity Symbol Detectors with Parallel Structures. In: IEEE GLOBECOM, pp. 704–708 (1990)

[22] Kawaji, S., Maeda, T., Matsunaga, N.: Fuzzy Control Using Knowledge Acquired from PD Control. Industrial Electronics, Control and Instrumentation 2, 1549–1554 (1991)

[23] Gwaltney, D., King, K., Smith, K., Ormsby, J.: Implementation of Adaptive Digital Controllers on Programmable Logic Devices (2002), https://www.researchgate.net/publication/23892297_Implementation_of_Adaptive_Digital_Controllers_on_Programmable_Logic_Devices

[24] Taur, J., Tao, C.: Design and Analysis of Region-Wise Linear Fuzzy Controllers. IEEE Trans. on Systems, Man and Cybernetics 27(3), 526–532 (1997)

[25] Macia, N.F., Thaler, G.J.: Modeling and Control of Dynamic Systems. Delmar Cengage Learning, Albany (2004)

[26] Barolli, L., et al.: A CAC scheme for multimedia applications based on fuzzy logic. In: Proc. Int. C. Advanced Inf. Networking & Apps., Taipei, Taiwan (2005)

[27] Ascia, G., Catania, V., Panno, D.: An evolutionary management scheme in high-performance packet switches. IEEE/ACM Trans. Networking 13(2), 262–275 (2005)

[28] Lin, H.-P., Juang, R.-T., Lin, D.-B.: Validation of an improved location-based handover algorithm using GSM measurement data. IEEE Trans. Mobile Comput. 4(5), 530–536 (2005)

[29] Razib, A., Dick, S., Gaudet, V.: Design of a High-Speed Fuzzy Logic Controller Based on Log-Domain Arithmetic. In: Proceedings, ISMVL 2009, Naha, Okinawa, Japan, May 21-23, pp. 139–144 (2009)

Fuzzy Clustering with Prototype Extraction for Census Data Analysis

Oleg Chertov and Marharyta Aleksandrova

Applied Mathematics Department, National Technical University of Ukraine "Kyiv Polytechnic Institute",
37 Peremohy Prospekt, 03056 Kyiv, Ukraine
chertov@i.ua, rita.v.aleksandrova@gmail.com

Abstract. Not long ago primary census data became available to publicity. It opened qualitatively new perspectives not only for researchers in demography and sociology, but also for those people, who somehow face processes occurring in society.

In this paper authors propose using Data Mining methods for searching hidden interconnections in census data. A novel clustering-based technique is described as well. It allows determining factors which influence people behavior, in particular decision-making process (as an example, a decision whether to have a baby or not). Proposed technique concerns contrast mining as it is based on dividing the whole set of respondents on two contrasting groups. The first group consists of those, who possess a certain feature (for instance, has a baby) unlike members of the second group. We propose define clustering based subgroups out of the first group and their prototypes out of the second one. By means of analyzing subgroups' and their prototypes' characteristics it is possible to identify which factors influence the decision-making process. Authors also provide an experimental example of the described approach usage, which additionally shows that fuzzy clustering provides more accurate results than hard clustering techniques.

1 Introduction

Nowadays there exist two considerable global tendencies. The first trend is that the amount of digital information yearly produced by humanity increases significantly. Thus, for the last five years it grew up by 9 times and comprised almost $1.8 \cdot 10^{12}$ Gb [1]. The second trend is related to the fact that not only aggregated (consolidated) data but also primary ones become more available. Researchers can relatively easy get access to the great variety of primary data such as information on patients' hospital treatment (so-called Clinical Data Repositories [2]), electronic commerce results in big automated collections of consumer data, microfiles with large census data samples etc.

Usually, the access is given not to the complete primary data sets but to the data samples (microfiles). In addition, values of some attributes are masked somehow, or even absolutely unavailable. These restrictions are necessary to provide the published

R.R. Yager et al. (Eds.): *Soft Computing: State of the Art Theory*, STUDFUZZ 291, pp. 289–313.
DOI: 10.1007/978-3-642-34922-5_20

data anonymity, which is a subject to special legal regulation in different countries (e.g., see the Health Insurance Portability and Accountability Act of 1996 (HIPAA) [3]; the Patient Safety and Quality Improvement Act of 2005 (PSQIA) [4] concerning the health data protection in the USA; Directive on privacy and electronic communications [5] about electronic commerce in the EU; the State Statistics Law [6] about providing confidentiality of the primary statistical information in Ukraine).

However, the amount of available primary data keeps growing. The IPUMS project is the most significant example. It started in 1992 [7] with the main goal to collect and distribute census data for researchers from around the world. Within it, more than 397 million person records collected from 185 censuses held in 62 countries (at the moment this paper is being written) are accessible. The social importance of results, which can be obtained through such data analysis, can scarcely be overestimated. Still, such analysis requires using appropriate tools.

2 Related Works

2.1 Methods for Census Data Analysis

Statistical and OLAP databases are the main sources of population census data [8]. Different statistical methods are actively used for analyzing it, among them analysis of variance (ANOVA), regression analysis, log-linear analysis, nonparametric approaches [9]. While using statistical methods it is possible to achieve important results, but this approach requires prior knowledge or at least assumptions about some patterns or interconnections existence.

2.2 Basic Data Mining Techniques

As opposite to statistical methods intelligent data analysis (*Data Mining*) makes it possible to discover such patterns, which even wasn't suspected to exist. That is why Data Mining refers to extracting ("mining") knowledge from large amounts of data [10].

Among the most widely used Data Mining techniques it is possible to define [11]: clustering, classifying, nearest neighbor prediction, decision trees, neural networks, and association rules.

Clustering is a process of grouping related points in a given set on the basis of having similar characteristics [12]. Thus, clustering discovers natural accumulations in data sets.

Modern clustering algorithms can be divided in two groups: on-line and off-line methods [13]. On-line methods use every new data point for cluster centers identification. Thus, the system learns while adding new elements. The other approach computes cluster centers only once and they can't be changed lately. Considering static nature of the census data it is logical to use off-line methods.

Commonly used off-line algorithms are k-means clustering, fuzzy c-means clustering, mountain clustering, and subtractive clustering [13]. First two algorithms require prior knowledge of clusters number. Both mountain and subtractive clustering techniques implement the same algorithm. The only major difference is that the first one examines every possible point in the data space (bounded by minimum and maximum values in each dimension) in order to discover cluster centers, and the second one goes only over points of the clustered data set. The latter approach significantly speeds up the algorithm's performance.

2.3 Contrast Data Mining

As it was said in [14] "contrast data mining is concerned with the mining of patterns and models that contrast two or more datasets".

It is natural for humanity to contrast different things and notions: day and night, good and evil, true and false. "Without darkness there would be no concept of light. In order to realize light itself there must be its antipode – darkness". This aphorism belongs to Lion Feuchtwanger and represents philosophical view on the nature of contrasting and its importance in cognitive process. Probably this confrontation takes its origin from the very beginning of mankind evolution. We can see it, for example, in fairy tale characters, which are bad and evil or kind and just. This tendency continues in modern culture as well (cinema and literature keep contrasting good and evil characters).

In the late 90s and early 00's a new direction based on the similar contrasting of two or more datasets have occurred in Data Mining research field. It was successfully used for solving problems of different nature and thus gained a wide popularity. Papers [15, 16] provide a comprehensive study of contrast mining and its applications.

According to [15] data analysis techniques in this field can be divided into predictive and descriptive induction. Predictive induction mainly dials with labeled data and subgroup discovery trying to predict class value of previously unseen examples. The latter approach works with contrast sets introduced by S.D. Bay and M.J. Pazzati in [17] and emerging patterns proposed by G. Dong and J. Li in [18]. Although until recently this techniques have evolved separately mainly in machine learning and data mining communities respectively, Novak et al. in their paper [15] proved equivalence of notions "subgroup discovery", "contrast sets" and "emerging patterns". Authors also marked out some approaches which are closely related to contrast mining. Among them fundamental rule changes [19], closed sets [20] and exception rule [21] mining.

Despite the fact that contrast mining is a relatively new approach it has been developed intensively by researchers from different countries. Dong and Bailey in their paper [16] mentioned that scientists have already developed more powerful constructs based on contrasting patterns. Among them disjunctive emerging patterns [22], fuzzy emerging patterns [23], contrast inequalities [24], contrast functions [25] and emerging cubes [26].

Theoretical researches in contrast mining also concern developing contrast patterns mining algorithms. The basic ones are STUCCO [17] and border based algorithm [18].

Among most prominent results in contrast mining application we can mention classification [27] and clustering [28] algorithms based on contrast patterns notion, researches in bioinformatics [29], blog community [30] and image [31] analysis.

2.4 Data Mining in Census Analysis

Intelligent data analysis techniques are widely used in different scopes of human activities: education [32], medicine [33], banking [34], marketing [35], and so on. Still, Data Mining methods are almost not used for census data analysis. There are only few works dealing with intelligent analysis of census data.

References [36-38] describe SPADA (*spatial pattern discovery algorithm*) system, which was designed for discovering spatial association rules in census data. Proposed method of rules discovery is based on a multi-relational data mining approach and uses representation and reasoning techniques developed in inductive logic programming.

CHARM, an algorithm for mining all frequent closed itemsets is proposed in [39]. CHARM is an alternative to Apriori-inspired algorithms [40, 41]. It was tested on census databases.

In contrast mining researches census data was primary used for patterns discovering algorithms testing. In [17] authors tried to find differences between people holding PhD and bachelor degrees analyzing adult census data and STUCCO algorithm. In [18] dissimilarities between Texas and Michigan census data were investigated. Authors didn't provide any significant interpretation of the obtained contrast patterns.

Paper [42] discusses special requirements that occur for subgroup mining of spatial census data. Authors also describe subgroup mining system (SubgroupMiner), which provides multi-relational hypotheses support, efficient data base integration, discovery of causal subgroup structures, and visualization based interaction options.

Summing up we can say that data mining techniques are poorly used in census data analysis and mainly dial with consolidated data. We believe that using primary data can give qualitatively new and more interesting results.

Despite the fact, that contrast mining was previously applied to census data analysis, the purpose of such investigations didn’t concern searching factors, which can stimulate natural moving of objects from one contrasting set to another. We are also not aware of any works dealing similar tasks for other demographic or social data. Still this problem remain actual, especially when it is necessary to find ways of encouraging people to make important for the society decisions, such as to increase number of family members, to migrate to another region and so on.

Basic concept of this work was presented on the World Conference on Soft Computing, San-Francisco, USA in May 2011 [43]. We presented novel Influence Search Algorithm, which uses hard clustering as one of the steps. With its help we achieved some interesting results:

- providing financial support or cheep housing loans increases birth-rate;
- providing financial support to the youngest age group representatives with high education level probably won't contribute to their desire having babies, because

most highly educated couples decide to have a baby after both spouse are 30 years old;

- if we want couples from the oldest age group to have a baby, we should actively encourage them materially, a lot of families from this group lack of money or own dwelling; besides, if the first child is not born before one of spouse reaches 40 years, the probability of his appearance reduces significantly.

Still world around us is not strict and all concepts are usually fuzzy. Often objects belong to several groups with some membership grade. Thus the main scope of this paper is to investigate fuzzy clustering application within Influence Search Algorithm.

3 Problem Formulation and Solution

3.1 Microfile Structure

A census microfile is a set of records of two types [44]: houschold rccords and person records. Each record contains certain attributes. For example, among household attributes there are state code, building size, number of rooms, type of unit (housing unit, institutional group quarters, noninstitutional group quarters) so on; person records contain such information as sex, age, marital status, race. All attributes are numerically coded.

Each person record has a field which determines this person's relationship with a householder. This field can take different values: "husband/wife", "natural born son/daughter", "uncle/aunt" so on.

3.2 Paper Purpose

The main purpose of this paper is to expand proposed approach for identifying factors, which stimulate or vice versa do not contribute to the decision making process concerning respondents' life arrangement, with fuzzy clustering.

During the experiment we analyzed population census microfile, as it contains all necessary socio-economical data. Clustering was chosen as an analysis technique, because it allows discovering natural data accumulations, thus, it gives an opportunity to identify which parameters consolidate certain people groups. We used subtractive clustering algorithm to identify cluster number, because it doesn't require prior knowledge of this parameter and is relatively fast [45]. Note, that in our case it is almost impossible to define clusters number beforehand. FCM algorithm [46] was chosen as a fuzzy clustering technique.

3.3 Subtractive Clustering Algorithm

Before using subtractive algorithm all data points must be rescaled in order to fall within a unit hypercube [47]. After this potential $P_i^{(1)}$ of each data point x^i is calculated by (1).

$$P_i^{(1)} = \sum_{j=1}^{n} \exp\left(-\frac{\left\| x^i - x^j \right\|^2}{(r_a/2)^2} \right), \tag{1}$$

where r_a is a positive constant called cluster radius, n is a number of data-points in the clustered set.

A point with the highest potential is considered to be a first cluster center c_1. Then all potentials are recalculated by (2):

$$P_i^{(2)} = P_i^{(1)} - P_{c_1} \exp\left(-\frac{\left\| x^i - c_1 \right\|^2}{(r_a \eta/2)^2} \right), \tag{2}$$

where P_{c_1} is a potential of the first cluster center, η is a quash factor. By means of this recalculating influence of the first cluster center is excluded, as potentials of all close to it data points are reduced significantly.

Data point with the next highest potential $P_k^{(2)}$ is tested to be the second cluster center.

In general, after the m-th cluster center has been obtained potentials of all data points are recalculated by (3):

$$P_i^{(m+1)} = P_i^{(m)} - P_{c_m} \exp\left(-\frac{\left\| x^i - c_m \right\|^2}{(r_a \eta/2)^2} \right), \tag{3}$$

where c_m is the m-th cluster center.

Point c_{m+1}^* with the highest potential $P_l^{(m+1)}$ is accepted as an $(m+1)$-th cluster center if

$$P_l^{(m+1)} > \overline{\varepsilon} P_{c_1}, \tag{4}$$

where $\overline{\varepsilon}$ is an accept ratio – the potential of the cluster center candidate as a fraction of the first cluster center potential, above which data point c_{m+1}^* is accepted as a cluster center.

Else if condition (5) holds point c_{m+1}^* is rejected as a cluster center and the clustering process ends.

$$P_l^{(m+1)} < \underline{\varepsilon} P_{c_1}, \tag{5}$$

where $\underline{\varepsilon}$ is a reject ratio.

Otherwise the following condition is verified

$$\frac{d_{\min}}{r_a} + \frac{P_l^{(m+1)}}{P_{c_1}} \geq 1, \tag{6}$$

where $d_{\min}$ is the minimal distance between c_{m+1}^* and all previously defined cluster centers.

If (6) holds c^*_{m+1} is accepted as a cluster center and the algorithm starts next iteration. Else c^*_{m+1} is rejected and its potential is set to 0; data point with the next highest potential on the current iteration is re-tested.

Quash factor, accept and reject ratios are specified by the user. Their default values in MATLAB framework are 1.25, 0.5 and 0.15 respectively. Cluster radius is also an input argument. After all centers are defined the origin set can be divided into appropriate number of clusters by referring each point to that cluster, which center is the closest to it.

3.4 Fuzzy C-Means Clustering Algorithm

Fuzzy c-means clustering algorithm (FCM) [46] assigns each data point to every cluster with a certain membership grade, which is represented by a value within unit segment [0, 1].

Data partitioning is represented with a membership matrix *U*. Matrix columns correspond to data points and its rows – to clusters. Equation (7) represents the condition imposed on membership grades: sum of data point belongingness degrees to all clusters is always equal to unity.

$$\sum_{i=1}^{c} u_{ij} = 1, \quad \forall j = 1,\ldots,n, \tag{7}$$

where c is number of clusters, n is number of data points.

At the beginning of the algorithm membership matrix is initialized with random values within [0; 1] holding condition (7).

After that centers of fuzzy clusters are calculated by (8):

$$c_i = \sum_{j=1}^{n} u_{ij}^m x_j \Big/ \sum_{j=1}^{n} u_{ij}^m, \tag{8}$$

with c_i as a center of the i-th cluster, $m \in [1;\infty)$ as weighting exponent and x_j as a data point.

Then cost (objective) function J of the data dissimilarity is calculated by (9):

$$J = \sum_{i=1}^{c} J_i = \sum_{i=1}^{c} \sum_{j=1}^{n} u_{ij}^m \left\| c_i - x_j \right\|^2 . \tag{9}$$

Here clustering process finishes if one of the following conditions holds:

- J is less then a certain tolerance value;
- its improvement over previous iteration is below a certain threshold (minimum amount of improvement).

In practice other stopping conditions can also be used, for example, if number of iterations exceeds some predefined value (MATLAB framework).

If algorithm doesn't stop, values of the membership matrix *U* are recalculated by (10), cluster centers are updated according to (8) and new iteration starts.

$$u_{ij} = 1 \Bigg/ \sum_{k=1}^{c} \left(\frac{\|c_i - x_j\|}{\|c_k - x_j\|} \right)^{2/(m-1)} \tag{10}$$

So, fuzzy c-means clustering algorithm (consider realization implemented in MATLAB) accepts following input arguments: a set of data points, cluster number, weighting exponent, maximum number of iterations and minimum amount of improvement. Default values of the latter three arguments are 2, 100 and 10^{-5} respectively.

Considering the fact that initially matrix U contains random values, the algorithm results may vary. Hard clustering techniques can be also used beforehand in order to identify number of clusters. FCM returns membership matrix and found cluster centers.

3.5 Novel Influence Search Algorithm Based on Clustering and Prototype Definition

We proposed a novel Influence Search Algorithm for solving a problem posed in the current section. This algorithm consists of the following steps.

1. Separate two contrasting groups N_1 and N_2 out of the origin dataset. The first group should contain records about those respondents who possess a certain characteristic, and the second one – records about respondents who do not possess it. Additional restrictions can be also imposed on the group definition process (according to the results of the problem domain analysis).
2. Identify those attributes, which can potentially influence the chosen characteristic presence. Mark out attributes for clustering, i.e., attributes which are numerical or can be compared by numbers. Basing on the problem specific define invariant parameters for groups N_1 and N_2.
3. Cluster group N_1, divide it on subgroups.
4. Identify range of values for each invariant parameter corresponding to the subgroups bounds.
5. Using obtained values ranges define subgroups prototypes out of the group N_2.
6. Compare characteristics of clustering based subgroups and their prototypes; summarize results.

4 Experimental Results

The task of the experiment was to identify which factors influence human desire to have a baby.

We took a microfile with 5-percent sample of the California census data for 2000, which contains records about 610369 family households (we ignored subfamilies as number of households with subfamilies comprise only 3.6% of the initial sample) [44].

Following attributes were considered:

- home ownership,
- type of building,
- number of vehicles available,
- commercial business on property,
- spouse age,
- spouse education,
- spouse ancestry,
- class of worker for each spouse,
- husband's total income (in 1999).

Such parameter as wife's total income wasn't considered as a lot of women go on a maternity leave, thus, in our case this parameter can't be used as a family state factor.

We separated two contrasting groups out of the origin set of families. Group N_1 contains families with one or two children aged from 0 to 2 years, group N_2 is composed of families without children. Such restrictions on the children's age were imposed in order to track the change in family state from childless to a family with a little child (children).

To obtain reliable results we also imposed some additional restrictions on groups N_1 and N_2:

- all the families must be complete (presence of both spouse is obligate);
- both husband and wife must be without disabilities;
- spouse age must be within the most favorable period for having babies.

These conditions are quite relevant and obvious. For instance, it is clear that illness of the potential parents affects significantly their willingness and ability to have children.

In order to identify bounds of the most favorable age for having babies authors had to conduct additional researches. We calculated the number of families having a child or two children from 0 to 2 years depending on spouse age. Obtained results are shown on Fig. 1.

From the diagrams it is clearly seen that the age of the first babies bearing in families is almost normally distributed with average values 32 years for men and 30 years for women. As the interval of the most favorable age for having babies we took ranges 24-38 for men and 22-37 for women (values which correspond to 400 and more families). The size of the obtained sample is 3/4 of the initial set of families with a child or two children from 0 to 2 years.

Considering all imposed above restrictions we got groups N_1 and N_2 with 8299 and 12249 elements respectively.

The next step of the analysis was to identify parameters for clustering group N_1. We decided to use following parameters:

- spouse age,
- spouse education,
- total husband's income.

Such attributes as home ownership, class of worker and ancestry can't be used for clustering as they are qualitative, not quantitative characteristics and can be hardly compared by numbers.

As an invariant parameter for both contrasting groups we took spouse age, because all other parameters can change their values depending on state government policy (for example, young families can be provided with cheap education loans or material assistance).

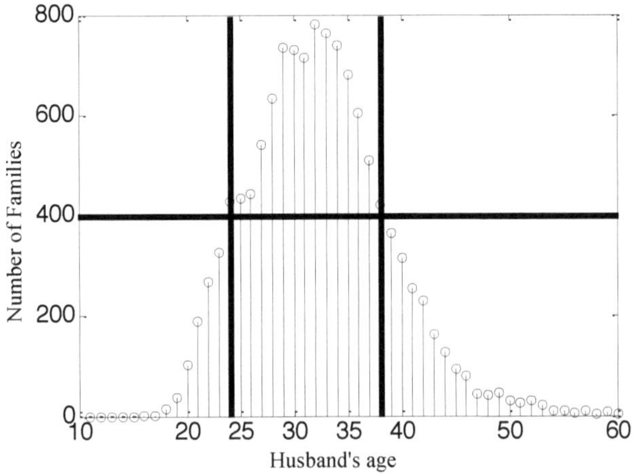

a)

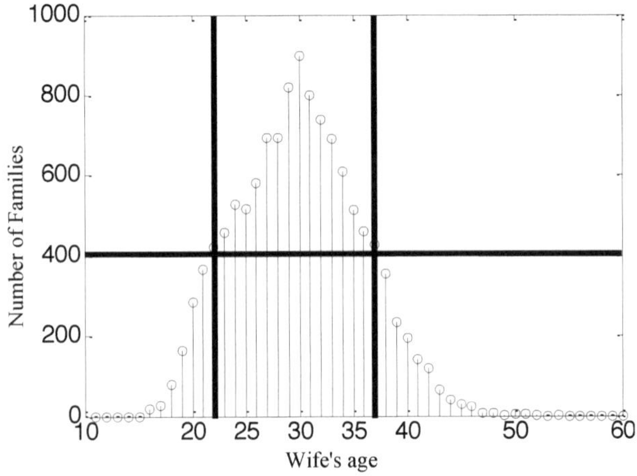

b)

Fig. 1. Husband's (a) and wife's (b) age distribution in families with little children

Using subtractive algorithm for clustering group N_1 on the mentioned above attributes (we used default values of input arguments and 0.5 as a cluster radius) resulted in 3 clusters. This value was taken as an input argument for the FCM (other arguments were taken with their default values). Because FCM itself doesn't perform data normalization we scaled data into a unit hypercube beforehand. It was done in order to eliminate strong influence of the variables with large ranges over those with narrow rage (compare, for example, variable "husband's total income" varying from $(-1)*10^4$ to $72*10^4$ and "education" with range interval 1–16).

As a result we obtained membership matrix and cluster centers (the latter was rescaled to the initial ranges). Cluster centers are given in Table 1 (education codes presented in Table 2). Because FCM algorithm considers data set as points in continuous space, it can result in fractional values of the cluster centers' coordinates, even if data is discrete itself. Unlike FCM subtractive clustering works only with data points of the investigated set. So each resulting cluster center corresponds to a point from the original set. Fractional coordinates of cluster centers obtained with FCM must be interpreted correctly. For example, husband's age in the first cluster center with value 30.94 means that most men from this cluster are of age close to 31. Still value 30.94 provides minimum of the objective function (9).

Table 1. Cluster Centers

Clusters	**1**	**2**	**3**
Husband's age	30.94	27.24	34.66
Wife's age	29.24	25.39	32.97
Husband's education	11.61	9.28	12.25
Wife's education	11.68	9.58	12.24
Husband's total income (in 1999), $*10^4	5.26	3.19	6.80

Table 2. Education Codes

1	No schooling completed	**9**	High school graduate
2	Nursery school to 4th grade	**10**	Some college, but less than 1 year
3	5th grade or 6th grade	**11**	One or more years of college, no degree
4	7th grade or 8th grade	**12**	Associate degree
5	9th grade	**13**	Bachelor's degree
6	10th grade	**14**	Master's degree
7	11th grade	**15**	Professional degree
8	12th grade, no diploma	**16**	Doctorate degree

Analyzing obtained cluster centers we can say that:

- cluster №2 contains the youngest couples with the lowest education level and the lowest incomes; relatively low income and education levels are partly a consequence of the fact that some respondents are still studying;
- cluster №3 contains the oldest couples with the highest education and income levels;

- cluster №1 is comprised of those couples whose age is the closest to the expected values of the age distribution; education and income have intermediate values between 2-d and 3-d clusters;
- in each cluster husband and wife have almost the same education levels.

After that all data points were divided into 3 clusters: a point was related to that cluster to which it belongs with the highest grade.

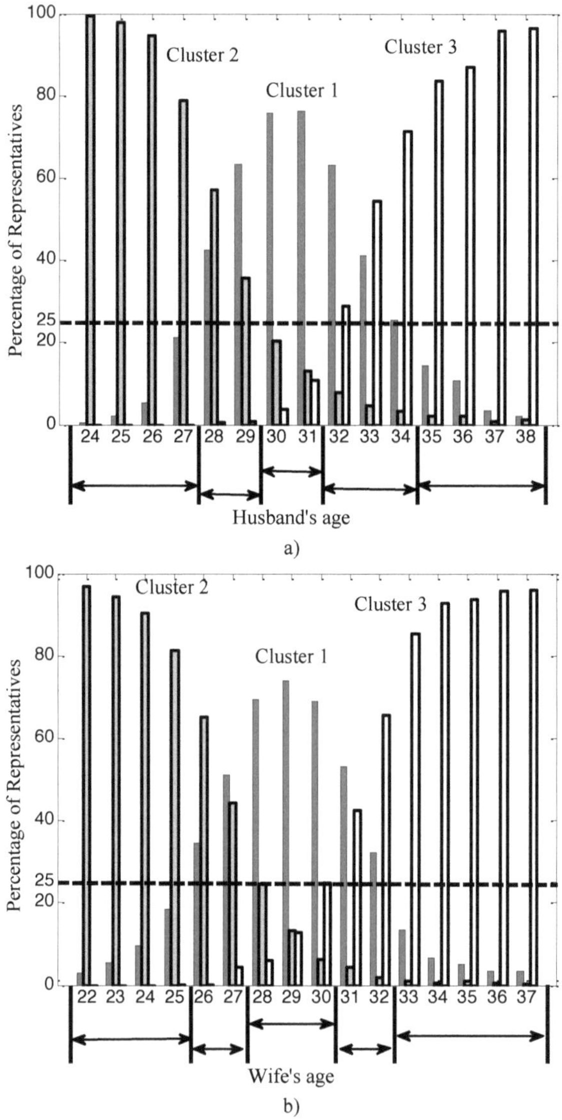

Fig. 2. Husband's (a) and wife's (b) age distribution through clusters for fuzzy clustering (FCM)

In order to divide group N_1 on subgroups we calculated the percentage of representatives of a certain age *through* clusters. Results are presented on Fig. 2. From it we can clearly see that almost all young couples belong to the second cluster, families with the oldest husbands and wives fall mostly within the third cluster, and a lot of families where spouse are of middle age correspond to the first cluster. It is also clearly seen that there are two cross-over areas between the first and the second, and the second and the third clusters, with high values of representatives in more than one cluster.

So, original group N_1 was divided into five subgroups in each invariant parameter with three subgroups corresponding to "pure" clusters and two crossovering ones. We assigned a certain age to a "pure" clustering group if it has more then 25% of representatives only in one cluster. Resulting subgroups' bounds are given in Table 3.

Table 3. Subgroup Bounds

Subgroup	Husband's age	Wife's age
1 – "young"	24-27	22-25
2 – "young-middle"	28-29	26-27
3 – "middle"	30-31	28-30
4 – "middle-old"	32-34	31-32
5 – "old"	35-38	33-37

For the comparison purposes we present the same figures (see Fig. 3) for clusters obtained with hard, namely subtractive algorithm. It is almost impossible to define age bounds using it, so in our previous work [43] we used slightly different technique. We calculated the percentage of representatives of the same age *within* each cluster and values corresponding to 80 or more percents of the appropriate number series maximum value were considered as cluster-defining bounds (Fig. 4). As a result we got age ranges 30-34 for men and 28-32 for women as the first cluster bounds, 27-30 and 24-31 as the second cluster bounds, 32-36 and 29-34 as the third cluster bounds. Resulting subgroups were more overlapping. Thus we've got an interesting result: *using fuzzy clustering gives us more distinct subgroups.*

Considering age of both husband and wife we got 25 subgroups. Distribution of families within subgroups and their prototypes is given in tables 4 and 5 with diagonal elements in bold, row maximums crossed and column maximums shadowed (values less than 1% are in italics). Note that absolute difference between corresponding elements in tables doesn't exceed 1.4%. Subgroups within bold frame were considered for more detailed investigation according to the next steps of the proposed algorithm. Subgroups were chosen by to following criteria:

- diagonals elements;
- maximal elements in rows and columns;
- all other elements with values greater or equal 5%.

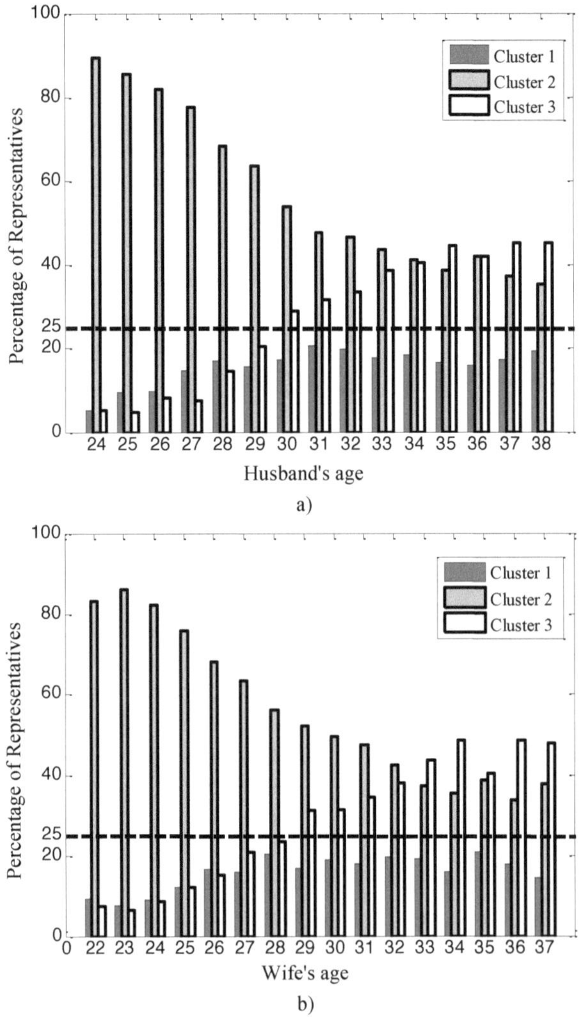

Fig. 3. Husband's (a) and wife's (b) age distribution through clusters for hard clustering (subtractive algorithm)

Chosen subgroups were organized in a sequence S : 1-1, 1-2, 2-2, 2-3, 3-3, 4-3, 4-4, 4-5, 5-5 (first digit represents husband age group, second – wife age group). This sequence covers 66.45% of group N_1 and 66.68% of group N_2 .

The next stage of the experiment was to compare characteristics of subgroups from N_1 and their prototypes from N_2 . After the appropriate calculations were held, it turned out that significant differences take place only for the following attributes: husband's total income, spouse education and ancestry, home ownership, type of building.

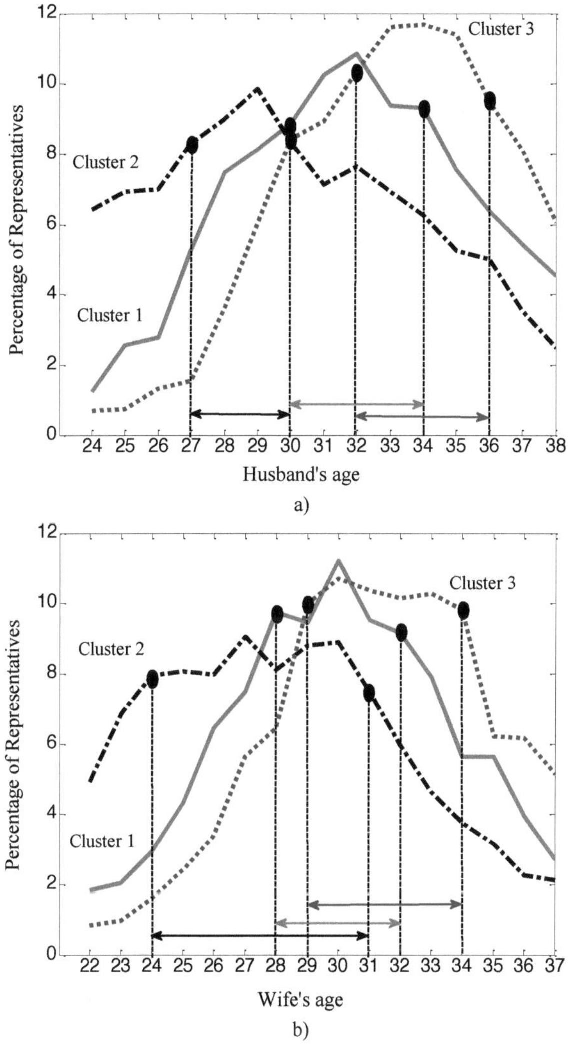

Fig. 4. Husband's (a) and wife's (b) age distribution within clusters for hard clustering (subtractive algorithm)

Table 4. Percentage of families within subgroups of N_1

Wife age group / **Husband age group**	**1**	**2**	**3**	**4**	**5**
1	**~~10.74~~**	4.63	2.55	*0.51*	*0.39*
2	3.66	**3.59**	~~6.06~~	1.43	*0.90*
3	1.77	2.63	**~~6.98~~**	3.33	1.86
4	1.60	2.13	7.18	**6.65**	~~8.38~~
5	*0.91*	1.19	4.13	4.41	**~~12.27~~**

Table 5. Percentage of families within subgroups of N_2

Wife age group / Husband age group	1	2	3	4	5
1	~~**9.89**~~	5.51	2.78	*0.60*	*0.63*
2	3.28	**4.53**	~~7.04~~	1.43	1.07
3	2.13	2.98	~~**7.52**~~	3.03	2.03
4	1.28	1.83	6.80	**6.00**	~~6.95~~
5	*0.74*	1.09	4.15	4.13	~~**12.44**~~

Fig. 5 shows husband's total income distribution within each subgroup and appropriate prototype. Corresponding income intervals are given in Table 6. Figure 6 reveals that families from contrasting groups N_1 and N_2 have practically the same income levels. While moving through the elements of the sequence S income levels tend to increase. Subgroups with younger couples (from 1-1 to 2-3) born children despite the fact, that they usually have slightly lower incomes than families from prototypes. Still families from the elder age groups require more money for having babies.

Fig. 7 shows both husband's and wife's education distribution. By analyzing it we can say that the families from the beginning of the sequence S born children much more often if they are less educated. Young couples (from 1-1 to 2-2) with high education level usually don't have babies. This trend becomes smoother while spouse age increases. Subgroups from 2-3 to 4-4 have almost the same education levels, and in subgroups 4-5 and 5-5 we can see the inverse dependency (families with children have slightly higher education levels).

It should be noted that prototypes of young subgroups are the source of replenishment for the elder subgroups. Indeed, with the course of time families with highly educated spouse can move to one of the N_1 subgroups depending on when the first baby (babies) appears.

Most families in California either own a dwelling with a mortgage or loan or rent it for the cash rent. In order to take into account influence of this parameter we calculated difference between the percentages of representative who own a dwelling or rent it in subgroups and their prototypes. Thus positive values indicate that parameter has a positive effect on desire to have a baby and vice versa. Absolute value indicates the effect strength. Obtained distribution is shown on Fig. 7a.

Table 6. Husband's Total Income Intervals (*10^4)

1	[-1; 0)	**7**	[8; 10)
2	[0; 1)	**8**	[10; 20)
3	[1; 2)	**9**	[20; 30)
4	[2; 4)	**10**	[30; 40)
5	[4; 6)	**11**	[40; 72]
6	[6; 8)		

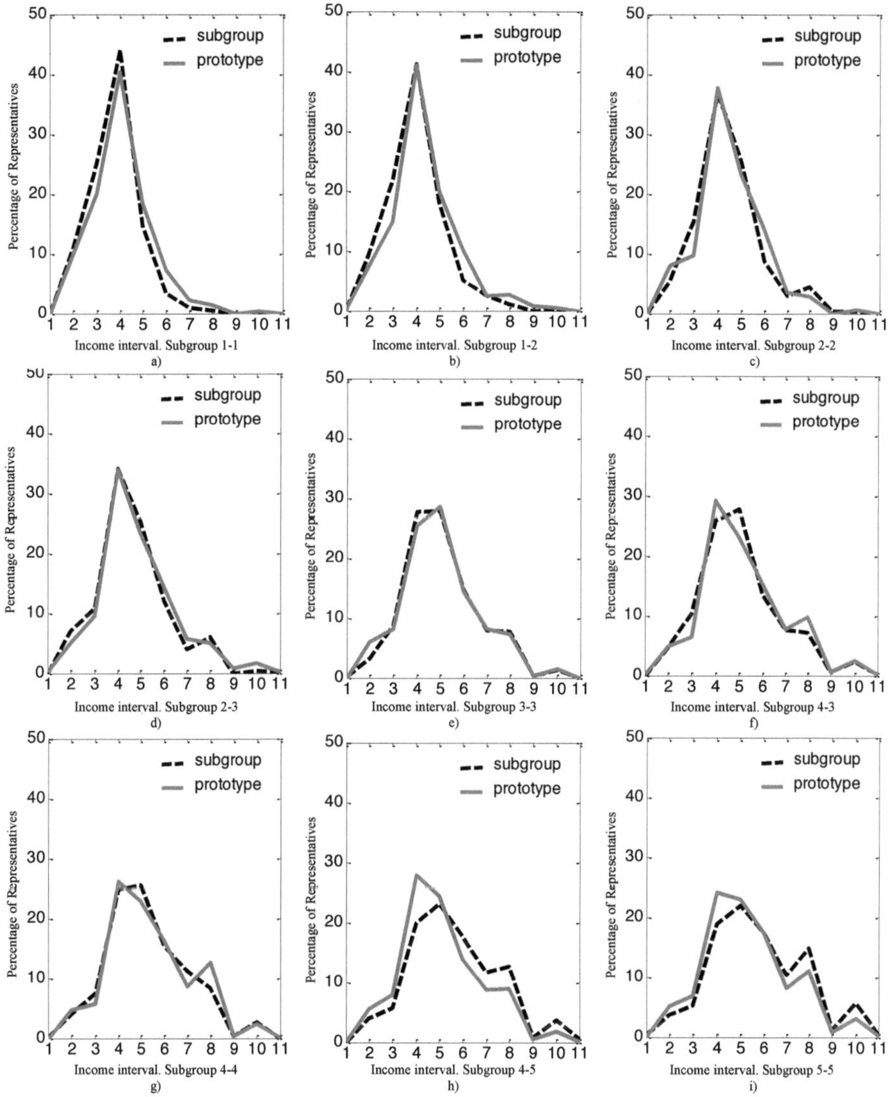

Fig. 5. Husband's income distribution in subgroups (dashed black line) and their prototypes (gray line)

From the figure we can see that prior to subgroup 2-2 the fact of having own dwelling doesn't increase the desire to have a baby. But starting from this point most families from prototypes lack own dwelling. The strongest effect is observed in subgroup 3-3, where people are of the most favorable age for having children.

Fig. 7b shows the same distribution for such values of the parameter "building type":

- a one-family house detached from any other house;
- a one-family house attached to one or more houses;
- a building with 2..50 apartments.

The general distribution tendency is similar to above described case. Still we can see that having detached house have strong positive effect in all subgroups. Also there is almost strong monotonous increase in intervals from 1-1 to 3-3 and from 4-3 to 5-5.

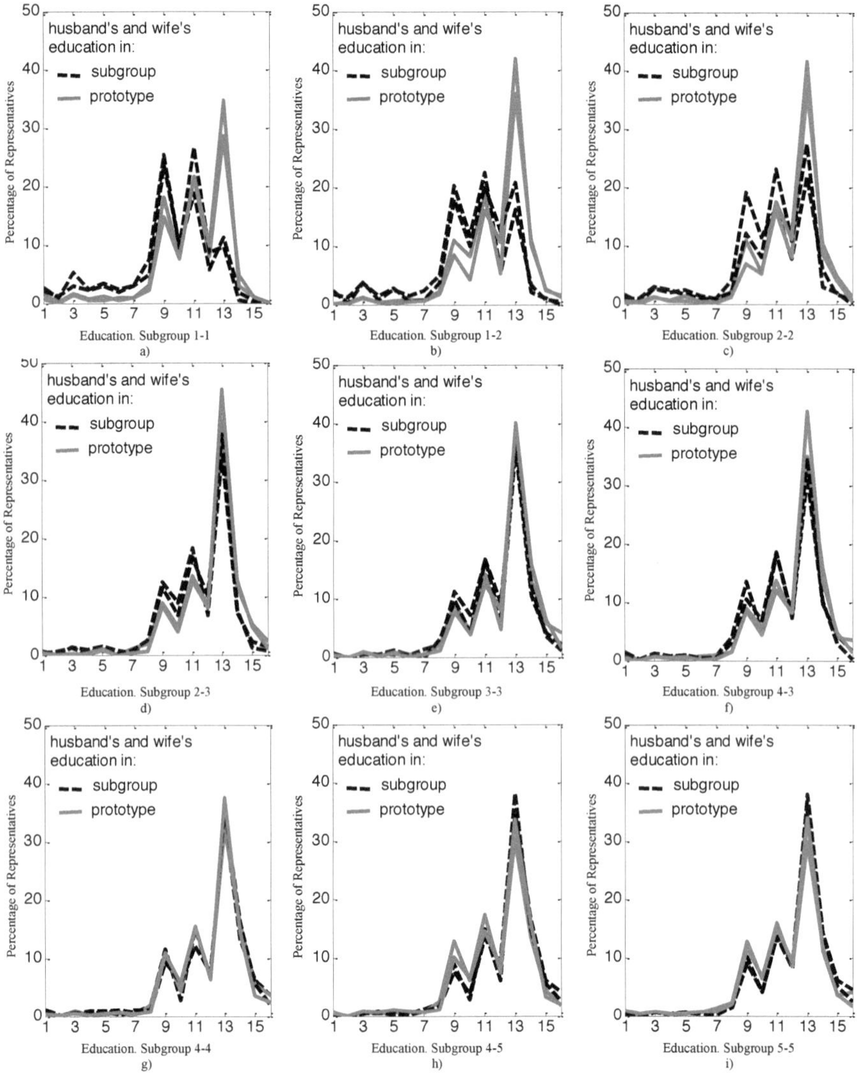

Fig. 6. Husband's and wife's education distribution in subgroups (dashed black line) and their prototypes (gray line)

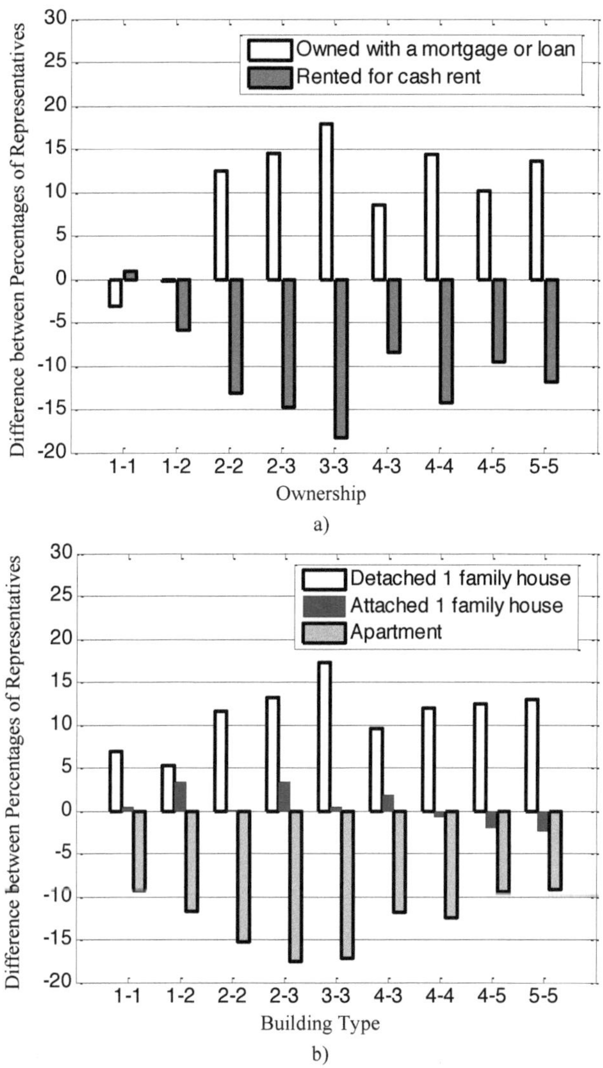

Fig. 7. Distribution of difference between percentage of representatives in subgroups and their prototypes for parameters ownership (a) and building type (b)

It is interesting to note that prior to subgroup 4-3 families living in attached houses are much more willing to born a baby than elder couples.

Figure 8 shows distributions of representatives within groups N_1 and N_2 by spouse ancestry (Table 7 reveals ancestry codes). As we can see this parameter has three dominant values:

- West Europe;
- Mexico;
- other (not South) Asia.

Table 7. Ancestry Codes

1	West Europe	**5**	Central America Islands	**9**	Australia
2	East Europe	**6**	North Africa and South Asia	**10**	Pasific
3	Mexico	**7**	other Africa	**11**	Afro-American
4	Latino	**8**	other Asia	**12**	Other American

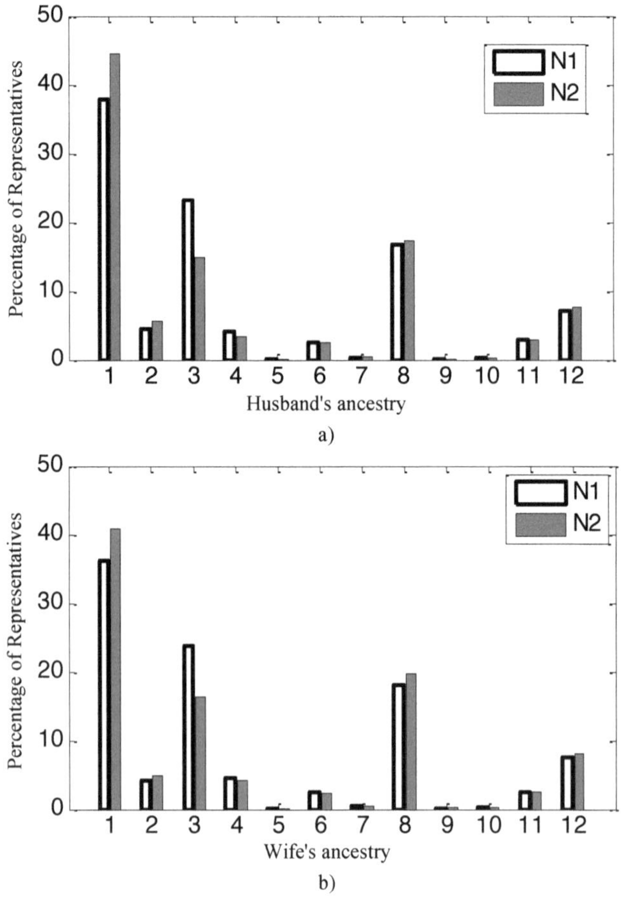

Fig. 8. Distribution of families by husband's (a) and wife's (b) ancestry

These values were examined in details. Once again we calculated the difference between percentage of representatives of a certain ancestry in each subgroup and its prototype. Resulting distributions are presented on Fig. 9.

Analyzing figure we can say that people with origin from West Europe are least inclined to have children. This effect goes monotonously from -15% difference for subgroup 1-1 to almost 0% for subgroup 5-5.

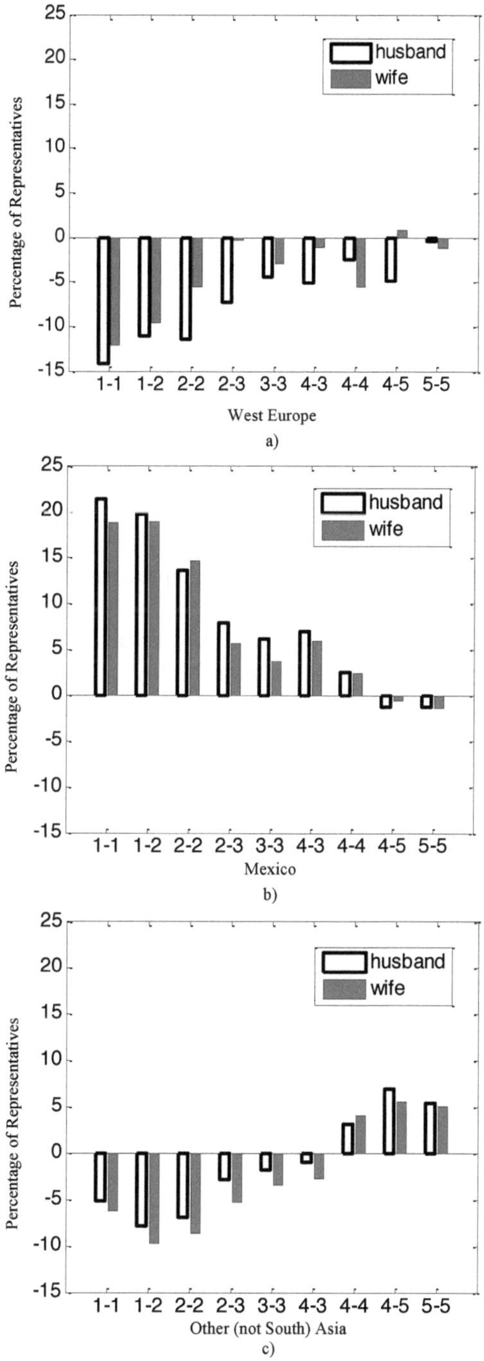

Fig. 9. Distribution of difference between percentage of representatives of a certain ancestry in subgroups and their prototypes: West Europe (a), Mexico (b) and not South Asia (c)

People with Mexican origins are vice versa most inclined to born children. This parameter has positive effect for almost all subgroups except the eldest 4-5 and 5-5, where appropriate values are slightly less than zero.

An interesting situation occurs for people with non South Asia ancestry. Prior to subgroup 4-4 parameter has negative effect with the strongest impact for subgroup 1-2. While moving from 4-4 to 5-5 we can see positive values with maximum for group 4-5.

Summing up experimental results we can supplement conclusions from our previous work [43] and say the following:

- indeed providing financial support contributes to the birth rate increase but most people lack not money but own separate dwelling;
- young age subgroups (1-1, 1-2) with low education level should be encouraged by providing cheep house rents (as we saw from the experiment this age group is less susceptible to ownership type); house can be detached or attached to another one;
- young families with high education level usually don't born children, so they shouldn't be considered before spouse become older;
- special attention should be paid to young couples with Mexican origins, because it is their best reproductive age;
- families from the middle age subgroups must be actively encouraged with cheep housing loans, because as it is seen from Fig. 7 these parameters have the strongest effect in this case; also it should be noted that this period is the most favorable for having babies in general, so investment will probably yield the greatest effect here;
- if spouse belong to the elder age subgroups they most likely lack own detached houses (note that attached houses have rather negative effect here); in this case special attention must be provided to people originated from not South Asia and West Europe.

5 Conclusions

Current paper shows that using Data Mining techniques for statistical data analysis, in particular for census data analysis, we can get results unreachable by other analysis methods. However, using only Data Mining may not be enough. Analysis algorithms which include Data Mining techniques as elements, just like the algorithm proposed in this paper, are of particular interest.

Developed Influence Search Algorithm is based on clustering a set which elements possess a certain feature, defining clustering based subgroups from it and subgroups prototypes out of the set with elements without this feature. Comparing characteristics of clustering subgroups and their prototypes we can give some recommendations regarding question how to "move" respondents from the second set to the first one (i.e., add this feature to respondents).

Methodologically this algorithm belongs to the contrast mining field. Nevertheless previous works in contrast mining are dedicated to other problems; they usually investigate some area, but not identify factors which can stimulate decision making process.

It was also shown that fuzzy clustering techniques are more effective than hard clustering and despite the fact that algorithm is *fuzzy* it helps determine more *strict* bounds.

Among prospects for further researches we can outline the following:

- joint analysis of such population influencing factors as natality and migration;
- providing similar analysis for different US states and countries;
- further development of the proposed approach with another Data Mining techniques, such as decision trees and association rules learning.

References

[1] Gantz, J., Reinsel, D.: The 2011 Digital Universe study: extracting value from chaos. IDC iview (June 2011), http://www.emc.com/collateral/demos/microsites/emc-digital-universe-2011/index.htm

[2] Mullins, I., et al.: Data Mining and clinical data repositories: insights from a 667,000 patient data set. Comput. in Biology and Medicine 36(12), 1351–1377 (2006)

[3] Public Law 104-191, 104th Congress. Health Insurance Portability and Accountability Act of 1996 (HIPAA) (August 21, 1996), http://aspe.hhs.gov/admnsimp/pl104191.htm

[4] Patient Safety and Quality Improvement Act of 2005 (PSQIA). Federal Register 73(266) (2001)

[5] Directive 2002/58/EC of the European Parliament and of the Council of 12 July 2002. Official J. of the European Communities L 201 (July 31, 2002)

[6] The Law of Ukraine On State Statistics as for 2009, March 5 (in Ukrainian), http://zakon1.rada.gov.ua/cgi-bin/laws/main.cgi?nreg=2614-12&p=1265575855780241

[7] Minnesota Population Center, University of Minnesota. Integrated Public Use Microdata Series International, https://international.ipums.org/international/

[8] Lenz, H.-J., Shoshani, A.: Summarizability in OLAP and statistical data bases. In: Proc. 9th Int. Conf. Scientific and Statistical Database Manage, SSDBM 1997, Olympia, WA, USA (1997)

[9] U.S. Census Bureau. Statistical Quality Standard E1: Analyzing Data, http://www.census.gov/quality/standards/standarde1.html

[10] Han, J., Kamber, M.: Data Mining: Concepts and Techniques, 2nd edn. Morgan Kaufmann Publishers, San Francisco (2006)

[11] Berson, A., Smith, S., Thearling, K.: An Overview of Data Mining Techniques (2005), http://www.stat.ucla.edu/~hqxu/stat19/DM-Techniques.pdf

[12] Colet, E.: Clustering and Classification: Data Mining Approaches (July 4, 2004), http://www.taborcommunications.com/dsstar/00/0704/101861.html

[13] Hammouda, K., Karay, F.: A comparative study of data clustering techniques, http://pami.uwaterloo.ca/pub/hammouda/sde625-paper.pdf

[14] Dong, G.: International Workshop on Contrast Data Mining and Applications (2011), http://www.cs.wright.edu/~gdong/ContrastDMWorkshop.pdf

[15] Novak, P.K., Lavrac, N., Webb, G.I.: Supervised Descriptive Rule Discovery: A Unifying Survey of Contrast Set, Emerging Pattern and Subgroup Mining. J. of Mach. Learning Research 10, 377–403 (2009)

[16] Dong, G., Bailey, J.: Overview of Contrast Data Mining as a Field and Preview of an Upcoming Book. In: 2011 IEEE 11th Int. Conf. on Data Mining Workshops, Vancouver, Canada (2011)
[17] Bay, S.D., Pazzani, M.J.: Detecting change in categorical data: mining contrast sets. In: Proc. 5th ACM SIGKDD Int. Conf. on Knowledge Discovery and Data Mining, KKD 1999, San Diego, CA, USA (1999)
[18] Dong, G., Li, J.: Efficient Mining of Emerging Patterns: Discovering Trends and Differences. In: Proc. 5th ACM SIGKDD Int. Conf. on Knowledge Discovery and Data Mining, KKD 1999, San Diego, CA, USA (1999)
[19] Liu, B., Hsu, W., Ma, Y.: Discovering the set of fundamental rule changes. In: Proc. of the 7th ACM SIGKDD Int. Conf. on Knowledge Discovery and Data Mining (KDD 2001), San Francisco, CA, USA, pp. 335–340 (2001)
[20] Garriga, G.C., Kralj, P., Lavrač, N.: Closed Sets for Labeled Data. In: Fürnkranz, J., Scheffer, T., Spiliopoulou, M. (eds.) PKDD 2006. LNCS (LNAI), vol. 4213, pp. 163–174. Springer, Heidelberg (2006)
[21] Daly, O., Taniar, D.: Exception rules in data mining. In: Encyclopedia of Information Science and Technology (II), pp. 1144–1148. Idea Group Reference (2005)
[22] Loekito, E., Bailey, J.: Fast Mining of High Dimensional Expressive Contrast Patterns Using Zero-suppressed Binary Decision Diagrams. In: Proc. of the 12th ACM SIGKDD Int. Conf. on Knowledge Discovery and Data Mining (KDD 2006), Philadelphia, PA, USA, pp. 307–316 (2006)
[23] García-Borroto, M., Trinidad, J.F.M., Carrasco-Ochoa, J.A.: Fuzzy emerging patterns for classifying hard domains. Knowledge and Inform. Syst. 28(2), 473–489 (2011)
[24] Duan, L., et al.: Mining contrast inequalities in numeric dataset. In: Int. Conf. on Web-Age Inform. Manage. (WAIM), Jiuzhaigou, China, pp. 194–205 (2010)
[25] Duan, L., Tang, C., Tang, L., Zhang, T., Zuo, J.: Mining Class Contrast Functions by Gene Expression Programming. In: Huang, R., Yang, Q., Pei, J., Gama, J., Meng, X., Li, X. (eds.) ADMA 2009. LNCS, vol. 5678, pp. 116–127. Springer, Heidelberg (2009)
[26] Nedjar, S., Cicchetti, R., Lakhal, L.: Extracting semantics in OLAP databases using emerging cubes. Inform. Sci. 181, 2036–2059 (2011)
[27] Ramamohanarao, K., Bailey, J., Fan, H.: Efficient Mining of Contrast Patterns and Their Applications to Classification. In: Proc. of the 2005 3rd Int. Conf. on Intelligent Sensing and Inform. Process (ICISIP 2005), pp. 39–47. IEEE Computer Society, Washington, DC (2005)
[28] Fore, N., Dong, G.: CPC: A contrast pattern based clustering algorithm requiring no distance function. Department of Computer Science and Engineering, Wright State University, OH, USA, Tech. Rep. (2011)
[29] Li, J., Wong, L.: Identifying good diagnostic gene groups from gene expression profiles using the concept of emerging patterns. Bioinformatics 18(5), 725–734 (2002)
[30] Dong, G., Fore, N.: Discovering dynamic logical blog communities based on their distinct interest profiles. In: The First Int. Conf. on Social Eco-Informatics (SOTICS 2011), Barcelona, Spain, pp. 24–30 (2011)
[31] Kobyliński, Ł., Walczak, K.: Jumping Emerging Patterns with Occurrence Count in Image Classification. In: Washio, T., Suzuki, E., Ting, K.M., Inokuchi, A. (eds.) PAKDD 2008. LNCS (LNAI), vol. 5012, pp. 904–909. Springer, Heidelberg (2008)
[32] Encheva, S., Tumin, S.: Problem identification based on fuzzy functions. WSEAS Trans. on Advances in Eng. Educ. 6(9), 111–120 (2009)
[33] Gupta, A., Kumar, N., Bhatnagar, V.: Analysis of medical data using Data Mining and formal concept analysis. Proc. World Academy of Sci., Ing. and Technology 6 (2005)

[34] Nonyelum, O.: Potential value of Data Mining for customer relationship marketing in the banking industry. Adv. in Nat. Appl. Sci. 3(1), 73–78 (2009)
[35] Ngai, E.W.T., Xiu, L., Chau, D.C.K.: Application of Data Mining techniques in customer relationship management: a literature review and classification. Expert Syst. with Applicat. 36, 2592–2602 (2009)
[36] Malerba, D., Esposito, F., Lisi, F., Appice, A.: Mining spatial association rules in census data. Research in Official Statistics 1, 19–45 (2002)
[37] Malerba, D., Lisi, F., Appice, A., Sblendorio, F.: Mining census and geographic data in urban planning environments. In: Santini, L., Zotta, D. (eds.) Atti della Terza Conferenza Nazionale su Informatica e Pianificazione Urbana e Territoriale (INPUT 2003), Alinea Editrice, Firenze, Italy (2003)
[38] Appice, A., Ceci, M., Lanza, A., Lisi, F., Malerba, D.: Discovery of spatial association rules in geo-referenced census data: a relational mining approach. Intelligent Data Analysis 7, 541–566 (2003)
[39] Zaki, M., Hsiao, C.: CHARM: an efficient algorithm for closed itemset mining. In: Proc. 2002 SIAM Int. Conf. Data Mining, Arlington, VA, USA, pp. 457–473 (2002)
[40] Park, J., Chen, M., Yu, P.: An effective hash-based algorithm for mining association rules. In: Proc. 1995 ACM SIGMOD Int. Conf. Manage. Data, San Jose, CA, USA (1995)
[41] Brin, S., Motwani, R., Ullman, J., Tsur, S.: Dynamic itemset counting and implication rules for market basket data. In: Proc. 1997 ACM SIGMOD Int. Conf. Manage. Data, Tucson, AZ, USA (1997)
[42] Klosgen, W., May, M.: Census data mining – an application. In: Proc. 6th European Conf. Principles of Data Mining, Knowledge Discovery (PKDD 2002), pp. 65–79 (2002)
[43] Chertov, O., Aleksandrova, M.: Clustering with prototype extraction for census data analysis. In: Proc. World Conf. Soft Computing WConSC 2011, San Francisco, CA, USA (2011), http://arxiv.org/abs/1106.5122
[44] U.S. Census. 5-Percent Public Use Microdata Sample Files (2000), http://www.census.gov/Press-Release/www/2003/PUMS5.html
[45] Priyono, A., et al.: Generation of fuzzy rules with subtractive clustering. J. Technology 43, 143–153 (2005)
[46] Bezdek, J.C., Ehrlich, R., Full, W.: FCM: The fuzzy c-means clustering algorithm. Computers & Geosciences 10(2-3), 191–203 (1984)
[47] Chiu, S.L.: Fuzzy model identification based on cluster estimation. J. Intelligent Fuzzy Syst. 2, 267–278 (1994)

Author Index

MIX
Papier aus verantwortungsvollen Quellen
Paper from responsible sources
FSC® C105338

If you have any concerns about our products, you can contact us on
ProductSafety@springernature.com

In case Publisher is established outside the EU, the EU authorized representative is:
Springer Nature Customer Service Center GmbH
Europaplatz 3, 69115 Heidelberg, Germany

Printed by Libri Plureos GmbH
in Hamburg, Germany